Lecture Notes in Computer Science 2918

Edited by G. Goos, J. Hartmanis, and J. van Leeuwen

Springer
Berlin
Heidelberg
New York
Hong Kong
London
Milan
Paris
Tokyo

Samir R. Das   Sajal K. Das (Eds.)

Distributed Computing – IWDC 2003

5th International Workshop
Kolkata, India, December 27-30, 2003
Proceedings

 Springer

Series Editors

Gerhard Goos, Karlsruhe University, Germany
Juris Hartmanis, Cornell University, NY, USA
Jan van Leeuwen, Utrecht University, The Netherlands

Volume Editors

Samir R. Das
State University of New York
Computer Science Department
Stony Brook, NY 11794-4400, USA
E-mail: samir@cs.sunysb.edu

Sajal K. Das
The University of Texas at Arlington
Department of Computer Science and Engineering
P.O.Box 19015, Arlington, TX 76019, USA
E-mail: das@cse.uta.edu

Cataloging-in-Publication Data applied for

A catalog record for this book is available from the Library of Congress.

Bibliographic information published by Die Deutsche Bibliothek
Die Deutsche Bibliothek lists this publication in the Deutsche Nationalbibliografie;
detailed bibliographic data is available in the Internet at <http://dnb.ddb.de>.

CR Subject Classification (1998): C.2, D.1.3, D.2.12, D.4, F.2, F.1, H.4

ISSN 0302-9743
ISBN 3-540-20745-7 Springer-Verlag Berlin Heidelberg New York

Springer-Verlag is a part of Springer Science+Business Media

springeronline.com

© Springer-Verlag Berlin Heidelberg 2003
Printed in Germany

Typesetting: Camera-ready by author, data conversion by DA-TeX Gerd Blumenstein
Printed on acid-free paper   SPIN: 10977552   06/3142   5 4 3 2 1 0

General Chairs' Message

Within a short period of time, IWDC has been able to establish itself as a major international workshop in Distributed Computing. This year, the bulk of the papers and invited talks focused on wireless and mobile computing, sensor networks, and security, which is in tune with the way the world around us is unfolding. Thanks to the efforts of the publicity chairs, the program chairs and the various coordination chairs, we succeeded in making a solid connection with the international community. We were particularly happy to present before the attendees an array of invited speakers, each one being an expert in his own area. We were glad that the talks were enjoyable and illuminating, spawning thought-provoking questions.

This year, many people invested considerable time and effort to make the event a success. Our publicity chairs Debashis Saha and Rajkumar Buyya did an outstanding job of reaching out to the community. Our Asia-Pacific coordination chairs Kyung-Hyune Rhee and Eun-Yi Jung provided excellent support in reaching out to the regional institutions. The two program chairs Samir Das and Chandan Mazumdar worked very hard to put together a solid technical program that had both depth and breadth. Our special thanks to the publication chair Sajal Das and the program chair Samir Das who spent countless hours taking care of numerous issues beyond their own portfolios. All three tutorials addressed issues of current interest. We hope that our local arrangements for the event met everyone's expectations.

To run a workshop, we need money. Fundraising was not easy, and we are thankful to our finance chairs Nabendu Chaki and Avijit Kar for their efforts. Special thanks go to the industry chairs Rabinda N. Lahiri and Rajeev Shorey, all our co-sponsors, and the various co-ordination chairs – without your support this event would not have taken place. Finally, we thank all the authors presented their papers, the panelists who shared their knowledge and experiences, and all the attendees of IWDC 2003 for their contributions and participation.

October 12, 2003   Sukumar Ghosh
                                                                         Pradip K. Das

Program Chairs' Message

Welcome to the proceedings of the 5th International Workshop on Distributed Computing, IWDC 2003. This year we received 105 paper submissions, a significant increase over previous years. We received submissions from many countries around the world: India, United States, China, Australia, Korea, France, Italy, Japan, Canada, Taiwan, Pakistan, Brazil, Algeria, Portugal, Turkey, Finland and New Zealand – mentioned here in the decreasing order of number of submissions we received. About half of the papers received were from outside India, giving this workshop a true international flavor.

The 30 members of the technical program committee along with a team of more than 50 external reviewers worked hard on the reviews under a very strict timeline. At least 3 reviews were sought for every paper. Almost all the requested reviews were returned. At the end of the review period, the program chairs selected 32 papers for presentation in the workshop and inclusion in these proceedings. Because of the limitations of a single-track workshop, many good papers could not be accommodated. This made IWDC 2003 a highly selective workshop.

We were also fortunate to have an array of keynote and invited speakers – Sushil Jajodia (George Mason University), Ted Herman (University of Iowa), Amitava Bagchi (IIM Calcutta), Bishnu Pradhan (KReSIT, IIT Bombay, and Media Lab, Asia), Dharma Agrawal (University of Cincinnati) and Laxmi Bhuyan (University of California at Riverside). Their talks provided us with the unique opportunity to hear from the leaders of the field.

This year a poster session was designed for student papers. These papers went through a separate submission and review process. A separate book of proceedings will contain the accepted student papers.

We thank all authors for their interest in IWDC 2003, and all program committee members and external reviewers for their commitment in spite of a tight schedule and a high review load. Special thanks are due to our Publications Chair, Sajal Das, for his commitment to make the proceedings a high-quality, archival publication; and to Vishnu Navda, who created and administered the internal web site for the paper reviews, handled scores of administrative emails, and helped significantly in editing the proceedings you are now holding.

We hope that you will find the proceedings of IWDC 2003 program technically rewarding.

October 12, 2003   Samir R. Das
                                                                Chandan Mazumdar

Executive Committee

General Chairs:
Sukumar Ghosh, University of Iowa, USA
Pradip K. Das, Jadavpur University, India

Program Chairs:
Samir R. Das, SUNY at Stony Brook, USA
Chandan Mazumdar, Jadavpur University, India

Organizing Chairs:
Asim K. Pal, Indian Institute of Management, Calcutta, India
Suranjan Ghose, Jadavpur University, India

Publication Chair:
Sajal K. Das, Univ. of Texas, Arlington, USA

Tutorial Chair:
Ambuj Mahanti, Indian Institute of Management, Calcutta, India

Asia-Pacific Coordination Chairs:
Kyung-Hyune Rhee, PuKyong National University, Korea
Eun-Yi Jung, Choonhae College, Korea

Student Activity Chairs:
Ranjan Dasgupta, Technical Teachers' Training Institute, Eastern Region, India
Sankhayan Choudhury, Calcutta University, India

Industry Chairs:
Rabindra N. Lahiri, TCS Calcutta, India
Rajeev Shorey, IBM Solutions Research Lab, New Delhi, India

Publicity Chairs:
Debashis Saha, Indian Institute of Management Calcutta, India
Rajkumar Buyya, Univ. of Melbourne, Australia

Finance Chairs:
Avijit Kar, Jadavpur University, India
Nabendu Chaki, Calcutta University, India

IFIP/IEEE Coordination Chairs:
Augusto Casaca, Chair, TC6-IFIP
Ram G. Gupta, Chair, IEEE-AES-COM-LEOS Society Chapter, India

Online and Logistics Coordinator:
M. Scalem, Exec. Chair, CSI-IIMC Student Branch, IIM Calcutta

Student Track Coordinator:
CSI-IIMC Student Branch, IIM Calcutta, India

International Steering Committee:

Sukumar Ghosh	University of Iowa, USA, Chair
Man-Gon Park	Colombo Plan Staff College–Intergovernmental International Organization
Somprakash Bandyopadhyay	Indian Institute of Management Calcutta, India
Subhangsu Bandyopadhyay	Calcutta University, India
Pradip K. Das	Jadavpur University, India
Sajal K. Das	University of Texas, Arlington, USA
Sanjib C. De Sarkar	Indian Institute of Technology, Kharagpur, India
Shyamal Majumdar	Technical Teachers' Training Institute, Eastern Region, India
Prasanta K. Nandi	Bengal Engineering College (DU), India
Bhabani P. Sinha	Indian Statistical Institute, Kolkata, India
T. Srikanthan	Nanyang Technological University, Singapore

Program Committee

Chairs

Samir R. Das	SUNY at Stony Brook, USA
Chandan Mazumdar	Jadavpur University, India

Members

Dharma P. Agarwal	University of Cincinnati, USA
A.K. Aggarwal	University of Windsor, Canada
Aditya Bagchi	ISI Kolkata, India
Amitava Bagchi	IIM Calcutta, India
Shivaji Bandopadhyay	Jadavpur University, India
Prithviraj Banerjee	Northwestern University, USA
Anupam Basu	IIT Kharagpur, India
Dhruba Basu	Techno India, Salt Lake, Kolkata, India
Swapan Bhattacharya	Jadavpur University, India
Rupak Biswas	NASA, USA
Rajkumar Buyya	University of Melbourne, Australia
Nabendu Chaki	University of Calcutta, India
Mainak Chatterjee	University of Central Florida, USA
Partha S. Dasgupta	IIM Calcutta, India
Ajoy K. Dutta	University of Nevada, Las Vegas, USA
Arobinda Gupta	IIT Kharagpur, India
Mohan Gurusamy	National University of Singapore, Singapore
Anant K. Jain	Telecom Consultant, USA
Sanjay Jha	University of New South Wales, Australia
Sanjay Madria	University of Missouri-Rolla, USA
Shikharesh Majumdar	Carleton University, Canada
Archan Misra	IBM, USA
Asis Nasipuri	University of North Carolina, USA
Viktor K. Prasanna	University of Southern California, USA
Anup K. Sen	IIM Calcutta, India
Sandip Sen	University of Tulsa, USA
Rajeev Shorey	IBM Research Lab, India
Mukesh Singhal	University of Kentucky, USA
Bhabani P. Sinha	ISI Kolkata, India
Satish Tripathi	University of California, USA

External Reviewers

The following reviewers external to the program committee participated in the review process. We greatly appreciate their contributions.

Abhishek Karnik
Abhishek Roy
Anup Kumar Bandyopadhyay
Archana Mohanty
Arjita Ghosh
A. Sinha
Amitava Gupta
Cong Zhang
G. Sajith
Haitao Lin
Himadri Sekhar Paul
Hrushikesh Mohanty
Imran Pirwani
Isai Arasu Marichamy
Iti SahaMisra
Jaya Sil
Kalyan Basu
Krishna Paul
Loudon Blair
Lillykutty Jacob
Mahesh Marina
Mridul Sankar Barik
Nabendu Chaki
Paramartha Dutta
Parijat Prosun Kar
Partha Sarathi Dasgupta

Pradip K. Das
Rajib Mall
R.K. Ghosh
Romit RoyChoudhury
Sabyasachi Saha
Samrat Ganguly
Sanghamitra Bandyopadhyay
Sanjoy Kumar Saha
Sarmistha Neogy
Sethuraman J.
Shahabuddin Muhammad
Shamik Sural
Shweta Jain
Somprakash Bandyopadhyay
Sourav Pal
Sridhar K.N.
Stephane Airiau
Sudipta Rakshit
Sukumar Ghosh
Swapan Bhattacharya
Swarup Mandal
Ujjwal Maulik
Umesh Deshpande
Uttam Kumar Sarkar
Venkatesh
Vishnu Navda

Table of Contents

Session III: Parallel and Distributed Systems

Session IV: Wireless and Mobile Networking

Keynote Talk II

Session V: Ad hoc and Sensor Networks

Invited Talk I

Session VI: Learning and Optimization

Invited Talk II

Session VII: Optical Networking

Scalable Group Rekeying
for Secure Multicast: A Survey

Sencun Zhu and Sushil Jajodia

Center for Secure Information Systems
George Mason University, Fairfax, VA 22030-4444, USA
{szhu1,jajodia}@gmu.edu

Abstract. Many group key management approaches have been recently proposed for providing scalable group rekeying to support secure communications for large and dynamic groups. We give an overview of these approaches and show the research trends. Finally, we discuss some new research directions on this subject.

1   Introduction

Many multicast-based applications (e.g., pay-per-view, online auction, and teleconferencing) require a secure communication model. However, IP Multicast, the multicast service proposed for the Internet, does not provide any security mechanisms; indeed, anyone can join a multicast group to receive data from the data sources or send data to the group. Therefore, cryptographic techniques need to be employed to achieve data confidentiality. One solution is to let all members in a group share a key that is used for encrypting data. To provide backward and forward confidentiality [17] (i.e., a new member should not be allowed to decrypt the earlier communication and a revoked user should not be able to decrypt the future communication) this shared group key must be updated for every membership change and redistributed to all authorized members in a secure, reliable, and timely fashion. This process is referred to as *group rekeying*.

A simple approach for group rekeying is one in which the group key server encrypts and sends the updated group key individually to each member. This approach is not scalable because its costs increase linearly with the group size. For large groups with very frequent membership changes, scalable group rekeying becomes an especially challenging issue.

In recent years, many approaches for scalable group rekeying have been proposed [2, 3, 5, 10, 11, 16, 17]. Due to different research focuses and the diversity of performance metrics such as bandwidth overhead and rekeying latency, each approach has its own merits, and some of them are complementary to each other. Therefore, rather than comparing one scheme to another, we are more interested in showing the previous research trends and envisioning some future research directions. We studied these approaches chronologically and made the following observations:

S. R. Das, S. K. Das (Eds.): IWDC 2003, LNCS 2918, pp. 1–10, 2003.
© Springer-Verlag Berlin Heidelberg 2003

- The research interests have been moving from stateful protocols to stateless protocols. Stateless protocols such as Subset Difference [10, 7] are more attractive than stateful protocols such as LKH [16, 17] in applications where users may go offline frequently or have high loss rates. In a stateless protocol, a legitimate user can readily extract the new group key from the received keying materials despite the number of previous group rekeying operations he/she has missed, whereas in a stateful protocol a user must receive all keys of interest in all previous rekeying operations to be able to extract the current group key.
- Reliable key distribution has also become a research hot spot recently. A group rekeying process usually includes two phases. The first phase deals with the key encoding problem. A group key must be encrypted by some key encryption keys (KEKs) before its distribution. The purpose of a key encoding algorithm is to minimize the number of encryptions, including those of the group key and of some KEKs. In the literature, many approaches (e.g., LKH, OFT, and Subset-Difference) work on this phase. The second phase deals with the key distribution problem (i.e., how to distribute the encryptions output from a key encoding algorithm to all the members reliably and timely in the presence of packet losses). Recently, researchers have proposed customized reliable multicast protocols for group key distribution (e.g., Proactive-FEC [18] and WKA-BKR [15]). Other schemes, such as ELK [11], integrate these two phases.
- Self-healing key distribution, which can also be thought of as belonging to the key distribution phase, has drawn much attention recently. A self-healing key distribution scheme [13, 20] provides the property that a user is able to recover the lost previous group keys without asking the key server for retransmission, which is important to prevent the key server from being overwhelmed by feedback implosion.
- Several optimization schemes [1, 14, 19] have been proposed to further reduce the rekeying communication overhead by exploiting the characteristics of group members such as topology, membership durations, and loss rates.

The reminder of this paper is organized as follows. We will discuss the existing work in more detail in Section 2 and then list some new research directions on scalable group rekeying in Section 3.

2   A Taxonomy of Group Rekeying Protocols

This section details several group rekeying schemes. Note that the approaches we will discuss do not cover the literature entirely. We only discuss some work on centralized group rekeying where a key server controls the group membership. We do not discuss contributory group rekeying, which are typically less scalable, and other tightly related issues, such as multicast source authentication.

2.1   Stateful Protocols

A stateful protocol is one in which a member must receive all keys of interest in all of the previous rekeying operations to be able to decrypt the new group key, unless he/she asks the key server for key retransmission. Most of the logical-key-tree-based group rekeying protocols in the literature (e.g., LKH [16, 17], OFT [2], and ELK [11]) are stateful protocols. We illustrate below the idea of logical key tree based on LKH and show why it is stateful.

The use of logical key trees for scalable group rekeying was independently proposed by Wallner et al. [17] and Wong et al. [16]. The basis for the LKH approach for scalable group rekeying is a logical key tree which is maintained by the key server. The root of the key tree is the group key used for encrypting data in group communications and it is shared by all users. Every leaf node of the key tree is a key shared only between an individual user and the key server, whereas the middle level keys are KEKs used to facilitate the distribution of the root key. Of all these keys, each user owns only those keys that lie on the path from his/her individual leaf node to the root of the key tree. As a result, when a user joins or leaves the group, all of the keys on his/her path have to be changed and re-distributed to maintain backward and forward data confidentiality.

An example key tree is shown in Fig. 1. In this figure, $K_{1\text{-}9}$ is the group key shared by all users, $K_1, K_2, \ldots, K_9$ are individual keys, and $K_{123}, K_{456}, K_{789}$ are KEKs known only by users who are in the sub-trees rooted at these keys. We next illustrate member joins and leaves through an example, based on group-oriented rekeying [16].

Join Procedure Suppose in Fig. 1 the root key was $K_{1\text{-}8}$ and K_{789} was K_{78} before user U_9 joined the group, and they are replaced with keys $K_{1\text{-}9}$ and K_{789} respectively when U_9 joins. To distribute these new keys to all users, the key server encrypts $K_{1\text{-}9}$ with $K_{1\text{-}8}$, K_{789} with K_{78}. In addition, the key server encrypts $K_{1\text{-}9}$, K_{789} with K_9. All of the encrypted keys are multicast to the group, and each user can extract the keys he/she needs independently.

Departure Procedure When user U_4 departs from the group, the keys K_{456} and $K_{1\text{-}9}$ need to be changed. Assume these keys are replaced with keys K'_{456} and $K'_{1\text{-}9}$ respectively. Now the key server encrypts $K'_{1\text{-}9}$ with K_{123}, K'_{456} and K_{789} separately, encrypts K'_{456} with K_5 and K_6 separately, and then multicasts these five encrypted keys to the group.

Consider user U_7 in Fig. 1. During the group rekeying for adding U_9 into the group, U_7 must receive K_{789}; otherwise, he/she will not be able to decrypt the new group key $K'_{1\text{-}9}$ during the group rekeying for revoking user U_4. Therefore, LKH is a stateful rekeying protocol. LKH is very efficient and hence scalable for group rekeying when compared to a unicast-based naive approach. Let N be the group size and d be the degree of the key tree; then the communication cost for rekeying is $O(log_d N)$, whereas the naive approach requires a communication cost of $O(N)$.

However, for a large group with very dynamic memberships, LKH may not scale well [12] because it performs a group rekeying for every membership change.

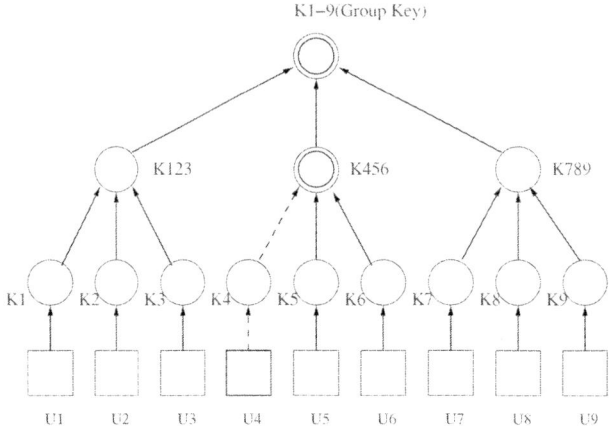

Fig. 1. An example of a logical key tree. The root key is the group key and a leaf key K_i is an individual key shared between the key server and a user U_i.

To reduce the frequency of group rekeying operations, researchers [12, 18] have proposed to use batched rekeying instead of individual rekeying. Batched rekeying can be done in a periodic fashion, so that the rekeying frequency is decoupled from the membership dynamics of a group and hence the processing overhead at the key server can be reduced. In addition, using batched rekeying can reduce the overall bandwidth consumption significantly. This is because every compromised key in the key tree due to member joins or leaves only needs to be updated once despite the number of joins and leaves. For example, in Fig. 1 when users U_4 and U_6 both depart from the group during the same rekeying period, $K_{1\text{-}9}$ and K_{456} only need to be changed once.

2.2   Stateless Protocols

A stateless rekeying protocol is one that allows a legitimate group member to obtain the group key from the received rekeying materials, despite the number of previous rekeying operations it has missed. The statelessness property is very desirable for multicast applications in which group members go offline frequently or experience high packet loss rates. A simplest stateless protocol, which we refer to as *flat-tree rekeying*, is one in which the key server encrypts the group key with the individual key of each member and then multicasts all the encryptions. Every member uses his/her individual key to decrypt one of the encryptions to obtain the group key. However, this scheme does not scale well with the group size.

Naor et al. [10] propose a more scalable stateless protocol called *subset-difference rekeying* (SDR). In SDR, the key server maintains a logical binary key tree and maps every member to a leaf node of the key tree. Let V_i and V_j

denote two vertices in the key tree and V_i is an ancestor of V_j. Let S_{ij} be a sub-set, then S_{ij} can be thought as the set of users in the sub-tree rooted at node V_i minus the set of users in the sub-tree rooted at node V_j. Each subset is asso-ciated with a unique key that is only known to the members belonging to this subset. During a rekey operation, the key server partitions the current members of the group into a *minimal* number of such subsets and then encrypts the new group key with the unique key of each subset separately. Hence, the number of encrypted keys to be distributed to the users is the same as the number of subsets the method generates. A group member only needs to receive exactly one encrypted key in every rekeying operation, which is the new group key en-crypted with the key of a subset to which he/she belongs. Note that the SDR scheme falls back to the flat-tree rekeying scheme when every subset contains one member.

In SDR the average number of subsets is $1.25r$ when there are r revoked users in the system. The communication complexity (i.e., the number of sub-sets) is independent of the group size N, which makes this scheme very scalable with the group size. However, because r always grows, the performance of SDR degrades with time. Note that in LKH the rekeying cost is determined by the group size N and the number of users being revoked since the *previous* rekey-ing operation; thus, LKH does not incur this performance degradation. This comparison suggests that SDR is suitable for applications in which the number of revoked users r is relatively small, particularly when $r \ll N$, whereas LKH seems to be preferable for applications that have a large number of revoked users. Chen and Dondeti [4] further show that SDR incurs smaller bandwidth overhead than LKH for rekeying of large batches. On the other hand, in SDR each user stores $0.5log^2N$ keys, whereas in LKH each user stores $logN$ keys. Halevy and Shamir [7] present a variant of SDR, which allows a tradeoff between user storage and communication cost.

2.3 Reliable Key Distribution

During a group rekeying based on a rekeying algorithm such as LKH, the key server first updates the compromised keys, including the group key and a fraction of KEKs; it then encrypts the updated keys with appropriate noncompromised KEKs; finally it multicasts all of the encrypted keys to the group. The purpose of a reliable key distribution protocol is to ensure that most of the group members receive the keys of interest in a *reliable* and *timely* fashion in the presence of packet losses in a network.

Sending every packet multiple times increases the probability that every user receives his/her keys of interest. However, Setia et al. [15] show that both the proactive FEC approach [18] and the WKA-BKR approach [15] incur a much smaller bandwidth overhead than the multi-send approach. A key server can employ proactive FEC, WKA-BKR, or their hybrid WFEC-BKR [20] for reliable distribution of the encrypted keys output from a key encoding algorithm such as LKH, OFT, or SDR. We below introduce in more detail the proactive FEC

and WKA-BKR approaches and another reliable key distribution protocol called
ELK [11].

Proactive FEC-based Key Delivery In the proactive FEC-based approach [18],
the key server packs the encrypted keys into packets of s_k keys. These packets are
divided into FEC blocks of k packets. The key server then generates $\lceil (\rho - 1)k \rceil$
parity packets for each block based on Reed Solomon Erasure (RSE) correcting
codes [8], where $\rho \geq 1$ is the proactivity factor. A user interested in the packets
from a certain block can recover all of the original packets in the block as long
as he/she receives any k out of $\lceil k\rho \rceil$ packets from the block. If a user does not
receive a packet that contains the encrypted keys of interest, but receives $t(t < k)$
packets from the block that contains this packet, he/she will ask the key server
for retransmission of $k - t$ new parity packets. The key server collects all of the
retransmission requests, and then for each block it generates and transmits the
maximum number of new parity packets required by users. The retransmission
phase continues until all of the users have successfully received their keys.

WKA-BKR The WKA-BKR scheme [15] uses the simple packet replication technique which sends every packet multiple times, but it exploits two properties of
logical key trees. First, the encrypted keys may have different replication weights,
depending on the number of users interested in them and the loss rates of these
users. For example, in LKH the keys closer to the root of the key tree are needed
by more members than are the keys less closer to the root, and in SDR some
subsets cover a larger number of users than other subsets. If a key is needed by
more users and these users have a higher loss rate, it should be given a higher
degree of replication so that most of these users will receive the key reliably and
timely. Hence, in this scheme the key server first determines the weight w_i for
each encrypted key K_i based on the users interested in that key. It then packs
the keys that have the same weight $\lfloor w_i \rfloor$ into the set of packets p_i. When broadcasting the packets, the key server sends packets in p_i $\lfloor w_i \rfloor$ times. This process
is called weighted key assignment (WKA). Second, during the retransmission
phase, since each user requesting a retransmission only needs a small fraction
of keys to decode the current group key, there is no need for the key server
to retransmit the entire packet sent in the previous round that contained the
requested key. Instead, the key server repackages the keys that need to be retransmitted into new packets before retransmitting them. This process is called
batched key retransmission (BKR). The WKA-BKR scheme has been shown to
have a lower bandwidth overhead than the other schemes in most scenarios.

ELK Unlike LKH, OFT, WKA-BKR, or proactive FEC, which provide either
key encoding or key distribution, ELK [11] integrates the key encoding phase
and the key distribution phase. In ELK, the key server maintains a binary key
tree that is similar to the one in OFT [2]. During a group rekeying, the key server
not only generates new keys but also derives some "hint" information from these
new keys. The new keys are distributed alone without replication, whereas the
hints are usually replicated and distributed together with data packets. When

a member loses a part of keys of his/her interest but receives the corresponding hint, he/she can recover the lost keys by brute-forcing a small key space (e.g., 16 bits). On the other hand, a revoked user must brute-force a larger key space (e.g., 44 bits) to recover the new group key. Thus, ELK achieves reliability and moderate security through a tradeoff of communication overhead with member computation. Note that ELK is also a stateful rekeying protocol.

2.4   Self-Healing Key Distribution

The reliable key delivery protocols discussed in Section 2.3 work well for scenarios where a user experiences random packet losses. However, a user might have to request multiple packet retransmissions until he/she finally receives the encrypted key of interest. Hence, there is no guarantee that he/she will receive the group key before the next group rekeying event. This is especially true for receivers that are experiencing intermittent burst packet losses. Another similar scenario arises when a user is offline (while still a member of the group) at the time of group rekeying. If the user receives data that was encrypted using a group key that he/she has not received, he/she will need to obtain that group key.

A self-healing key delivery protocol allows a user to obtain missing group keys on his/her own without requesting a retransmission from the key server. This is accomplished by combining information from the current key update broadcast with information received in previous key update broadcasts. In this section, we discuss two such schemes.

Zhu et al. [20] propose a self-healing key distribution scheme for SDR. The key idea is to bind the ability of a user to recover a previous group key to his/her membership duration. They use a one-way hash key chain such that revoked members and new members cannot collude to recover the previous keys they should not know. The communication overhead of this protocol is very small. To allow a legitimate member to recover s previous group keys, the additional encryptions to be distributed are at most $3s$.

Staddon et al [13] propose a self-healing key distribution protocol, using polynomial-based secret sharing techniques to achieve broadcast overhead of $O(t^2m)$ key sizes, where m is the number of sessions over which self-healing is possible and t is the maximal allowed number of revoked nodes in the m sessions. The protocol is also stateless in addition to self-healing. However, this scheme also has several disadvantages that may discourage its deployment for some applications. First, in this scheme, an application is pre-divided into m sessions, and the key server initiates a group rekeying at the beginning of each session. Thus, this scheme cannot be used for networks which demand immediate user revocation due to security requirement. Second, t, the maximal number of compromised users during these m sessions, must be pre-determined and must not be exceeded; otherwise, the security of this scheme will be broken. Third, the broadcast size becomes very large even for a small t and m.

2.5   Optimization

Moyer et al. [9] suggest to keep the key tree balanced in LKH, so that the rekey cost is fixed to be $log(N)$ for a group of size N. However, Selcuk et al. [14] show that it may be beneficial to use an unbalanced key tree in scenarios where group members have different membership durations. Their idea is to organize the key tree with respect to the compromise probabilities of members, in a spirit similar to data compression algorithms such as Huffman and Shannon-Fano coding. Basically, the key server places a member that is more likely to be revoked closer to the root of the key tree. If the key server knows in advance or can make a good guess of the leaving probability of each member, this probabilistic organization of the LKH tree could lead to a smaller communication overhead than that in a balanced-key-tree case. Banerjee and Bhattacharjee [1] show that organizing members in a key tree according to their topological locations could also be beneficial, if the multicast topology is known to the key server.

Zhu et al. [19] present two performance optimizations that are applicable to group key management schemes based on the use of logical key trees. These optimizations involve simple modifications to the algorithms and data structures used by the key server to maintain the logical key tree for a group. The first optimization exploits the temporal patterns of group member joins and leaves. The main idea is to split the logical key tree into two partitions—a short-term partition and a long-term partition. When a member joins the group, the key server initially places it in the short-term partition. If the member is still in the group after a certain time threshold, the key server then moves it from the short-term partition to the long-term partition. The second scheme exploits the loss probabilities of group members. The key server maintains multiple key trees and places members with similar loss rates into the same key tree. Thus, the key server separates keys needed by high loss members from those needed by low loss members when it packs keys into packets. These two optimizations are shown to reduce communication overhead over the onekey-tree scheme for applications with certain member characteristics.

Summary Generally speaking, stateful protocols are more bandwidth efficient than stateless protocols, whereas stateless protocols are more preferable in the presence of high packet loss or when members go offline frequently. Both stateful and stateless protocols need a key distribution mechanism to distribute a new group key to all members reliably and in a timely fashion, and a self-healing key distribution mechanism to allow a member to recover a certain number of previous group keys on his/her own.

3   New Research Directions

There are mainly two directions for future research, based on different principles.

First, we can try to design a "perfect" rekeying scheme. Ideally, a scalable group rekeying scheme should not only have the communication overhead close to that of a stateful protocol such as LKH or OFT but have embedded reliability and self-healing mechanisms as well. If we cannot design such a perfect scheme, the alternative seems to be to augment a stateful protocol with some lightweight plugins. For example, we have seen that a customized reliable multicast scheme such as Proactive FEC or WKA-BKR can be used to increase the reliability of key delivery for LKH, but we do not know if there is a self-healing scheme that can be used for LKH. We note that the scheme in [20] has been designed to add reliability and self-healing to the stateless protocol SDR.

Second, we can try to apply or adapt the current schemes to new applications such as ad hoc networks or sensor networks. The constrained energy, limited bandwidth and computational capability, high link error rates, mobility, and network partition as a result of node failure or movement make group rekeying much more challenging in these applications than in IP multicast. The SDR with added self-healing seems to work well under these constraints if the number of revoked nodes in the system is small. However, if the number of revoked nodes could be very large, none of these schemes would work efficiently. Research that investigates issues such as which scheme is the best in what networks with what kind of loss probability and mobility models would be very helpful. Carman, Kruus, and Matt have analyzed several approaches for key management and distribution in sensor networks [6], but these issues remain to be investigated in mobile ad hoc networks.

References

[1] S. Banerjee, B. Bhattacharjee. Scalable Secure Group Communication over IP Multicast. International Conference on Network Protocols (ICNP) 2001, Riverside, California, November 2001. 2, 8

[2] D. Balenson, D. McGrew, and A. Sherman. Key Management for Large Dynamic Groups: One-Way Function Trees and Amortized Initialization. IETF Internet draft (work in progress), August 2000. 1, 3, 6

[3] B. Briscoe. MARKS: Zero Side Effect Multicast Key Management Using Arbitrarily Revealed Key Sequences. In Proc. of First International Workshop on Networked Group Communication, NGC 1999. 1

[4] W.Chen and L.Dondeti. Performance comparison of stateful and stateless group rekeying algorithms. In Proc. of Fourth International Workshop on Networked Group Communication, NGC 2002. 5

[5] R. Canetti, J. Garay, G. Itkis, D. Micciancio, M. Naor, B. Pinkas. Multicast Security: A Taxonomy and Some Efficient Constructions. In Proc. of IEEE INFOCOM'99, March 1999. 1

[6] D. Carman, P. Kruus and B. Matt, Constraints and approaches for distributed sensor network security, NAI Labs Technical Report No. 00010, 2000. 9

[7] D. Halevy, A. Shamir. The LSD Broadcast Encryption Scheme. In Proc. of Advances in Cryptology - CRYPTO 2002. 2, 5

[8] A. Mcauley. Reliable Broadband Communications Using a Burst Erasure Correcting Code. In Proc. of ACM SIGCOMM'90, Philadelphia, PA, September 1990. 6

[9] M. Moyer, J. Rao and P. Rohatgi. Maintaining Balanced Key Trees for Secure Multicast. Internet Draft, draft-irtf-smug-key-tree-balance-00.txt, June 1999. 8

[10] D. Naor, M. Naor, and J. Lotspiech. Revocation and Tracing Schemes for Stateless Receivers. In Advances in Cryptology - CRYPTO 2001. Springer-Verlag Inc. LNCS 2139, 2001, 41-62. 1, 2, 4

[11] A. Perrig, D. Song, D. Tygar. ELK, a new protocol for efficient large-group key distribution. In Proc. of the IEEE Symposium on Security and Privacy 2001, Oakland, CA, May 2001. 1, 2, 3, 6

[12] S. Setia, S. Koussih, S. Jajodia, E. Harder. Kronos: A Scalable Group Re-Keying Approach for Secure Multicast. In Proc. of the IEEE Symposium on Security and Privacy, Oakland, CA, May 2000. 3, 4

[13] J. Staddon, S. Miner, M. Franklin, D. Balfanz, M. Malkin and D. Dean. Self-Healing Key Distribution with Revocation. In Proc. of the IEEE Symposium on Security and Privacy, Oakland, CA, May 2002. 2, 7

[14] A. Selcuk, C. McCubbin, D. Sidhu. Probabilistic Optimization of LKH-based Multicast Key Distribution Schemes. Draft-selcuk-probabilistic-lkh-01.txt, Internet Draft, January 2000. 2, 8

[15] S. Setia, S. Zhu and S. Jajodia. A Comparative Performance Analysis of Reliable Group Rekey Transport Protocols for Secure Multicast. In Performance Evaluation 49(1/4): 21-41 (2002), Special issue Proceedings of Performance 2002, Rome, Italy, September 2002. 2, 5, 6

[16] C. Wong, M. Gouda, S. Lam. Secure Group Communication Using Key Graphs. In Proc. of SIGCOMM 1998, Vancouver, British Columbia, 68-79. 1, 2, 3

[17] D. Wallner, E. Harder and R. Agee. Key Management for Multicast: Issues and Architecture. Internet Draft, draft-wallner-key-arch-01.txt, September 1998. 1, 2, 3

[18] Y. Yang, X. Li, X. Zhang and S. Lam. Reliable group rekeying: Design and Performance Analysis. In Proc. of ACM SIGCOMM 2001, San Diego, CA, USA, August 2001, 27-38. 2, 4, 5, 6

[19] S. Zhu, S. Setia, and S. Jajodia. Performance Optimizations for Group Key Management Schemes. In Proc. of the 23rd IEEE ICDCS 2003, Providence, RI, May 2003. 2, 8

[20] S. Zhu, S. Setia, and S. Jajodia. Adding Reliable and Self-Healing Key Distribution to the Subset Difference Group Rekeying Method for Secure Multicast. In Proc. of 5th International Workshop on Networked Group Communications (NGC 2003), Germany, September 2003. 2, 5, 7, 9

A Note on Modalities
for Nonconjunctive Global Predicates

Ajay D. Kshemkalyani

Computer Science Department, Univ. of Illinois at Chicago
Chicago, IL 60607, USA
ajayk@cs.uic.edu

Abstract. Global predicate detection is an important problem in distributed executions. A conjunctive predicate is one in which each conjunct is defined over variables local to a single process. Polynomial space and time algorithms exist for detecting conjunctive predicates under the *Possibly* and *Definitely* modalities, as well as under a richer class of fine-grained modalities based on the temporal interaction of intervals. However, it is much more difficult to detect nonconjunctive predicates under the various modalities because the entire state lattice may need to be examined. We examine the feasibility of detecting nonconjunctive predicates under the fine-grained temporal modalities using the interval-based approach. We gain some insightful observations into how nonconjunctive predicates can be decomposed, and into the relationships among the intervals (at different processes) in which the local variables have values that can satisfy the nonconjunctive predicate.

1 Introduction

Predicate detection in a distributed system is important for various purposes such as debugging, monitoring, synchronization, and industrial process control [5, 6, 7, 8, 10, 12, 14]. Marzullo et al. defined two modalities under which predicates can hold for a distributed execution [5, 12].

- *Possibly*(ϕ): There exists a consistent observation of the execution such that ϕ holds in a global state of the observation.
- *Definitely*(ϕ): For every consistent observation of the execution, there exists a global state of it in which ϕ holds.

For any predicate ϕ, three orthogonal relational possibilities hold: (i) *Definitely*(ϕ), (ii) ¬*Definitely*(ϕ) ∧ *Possibly*(ϕ), (iii) ¬*Possibly*(ϕ) [10]. The orthogonal set $\Re$ of 40 fine-grained temporal interactions between any pair of intervals [9] provides the basis for more expressive power than the *Possibly* and *Definitely* modalities for specifying any predicate. A mapping from the fine-grained interactions to the *Possibly* and *Definitely* modalities was given in [10].

A conjunctive predicate is of the form $\bigwedge_i \phi_i$, where ϕ_i is a predicate defined on variables local to process P_i, e.g., $x_i = 3 \land y_j > 20$, where x_i and y_j are local to P_i and P_j, resp.. Conjunctive predicates form an important class of

S. R. Das, S. K. Das (Eds.): IWDC 2003, LNCS 2918, pp. 11–25, 2003.

predicates. They have been studied in [2, 3, 6, 7, 8, 14]. The following result was shown in [10].

Theorem 1. *[10] For a conjunctive predicate* $\phi = \wedge_{i \in N} \phi_i$, *let* ϕ_i *denote the component of* ϕ *local to process* P_i *and let* N *denote the set of processes. The following results are implicitly qualified over a set of intervals, containing one interval from each process.*

- *Definitely(ϕ) holds if and only if* $\bigwedge_{(\forall i \in N)(\forall j \in N)} Definitely(\phi_i \wedge \phi_j)$
- *¬Definitely(ϕ) ∧ Possibly(ϕ) holds if and only if*
  - *$(\exists i \in N)(\exists j \in N)¬Definitely(\phi_i \wedge \phi_j) \bigwedge (\bigwedge_{(\forall i \in N)(\forall j \in N)} Possibly(\phi_i \wedge \phi_j))$*
- *¬Possibly(ϕ) holds if and only if* $(\exists i \in N)(\exists j \in N)¬Possibly(\phi_i \wedge \phi_j)$

By Theorem 1, given a conjunctive predicate ϕ defined on any number of processes, $Definitely(\phi)$ and $Possibly(\phi)$ can be expressed in terms of the $Possibly$ and $Definitely$ modalities on predicates defined over all pairs of processes. Also, from the results in [10], the $Possibly$ and $Definitely$ modalities on any general predicates defined over a pair of processes have been mapped into the set $\Re$ of 40 orthogonal modalities between appropriately identified intervals on the two processes. Therefore, $Definitely(\phi)$ and $Possibly(\phi)$, where (conjunctive predicate) ϕ is defined over any number of processes, can be expressed in terms of the fine-grained orthogonal set $\Re$ of modalities over predicates defined over all pairs of processes.

Polynomial space and time algorithms exist for detecting conjunctive predicates under the $Possibly$ and $Definitely$ modalities [5, 12]. We have also designed polynomial complexity algorithms, to detect not just $Possibly$ and $Definitely$, but also the exact fine-grained relation between each pair of processes when $Possibly$ and $Definitely$ are true [2]. Two factors make the problem of detecting conjunctive predicates, even under the harder fine-grained modalities, solvable with polynomial complexities. First, each process can locally determine the local *intervals* or *durations* in which the local predicate is true. Second, as a result of the first factor, each process can identify alternating intervals when the truth value of the local predicate alternates. *Nonconjunctive* predicates are of the form $(x_i + y_j = 5) \wedge (x_i + z_k = 10)$ and $x_i + y_j + z_k = 10$. It is much more difficult to detect nonconjunctive predicates because the above two factors do not hold and the entire state execution lattice may need to be examined, leading to exponential complexity [1, 5, 12, 13].

The use of the interval-based approach to specify and detect conjunctive predicates under the rich class of modalities $\Re$ [2] prompts us to examine the feasibility of a similar interval-based approach for nonconjunctive predicates.

The following can be seen from [10]. (1) For a pair of processes, the mapping from the fine-grained set of interactions $\Re$ to the $Possibly/Definitely$ classification depends only on the intervals, and is independent of the "predicate type". (2) For more than 2 processes, a result similar to Theorem 1 can hold for nonconjunctive predicates *provided* the intervals can be first identified appropriately. The semantics of the intervals need to be defined and identified carefully for nonconjunctive predicates because the entire state lattice may need to be examined. In this paper, we examine how nonconjunctive predicates can be specified/ detected under various fine-grained modalities using the interval-based approach. We introduce the notion of a *composite interval* in an attempt to identify intervals at each process. We gain some insightful observations into how nonconjunctive predicates can be decomposed, and into the relationships among the intervals (at different processes) in which the local variables have values that can satisfy the nonconjunctive predicate under various modalities.

Section 2 gives the execution model. Section 3 states the objectives precisely and introduces the solution approach. Section 4 summarizes the results for conjunctive predicates. Section 5 presents the interval-based analysis for nonconjunctive predicates, and makes the main observations. Section 6 discusses the notion of minimal intervals for nonconjunctive predicates when detecting the *Definitely* modality. Section 7 gives the conclusions.

2   System Model and Background

We assume an asynchronous distributed system in which $n = |N|$ processes communicate by reliable message passing. To model the system execution, let $\prec$ be an irreflexive partial ordering representing the causality relation on the event set E. E is partitioned into local executions at each process. Each E_i is a linearly ordered set of events executed by process P_i. An event e at P_i is denoted e_i. The causality relation on E is the transitive closure of the local ordering relation on each E_i and the ordering imposed by message send events and message receive events [11].

A *cut* C is a subset of E such that if $e_i \in C$ then $(\forall e'_i)e'_i \prec e_i \implies e'_i \in C$. A *consistent cut* is a downward-closed subset of E in $(E, \prec)$ and denotes an execution prefix. The system state after the events in a cut is a global state; if the cut is consistent, the corresponding system state is a consistent global state. Each total ordering of $(E, \prec)$ is a *linear extension* that represents the global time ordering of events and globally observed states in some equivalent (isomorphic) execution. The global time interleaving of events is different in each such isomorphic execution, but all these executions have the same partial order. The *state lattice* of an execution represents all possible global states that can occur. There is a bijective mapping between the set of all paths in the state lattice and the set of all linear extensions of the execution, for a given execution. We assume that only consistent global states are included in the state lattice.

An interval at process P_i is identified by the (totally ordered) subset of adjacent events of E_i that occur in that interval. An interval of interest at a process is a duration in which the local predicate is true (for conjunctive predicates) or in which the local values may potentially satisfy the global predicate (for nonconjunctive predicates). Henceforth, unless otherwise specified, references to intervals will implicitly be to intervals of interest. For a nonconjunctive predicate, the intervals need to be identified carefully, based on appropriate semantics.

3 Objectives and Approach

In this paper, we examine how nonconjunctive predicates can be specified and detected under the various fine-grained modalities using the interval-based approach. We ask the following questions.

1. What semantics can be used to identify the intervals at each process when ϕ is a nonconjunctive predicate, in order to apply the interval-based approach [4, 10]? An example predicate is $x_i - y_j < 10$.
2. How can Theorem 1 be extended to nonconjunctive predicates? When ϕ is defined over more than two processes, can it be reexpressed in terms of predicates over pairs of processes such that Theorem 1 can then be used in some form? Example predicates are $(x_i + y_j = 5) \wedge (x_i + z_k = 10)$ and $x_i + y_j + z_k = 10$.

Knowing the inherent difficulty in dealing with nonconjunctive predicates, we do not expect startling new results. Rather, we hope to make some insightful observations into how nonconjunctive predicates can be decomposed, and into the relationships among the intervals (at different processes) where the local variables have values that can satisfy the nonconjunctive predicate. By answering these questions, we can specify/detect not just the *Possibly* and *Definitely* modalities but also the fine-grained modalities of $\Re$, on nonconjunctive predicates.

Based on Theorem 1 for conjunctive predicates, we reexpress the definition of *Possibly* and *Definitely* in terms of intervals when $n > 2$ processes.

Definition 1. *Let $\mathcal{I}$ be a set of intervals, containing one interval per process, such that during these intervals, the local predicates are true, (or more generally, the local variables using which global predicate ϕ is defined have values that may satisfy ϕ).*

- *Possibly(ϕ): (For some set $\mathcal{I}$ of intervals,) there exists a linear extension of $(E, \prec)$ such that for each pair of intervals X and Y in $\mathcal{I}$, the string $[min(X), max(X)]$ overlaps with the string $[min(Y), max(Y)]$.*
- *Definitely(ϕ): (For some set $\mathcal{I}$ of intervals,) for every linear extension of $(E, \prec)$, for each pair of intervals X and Y in $\mathcal{I}$, the string $[min(X), max(X)]$ overlaps with the string $[min(Y), max(Y)]$.*

This alternate definition cannot be applied to nonconjunctive predicates unless the semantics of the interval at each process is known, and the intervals can be identified somehow. To understand how the differences between conjunctive and nonconjunctive predicates affect the identification of intervals at each process, we first identify and discuss the salient features that lead to such differences.

Method of decomposing predicate. For conjunctive predicates, the global predicate can be simply decomposed as the conjunct of the local predicates. For nonconjunctive predicates, there are several choices for decomposing the global predicate. The most natural form is the disjunctive normal form

(DNF) where each disjunct is a conjunct over variables at all processes, and each disjunct can be satisfied by a set of "subintervals", one per process. Even a simple predicate defined over two processes, like $x_i + y_j = 10$ or $x_i + y_j > 10$ can lead to an infinite number of disjuncts. As one of our goals is to determine how a predicate over more than two processes can be reexpressed in terms of predicates over pairs of processes at a time, this reexpression needs to be done carefully.

Adjacency of local intervals. For conjunctive predicates, each local interval can be locally determined to be the maximum duration in which the local predicate is *true*, irrespective of values of the variables used to define the local predicate. Two such intervals can never be adjacent (otherwise they would form a larger interval). For nonconjunctive predicates, two adjacent intervals may be such that the values of the local variables can potentially satisfy global predicate ϕ. For the predicate $x_i + y_j = 10$, $x_i = 3 \wedge y_j = 7$ and $x_i = 4 \wedge y_j = 6$ correspond to two different disjuncts in the DNF expression. As such, the intervals in which x_i is 3 and in which x_i is 4 may be adjacent.

Composite intervals. A *composite interval* is defined as an interval containing multiple adjoining intervals which we term as *subintervals*. In each of the subintervals, the local variables using which the global predicate is defined may or may not satisfy that predicate under the specified modality, depending on which subintervals at other processes these subintervals overlap.
Composite intervals are not relevant to conjunctive predicates because in any interval in which the local predicate is true, the varying values of local variables do not matter. For nonconjunctive predicates, composite predicates are relevant because different sets of subintervals, each set containing one subinterval from each process, can cause predicate ϕ (or some disjunct(s) of ϕ when it is expressed in DNF) to be *true* in the desired modality.

We analyze the above features for the 8 combinations obtained by the choices: (i) conjunctive or nonconjunctive ϕ, (ii) *Possibly* or *Definitely* modalities, and (iii) ϕ being defined on two or more than two processes. This gives a better insight into the use of the interval-based approach for detecting nonconjunctive predicates under the various modalities of $\Re$.

4   Conjunctive Predicates

There are 4 independent combinations to consider for conjunctive predicates.

- *Possibly*(ϕ), conjunctive predicate, $n = 2$. If *Possibly*(ϕ) holds, there is some linear extension in which some pair of intervals X and Y overlap.
- *Definitely*(ϕ), conjunctive predicate, $n = 2$. If *Definitely*(ϕ) holds, a common pair of intervals X and Y overlap for each linear extension. If X overlaps Y in only some linear extensions and a disjoint interval X' overlaps Y in only all the other linear extensions, then from the properties of linear extensions, there must exist a linear extension in which neither X nor X' overlaps Y [10].

– $Possibly(\phi)$, conjunctive predicate, $n > 2$. As shown in [10], there is some
linear extension in which some set of intervals, containing one interval per
process, overlap pairwise.
– $Definitely(\phi)$, conjunctive predicate, $n > 2$. As shown in [10], a common
set of intervals, containing one interval per process, overlap pairwise for each
linear extension.

For all these four cases, we have the following results using our approach.

1. **Decomposing global predicate.** In CNF, each conjunct is defined on
   variables local to a single process.
2. **Adjacency of local intervals.** Two intervals cannot be adjacent at a process.
3. **Composite intervals.** Not relevant because intervals cannot be adjacent
   at a process.

5   Nonconjunctive Predicates and Intervals

5.1   Nonconjunctive Predicates on Two Processes

$Possibly(\phi)$, nonconjunctive predicate, $n = 2$. For some linear extension,
the local interval at one process overlaps the local interval at the other process
and the values of the local variables in these intervals satisfy the predicate.

1. **Decomposing global predicate.** The predicate can be reexpressed in
   DNF. For the *Possibly* modality to be satisfied, it is sufficient if any one
   disjunct is satisfied in some linear extension.
2. **Adjacency of local intervals.** Intervals can be adjacent because when
   the predicate is expressed in DNF, two adjacent intervals at a process may
   satisfy two different disjuncts.
3. **Composite intervals.** Even though local intervals may be adjacent, it is not
   necessary to consider composite intervals because for the *Possibly* modality,
   any one interval (without subintervals) at a process may be used to satisfy
   the modality of the predicate.

Once the local intervals, one per process, are identified, the fine-grained modality
for *Possibly* can be determined by using the tests and mappings from [10].

$Definitely(\phi)$, nonconjunctive predicate, $n = 2$. The predicate can be
reexpressed in DNF. For the *Definitely* modality to be satisfied, it is sufficient
if for every linear extension, some disjunct is satisfied. As it is sufficient that
different disjuncts can be satisfied, it is necessary to consider composite intervals.
Examples 1(a,b): Examples of composite intervals are given in Figure 1. Consider the predicate $Definitely(x = y)$. Two slightly differing executions are
shown. The state lattices are labeled using event numbers at the two processes.
In Figure 1(a), $Definitely(x = 1 \wedge y = 1)$ is *false*. Also, $Definitely(x = 2 \wedge y = 2)$

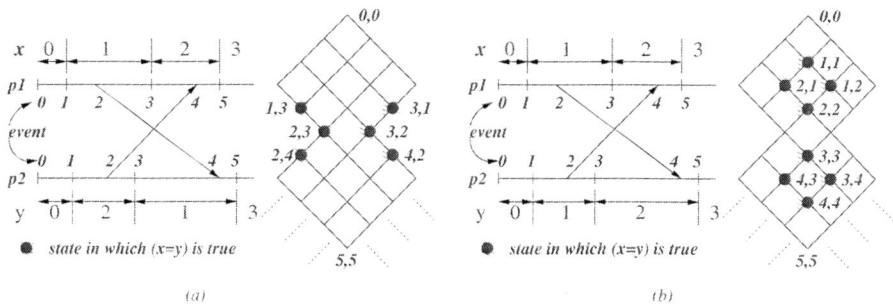

Fig. 1. Examples to show composite intervals for nonconjunctive predicates, for $n = 2$ processes

is *false*. However consider the following two composite intervals at P_1 and P_2, respectively. At P_1, the composite interval contains the subintervals when $x = 1$ and $x = 2$ in that order. At P_2, the composite interval contains the subintervals when $y = 2$ and $y = 1$ in that order. From the state lattice of this execution, observe that $Definitely(x = y)$ is *true* because exactly one of the following two states must hold in every execution: $x = 1 \wedge y = 1$, or $x = 2 \wedge y = 2$. The execution in Figure 1(b) differs only in that y first takes the value of 1 and then the value of 2. For this execution, the states in which $x = y$ are marked in the corresponding state lattice diagram. As can be observed from the lattice, $Definitely(x = y)$ is *false*. This example shows that the composite intervals need to be identified carefully after examining the state lattice.

Examples 2(a,b): Figure 2 shows two example executions with their corresponding state lattice diagrams. There are no messages exchanged in these executions. The state lattices are labeled so as to show only the values of the variables x and y. To detect $Definitely(x = y)$, observe that one has to seek recourse to examining the state lattice, and then determining the composite intervals. In Figure 2(a), $Definitely(x = y)$ is *true*. At P_1, the composite interval contains the subintervals when $x = 1, 2, 3, 4$. At P_2, the composite interval contains the subintervals when $y = 4, 3, 2, 1$. In Figure 2(b), $Definitely(x = y)$ is *false*.

These simple examples indicate that it is necessary to consider the state lattice to determine the composite intervals for the $Definitely$ modality.

1. **Decomposing global predicate**. The predicate can be reexpressed in DNF. For the $Definitely$ modality to be satisfied, it is sufficient if for every linear extension, some disjunct is satisfied.
2. **Adjacency of local intervals**. Intervals can be adjacent because when the predicate is expressed in DNF, two adjacent intervals at a process may satisfy two different disjuncts.

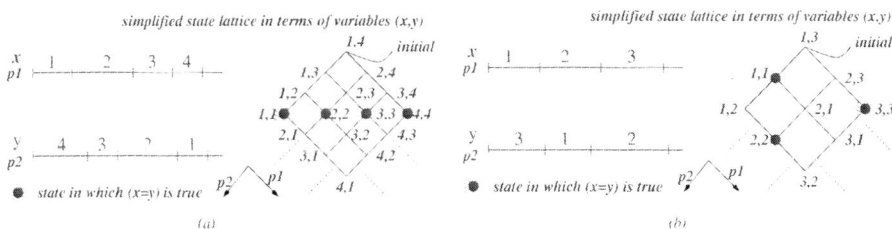

Fig. 2. Further examples for nonconjunctive predicates, for $n = 2$ processes

3. Composite intervals. As it is sufficient that different disjuncts can be satisfied in different linear extensions, it is necessary to consider composite intervals.

The identification of the semantics of the intervals for $Definitely(\phi)$ and the exact fine-grained relation(s) between the intervals will be dealt with in Section 6.

5.2   Nonconjunctive Predicates on More than Two Processes

The predicate can be reexpressed in DNF. For the $Definitely$ modality and $n = 2$, it was necessary to consider composite intervals, as also illustrated in Examples 1 and 2. For the $Definitely$ modality and $n > 2$, the same reasoning shows that because it is sufficient that different disjuncts can be satisfied in different linear extensions, therefore it is necessary to consider composite intervals. As the fine-grained modalities $\Re$ are expressed on processes pairwise, let us first try to adapt Definition 1 to nonconjunctive predicates. For $n > 2$, multiple pairs of processes need to be considered. In every linear extension, when the composite intervals pairwise overlap, only some combination(s) of the subintervals will actually overlap. For each pair of processes, there will exist multiple pairs of subintervals. However, if for each pair of processes, one of the pairs of subintervals overlap in each linear extension, that does not imply that there is a set of subintervals, one from each process, that will *collectively* overlap with each other in every linear extension. Thus for $Definitely(\phi)$, reexpressing ϕ as a conjunction of predicates on pairwise processes, the following *incorrect* Definition 2 would result.

Definition 2. *(Incorrect definition of $Definitely(\phi)$ when reexpressing ϕ as conjunction of predicates on pairwise processes:) (For some set $\mathcal{I}$ of composite intervals), for every linear extension, for each pair of composite intervals X and Y, one of the subinterval strings $[min(X'_i), max(X'_i)]$ overlaps with one of the subinterval strings $[min(Y'_j), max(Y'_j)]$.*

The above definition does not guarantee that if $[min(X'_i), max(X'_i)]$ overlaps with $[min(Y'_j), max(Y'_j)]$, and $[min(Y'_j), max(Y'_j)]$ overlaps with $[min(Z'_k),$-

$max(Z'_k)]$, then $[min(X'_i), max(X'_i)]$ overlaps with $[min(Z'_k), max(Z'_k)]$; instead, $[min(X''_i), max(X''_i)]$ may overlap with $[min(Z'_k), max(Z'_k)]$.

An argument analogous to the above shows that the same conclusion holds for $Possibly(\phi)$ if ϕ is reexpressed as a conjunction of predicates on pairwise processes, as in Definition 3.

Definition 3. *(Incorrect definition of $Possibly(\phi)$ when reexpressing ϕ as conjunction of predicates on pairwise processes:) (For some set $\mathcal{I}$ of composite intervals), for some linear extension, for each pair of composite intervals X and Y, one of the subinterval strings $[min(X'_i), max(X'_i)]$ overlaps with one of the subinterval strings $[min(Y'_j), max(Y'_j)]$.*

This unfortunately implies two negative results.

- The manner in which the global predicate is decomposed pairwise over processes must ensure that the same subinterval at any process is considered when determining overlaps with subintervals at other processes. Consider the predicate $x_i = y_j = z_k$. In DNF, this would be expressed as

$$(x_i = 1 \land y_j = 1 \land z_k = 1) \lor (x_i = 2 \land y_j = 2 \land z_k = 2) \lor \ldots\ldots$$

If this is reexpressed in terms of predicates over pairs of processes, as

$$((x_i = 1 \land y_j = 1) \land (x_i = 1 \land z_k = 1) \land (y_j = 1 \land z_k = 1)) \bigvee$$

$$((x_i = 2 \land y_j = 2) \land (x_i = 2 \land z_k = 2) \land (y_j = 2 \land z_k = 2)) \bigvee \ldots$$

then care must be taken to consider each disjunct separately.
Example 3(a): To detect ψ: $Definitely(x = y = z)$, if the predicate were reexpressed by splitting pairwise as per Definition 2, as

$$\psi' : Definitely(x = y) \land Definitely(y = z) \land Definitely(x = z),$$

then ψ' would be *true* in the execution of Figure 3(a). However, ψ' is not equivalent to ψ, and in this example, ψ is *false*.
Example 3(b): To detect ψ: $Possibly(x = y = z)$, if the predicate were reexpressed by splitting pairwise as per Definition 3, as

$$\psi' : Possibly(x = y) \land Possibly(y = z) \land Possibly(x = z),$$

then ψ' would be *true* in the execution of Figure 3(b). $Possibly(x = y = z = 1)$ is *false*; $Possibly(x = y = z = 2)$ is also *false*. ψ' is not equivalent to ψ. Definition 3 implicitly assumed that composite intervals with their subintervals are considered. But for the $Possibly$ modality, it is sufficient to ensure that *only one* subinterval from each process is considered, when checking for pairwise overlap (matching) between each pair of processes. However, to determine which subinterval from a process should be considered, one has to inevitably consider the global predicate (i.e., each disjunct of the global predicate) defined across all the processes, and hence the global state lattice. This suggests that the use of composite intervals with their subintervals is not useful for detecting $Possibly$ modality.

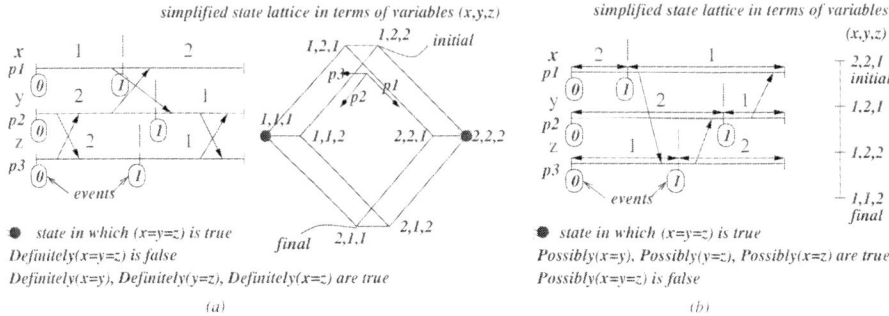

Fig. 3. Examples to show composite intervals, for $n = 3$ processes. Simplified state lattices that indicate only the values of the variables x, y, and z are shown for each execution. (a): $Definitely(x = y = z)$ is *false* although $Definitely(x = y)$, $Definitely(y = z)$, and $Definitely(x = z)$ are *true*. (b): $Possibly(x = y = z)$ is *false* although $Possibly(x = y)$, $Possibly(y = z)$, and $Possibly(x = z)$ are *true*

- For the $Definitely$ modality, one needs to inevitably examine the state lattice to determine whether multiple such sets of subintervals (one subinterval per process in each set) exist such that in each linear extension, there is mutual overlap between each pair of subintervals in at least one such set of subintervals. The subintervals at a process identify the composite interval at that process.

  Example 4: Figure 4 shows an example execution along with its state lattice diagram labeled using variable values. The predicate of interest here is $Definitely(x + y + z = 3)$. This is *true*, but can be determined only by examining the lattice and observing that each execution must necessarily pass through one of the 10 states marked in the lattice.

$Possibly(\phi)$, nonconjunctive predicate, $n > 2$.

1. **Decomposing global predicate**. The predicate can be reexpressed in DNF. For the $Possibly$ modality to be satisfied, it is sufficient if for *some* linear extension, *some* disjunct is satisfied. Consider each (instantiated) disjunct separately. If some disjunct is *true* in some global state, that is equivalent to there being pairwise overlap between the intervals, one per process, for which the local variable values in that state hold. $Possibly(\phi)$ is *true*, and the interval at each process that can satisfy that disjunct is identified by the duration in which the local variable values of that state persist.

   In this analysis, it is essential that ϕ must not be reexpressed as a conjunction of predicates on pairwise processes.

2. **Adjacency of local intervals**. Can be adjacent.

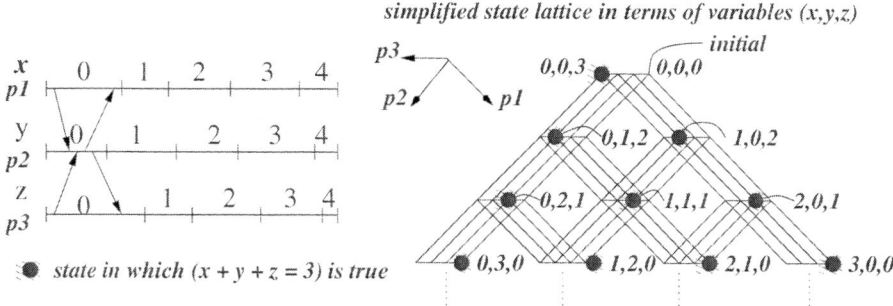

Fig. 4. Example to show composite intervals for nonconjunctive predicate $Definitely(x + y + z = 3)$, for $n = 3$ processes

3. Composite intervals. Not necessary and not useful.

Once the intervals (simple, not composite) are identified, Theorem 1 which was for conjunctive predicates can also be used along with the mapping from [10] to find the fine-grained modality for each pair of intervals corresponding to the disjunct that is satisfied.

$Definitely(\phi)$, nonconjunctive predicate, $n > 2$.

1. **Decomposing global predicate.** The predicate can be reexpressed in DNF. For the $Definitely$ modality to be satisfied, it is sufficient if for *every* linear extension, *some* disjunct is satisfied. This requires the state lattice to be examined.
2. **Adjacency of local intervals.** Can be adjacent.
3. **Composite intervals.** As in the case for $n = 2$, it is sufficient that different disjuncts can be satisfied in different linear extensions. Hence, it is necessary to consider composite intervals.

Section 6 shows how to identify the intervals and the fine-grained modalities between the intervals, for the $Definitely$ modality.

6   Minimal Intervals for $Def(\phi)$, for Nonconjunctive ϕ

For a conjunctive predicate, the interval at each process can span a duration in which local variables may take on multiple values, all of which satisfy the local predicate. More importantly, each interval can be determined locally. The following observation about the local intervals, one at each process, can be made. For each process P_i, let there be some contiguous range of events $D_i = [e_i^{x_i}, e_i^{x_i'}]$, such that the global predicate is *true* in at least one global state containing the local state after *each* event in D_i. Then the global predicate is necessarily *true* in *every* global state such that for each P_i, the local state is the state after some event in D_i. This is due to the conjunctive nature of the predicate. In terms of the state lattice, these states form a dense "convex" region. If $Definitely(\phi)$ is *true*, then every path in the lattice must pass through this region.

This observation can now be formally extended to nonconjunctive predicates.

Definition 4. (States in a sublattice in which ϕ is satisfied:)

1. *Let the set of states in the sublattice defined by the various D_i, one per process, be denoted $\prod_i D_i$.*
2. *Given the set of states $\prod_i D_i$ in a sublattice, the subset of states in which ϕ is true is denoted $S(\prod_i D_i)$. In general, $S(\prod_i D_i) \subseteq \prod_i D_i$ but for conjunctive predicates, $S(\prod_i D_i) = \prod_i D_i$.*

Thus, $S(\prod_i D_i)$ represents those states in the "convex" region in the lattice, where ϕ is *true*. Any equivalent execution must pass through at least one state in $S(\prod_i D_i)$ if $Definitely(\phi)$ is *true*.

Example 5(a): Consider the state lattice of an execution, shown in Figure 5(a). If the predicate ϕ is *true* in the states marked, and if ϕ is conjunctive, then ϕ is also necessarily *true* in all (consistent) states (v, w), where $1 \le v \le 7$ and $1 \le w \le 6$. Hence, $D_1 = [e_1^1, e_1^7]$ and $D_2 = [e_2^1, e_2^6]$. However, if ϕ is nonconjunctive, then the predicate can be *true* in only the 11 states $S(D_1 \times D_2)$ marked.

If $Definitely(\phi)$ holds for a nonconjunctive predicate, the identification of intervals at each process is useful for determining the fine-grained modality between each pair of processes. The sequence of events D_i identifies the interval at process P_i, provided that property DEF-SUBLATTICE is satisfied.

Property 1. (Property DEF-SUBLATTICE($\prod_i D_i$):) Every equivalent execution must pass through at least one state in which ϕ is *true*, among the states in $\prod_i D_i$.

The pairwise orthogonal relation between the intervals can be specified by considering the set of intervals identified by each D_i, one per process P_i. Observe that these intervals can be refined further to get more specific information on the fine-grained modalities that can hold when $Definitely(\phi)$ is *true*.

Example 5(a) contd.: ϕ is *true* in only 11 of the 42 states of the sublattice $D_1 \times D_2$. Each of these 11 states corresponds to a potentially different disjunct.

For each possible disjunct of ϕ, one of which will necessarily become *true* in each equivalent execution through $\prod_i D_i$, the pairwise orthogonal relations can be determined by executing DISJUNCT-FINE-GRAIN($S(\prod_i D_i)$), shown in Figure 6.

The smaller the set of states $S(\prod_i D_i)$, the smaller is $\mathcal{O}^*$ likely to be, and the more precise the information about the fine-grained modalities. *Minimal intervals are more useful because they can more accurately pinpoint the possible pairwise orthogonal relations.*

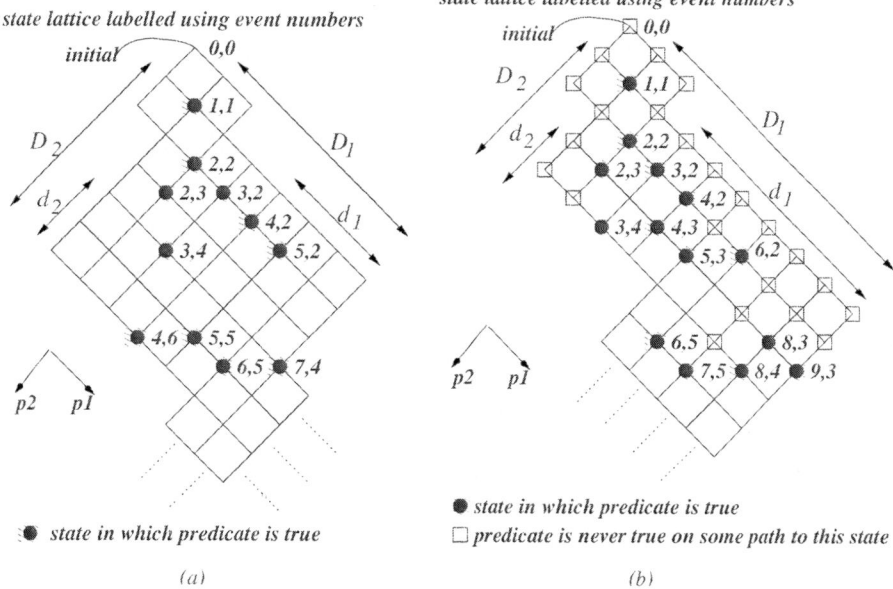

Fig. 5. Two examples to show minimal intervals for the *Definitely* modality for nonconjunctive predicates

Example 5(a) contd.: $D_1 = [e_1^1, e_1^7]$ and $D_2 = [e_2^1, e_2^6]$ are not *minimal*, in the sense that they have subintervals $d_1 = [e_1^4, e_1^7]$ and $d_2 = [e_2^4, e_2^6]$, resp., that satisfy the same property DEF-SUBLATTICE. Thus, every execution must pass through at least one state in which ϕ is *true* during the course of intervals d_1 and d_2.

Definition 5. (Minimal set of states in a sublattice, in which ϕ is satisfied:)

1. Let $S_{min} = S(\prod_i d_i)$ represent the set of states in which ϕ is true *in the minimal region satisfying* DEF-SUBLATTICE.
2. Let $S_{min}^* (\subseteq S_{min})$ be such that S_{min}^* represents the minimal subset of states in which ϕ is true *in the minimal region satisfying* DEF-SUBLATTICE.

Example 5(a) contd.: Every execution must pass through one state corresponding to $S_{min} = \{(e_1^3, e_2^4), (e_1^4, e_2^5), (e_1^5, e_2^5), (e_1^6, e_2^5), (e_1^7, e_2^4)\}$. There is a subset $S_{min}^* = \{(e_1^4, e_2^6), (e_1^5, e_2^5), (e_1^6, e_2^5), (e_1^7, e_2^4)\}$ of S_{min} such that every execution must pass through one state in S_{min}^*.

Example 5(b): In the example of Figure 5(b), ϕ is *true* in the 14 states shown. The example also shows each "reachable" state for which there is some path to that state such that the predicate is never *true* along that path.

DISJUNCT-FINE-GRAIN($S(\prod_i D_i)$)

1. For each global state ψ^k in $S(\prod_i D_i)$ (that satisfies predicate ϕ), do the following.
   (a) Identify I^k, the set of subintervals, one subinterval per process, corresponding to state ψ^k.
   (b) Using the tests and Theorem 1 [10], determine the orthogonal interaction $r_{i,j}^k$ $\in \Re$, between each pair of subintervals in I^k. Denote the set of such interactions as $\mathcal{O}^k$.
2. As at least one ψ^k state must occur in any execution, the possible interaction types are given by $\bigvee_k \mathcal{O}^*$, where $\mathcal{O}^* = \{ \mathcal{O}^k \mid \psi^k \in S(\prod_i D_i) \}$.

Fig. 6. Procedure DISJUNCT-FINE-GRAIN to identify possible pairwise fine-grained interaction types when $Definitely(\phi)$ holds

1. $D_1 = [e_1^1, e_1^9]$, $D_2 = [e_2^1, e_2^5]$, and $d_1 = [e_1^3, e_1^9]$, $d_2 = [e_2^3, e_2^5]$.
2. $S_{min} = \{(e_1^3, e_2^4), (e_1^4, e_2^3), (e_1^5, e_2^3), (e_1^6, e_2^5), (e_1^8, e_2^3), (e_1^7, e_2^5), (e_1^8, e_2^4), (e_1^9, e_2^3)\}$.
3. $S_{min}^* = \{ (e_1^3, e_2^4), (e_1^4, e_2^3), (e_1^5, e_2^3), (e_1^6, e_2^5), (e_1^7, e_2^5), (e_1^8, e_2^4), (e_1^9, e_2^3) \}$.

The fine-grained modalities for $Definitely(\phi)$ are given by DISJUNCT-FINE-GRAIN($S(\prod_i d_i)$). Finally, we remark that it is possible to devise an algorithm MIN-DISJUNCT-FINE-GRAIN($S(\prod_i d_i)$) that identifies S_{min}^*. Such an algorithm would also be exponential in the number of states examined.

7   Conclusions

This paper examined the feasibility of using intervals to determine the fine-grained modality of nonconjunctive predicates. Although it is known that detecting nonconjunctive predicates (under the $Possibly/Definitely$ modalities) involves examining an exponential number of states, nevertheless, this paper gave a better understanding of how the interval-based approach can be used to detect not just the $Possibly/Definitely$ but also the fine-grained modalities. Three parameters were used for the analysis – how to decompose the global predicate, adjacency of local intervals, and the use of composite intervals. The analysis showed how the interval-based approach can be used to determine nonconjunctive predicates under fine-grained modalities. This included how to identify the intervals for nonconjunctive predicates, and how Theorem 1 [10] based on intervals can be adapted/extended to nonconjunctive predicates.

Acknowledgements

This material is based upon work supported by the National Science Foundation under Grant No. CCR-9875617.

References

[1] Alagar, S., Venkatesan, S.: Techniques to Tackle State Explosion in Global Predicate Detection, IEEE Trans. Software Engg., 27(8): 704-714, 2001. 12

[2] Chandra, P., Kshemkalyani, A. D.: Algorithms for Detecting Global Predicates under Fine-Grained Modalities, Proc. ASIAN 2003, LNCS, Springer, 2003 (to appear). 12

[3] Chandra, P., Kshemkalyani, A. D.: Distributed Algorithm to Detect Strong Conjunctive Predicates, Information Processing Letters, 87(5): 243-249, September 2003. 12

[4] Chandra, P., Kshemkalyani, A. D.: Detection of Orthogonal Interval Relations, Proc. High-Performance Computing Conference, 323-333, LNCS 2552, Springer, 2002. 14

[5] Cooper, R., Marzullo, K.: Consistent Detection of Global Predicates, Proc. ACM/ONR Workshop on Parallel & Distributed Debugging, 163-173, May 1991. 11, 12

[6] Garg, V. K., Waldecker, B.: Detection of Weak Unstable Predicates in Distributed Programs, IEEE Trans. Parallel & Distributed Systems, 5(3):299-307, Mar. 1994. 11, 12

[7] Garg, V. K., Waldecker, B.: Detection of Strong Unstable Predicates in Distributed Programs, IEEE Trans. Parallel & Distributed Systems, 7(12):1323-1333, Dec. 1996. 11, 12

[8] Hurfin, M., Mizuno, M., Raynal, M., Singhal, M.: Efficient Distributed Detection of Conjunctions of Local Predicates, IEEE Trans. Software Engg., 24(8): 664-677, 1998. 11, 12

[9] Kshemkalyani, A. D.: Temporal Interactions of Intervals in Distributed Systems, Journal of Computer and System Sciences, 52(2): 287-298, April 1996. 11

[10] Kshemkalyani, A. D.: A Fine-Grained Modality Classification for Global Predicates, IEEE Trans. Parallel & Distributed Systems, 14(8): 807-816, August 2003. 11, 12, 14, 15, 16, 21, 24

[11] Lamport, L.: Time, Clocks, and the Ordering of Events in a Distributed System, Communications of the ACM, 21(7): 558-565, July 1978. 13

[12] Marzullo, K., Neiger, G.: Detection of Global State Predicates, Proc. 5th Workshop on Distributed Algorithms, LNCS 579, Springer-Verlag, 254-272, October 1991. 11, 12

[13] Stoller, S., Schneider, F.: Faster Possibility Detection by Combining Two Approaches, Proc. 9th Workshop on Distributed Algorithms, 318-332, LNCS 972, Springer-Verlag, 1995. 12

[14] Venkatesan, S., Dathan, B.: Testing and Debugging Distributed Programs Using Global Predicates, IEEE Trans. Software Engg., 21(2): 163-177, Feb. 1995. 11, 12

A Synchronous Self-stabilizing Minimal Domination Protocol in an Arbitrary Network Graph

Z. Xu, S. T. Hedetniemi, W. Goddard, and P. K. Srimani

Department of Computer Science, Clemson University
Clemson, South Carolina 29634 USA

Abstract. In this paper we propose a new self-stabilizing distributed algorithm for minimal domination protocol in an arbitrary network graph using the synchronous model; the proposed protocol is general in the sense that it can stabilize with every possible minimal dominating set of the graph.

1   Introduction

Most essential services for networked distributed systems (mobile or wired) involve maintaining a global predicate over the entire network (defined by some invariance relation on the global state of the network) by using local knowledge at each participating node. For example, a minimal spanning tree must be maintained to minimize latency and bandwidth requirements of multicast/broadcast messages or to implement echo-based distributed algorithms [8, 9, 1, 3]; a minimal dominating set must be maintained to optimize the number and the locations of the resource centers in a network [14]; an *(r,d)* configuration must be maintained in a network where various resources must be allocated but all nodes have a fixed capacity *r* [10]; a minimal coloring of the nodes must be maintained [15].

In this paper we propose a distributed algorithm to maintain a minimal dominating set in an arbitrary ad hoc network. Our algorithm is fault tolerant (reliable) in the sense that the algorithms can detect occasional link failures and/or new link creations in the network (e.g., due to mobility of the hosts) and can readjust the multi-cast tree. Our approach uses *self-stabilization* [5, 6] to design the fault-tolerant distributed algorithms.

The computation is performed in a distributed manner by using the mechanism of beacon messages. Mobile ad hoc networks use periodic beacon messages (also called ``keep alive" messages) to inform their neighbors of their continued presence. A node presumes that a neighboring node has moved away unless it receives its beacon message at stipulated interval. This beacon message provides an inexpensive way of periodically exchanging additional information between neighboring nodes. In our algorithm, a node takes action after receiving beacon messages (along with algorithm related information) from all the neighboring nodes. The most important contribution of the paper involves the analysis of the time complexity of the algorithms in terms of the number of *rounds* needed for the algorithm to stabilize after a topology change,

S. R. Das, S. K. Das (Eds.): IWDC 2003, LNCS 2918, pp. 26–32, 2003.

where a round is defined as a period of time in which each node in the system receives beacon messages from all its neighbors. The beacon messages provide information about its neighbor nodes synchronously (at specific time intervals). Thus, the synchronous paradigm used here to analyze the complexity of the self-stabilizing algorithms in ad hoc networks is very different from the traditional paradigm of an adversarial oracle used in proving the convergence and correctness of self-stabilizing distributed algorithms in general. Similar paradigms have been used in [2, 16, 12, 11].

2   System Model

We make the following assumptions about the system. A link-layer protocol at each node i maintains the identities of its neighbors in some list *neighbors(i)*. This data link protocol also resolves any contention for the shared medium by supporting logical links between neighbors and ensures that a message sent over a correct (or functioning) logical link is correctly received by the node at the other end. The logical links between two neighboring nodes are assumed to be bounded and FIFO. The link-layer protocol informs the upper layer of any creation/deletion of logical links using the *neighbor discovery protocol* described below.

Each node periodically (at intervals of t_b) broadcasts a *beacon* message. This forms the basis of the *neighbor discovery protocol*. When node i receives the beacon signal from node j which is not in its neighbors list *neighbors(i)*, it adds j to its neighbors list, thus establishing link (i,j). For each link (i,j), node i maintains a timer t_{ij} for each of its neighbors j. If node i does not receive a beacon signal from neighbor j in time t_b, it assumes that link (i,j) is no longer available and removes j from its neighbor set. Upon receiving a beacon signal from neighbor j, node i resets its appropriate timer.

When a node j sends a beacon message to any of its neighbors, say node i, it includes some additional information in the message that is used by node i to compute the cost of the link (i,j) as well as regarding the state of the node j, as used in the algorithm.

> **R1:** *if* $(| N[i] \cap S | = 1) \wedge (P(i) \notin N[i] \cap S)$
>   *then* $P(i):=j \in N[i] \cap S$
> **R2:** *if* $(| N[i] \cap S | = 0)$
>   *then if* $P(i) \neq i$
>   *then* $P(i) = i$
>   *else if* $(i < min(\{j \mid j \in N(i), P(j)=j\}))$
>   *then* $x(i):=1, P(i):=i$
>   *else* $P(i):=i$
> **R3:** *if* $(| N[i] \cap S | > 1) \wedge (x(i)=0) \wedge (P(i) \neq null)$
>   *then* $P(i):= null$
> **R4:** *if* $(| N[i] \cap S | > 1) \wedge (x(i)=1) \wedge (\forall j \in N(i)-S, P(j)=null)$
>   *then* $x(i):=0,$ *and* $P(i) = \begin{cases} S \cap N(i), \text{ if } |S \cap N(i)|=1 \\ null, \quad \text{otherwise} \end{cases}$

Fig. 1. Minimal Domination Protocol

The topology of the ad-hoc network is modeled by a (undirected) graph $G = (V,E)$, where V is the set of nodes and E is the set of links between neighboring nodes. We assume that the links between two adjacent nodes are always bidirectional. Since the nodes are mobile, the network topology changes with time. We assume that no node leaves the system and no new node joins the system; we also assume that transient link failures are handled by the link-layer protocol by using time-outs, retransmissions, and per-hop acknowledgments. Thus, the network graph has always the same node set but different edge sets. Further, we assume that the network topology remains connected. These assumptions hold in mobile ad hoc networks in which the movement of nodes is coordinated to ensure that the topology does not get disconnected. We also assume that each node is assigned a unique ID.

3   Minimal Dominating Set

Given an undirected graph $G= (V, E)$, a *dominating set* S is defined to be a subset of vertices such that $\forall v \in V\text{-}S$: N (i) $\cap S \neq \varnothing$ and a dominating set S is called *minimal* iff there does not exist another dominating set S' such that $S' \subset S$. Note that N (i) and N[i] respectively represent the open and the closed neighborhoods of the node i. In this section we present a synchronous model, self-stabilizing protocol for finding a minimal dominating set. Figure 1 shows the pseudo-code of the protocol that is executed at each node i, where $1\leq i \leq n$ (we assume nodes are numbered 1 through n). Each node i has two local variables: x(i), a Boolean flag (the value x(i)=0 indicates that $i \notin S$ while the value x(i) = 1 indicates that $i \in S$), and a pointer variable P(i). Note that *P(i)=i* indicates that node i is currently not dominated; P(i)=null indicates that the node i is dominated at least twice, i.e., $|N(i) \cap S| \geq 2$ and P(i)=j indicates that node i is dominated only by node j, i.e., $|N(i) \cap S| = \{j\}$.

4   Correctness Proof

Theorem 1: When the protocol stabilizes (terminates), the set $S = \{i \mid x(i) = 1\}$ gives a minimal dominating set of the network graph.

Proof: (a) Assume that the set S is not dominating when the protocol has terminated. Then, $\exists i \in V$, $S \cap N[i] = \phi$. Then, we have x(i)=0 and P(i)=i (by R2). Also, $\exists j \in S \cap N(i)$, $P(j)=j \wedge j < i$. If $|S \cap N[j]| = 1$, then by R1, node j has to move in the next round, a contradiction. If $|S \cap N[j]| > 1$, then by R3, j will move, again a contradiction. Hence, we have $|S \cap N[j]| = 0$. Since node j cannot make a move, then $\exists k \in S \cap N(j)$, $P(k)=k \wedge k < j$. Repeating the argument and noting there is only finitely many nodes, we reach a vertex v, where R2 will apply. This is contradiction to the hypothesis that the protocol is terminated. Therefore, S is dominating.

(b) Assume that S is dominating, but not minimal when the protocol terminates. Then $\exists i \in S$, such that $S' = S-\{i\}$ is a dominating set. Therefore, $\forall j \in N[i]$, $\exists k \in S-\{i\}$, $k \in N[j]$. If $x(j)=0$, then by R1 and R3, P(j) is either k or null. So R4 must apply on node i, a contradiction. Thus, S is minimal dominating set.   □

5   Convergence and Time Complexity

Lemma 1: If x(i) changes from 0 to 1 in a round, then any node $j \in N(i)$ cannot change its x(j) value in the same round

Proof: If x(j)=0, then by R2, only the smaller node between i and j is able to move; If x(j)=1, then by R2, node i can't move.   □

Lemma 2: If a node i changes its x-value from 0 to 1, then x(i) will not change again.

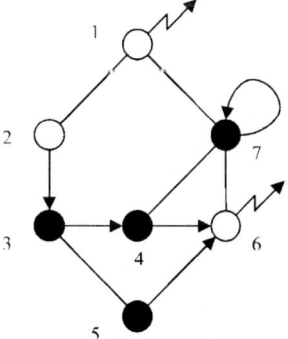

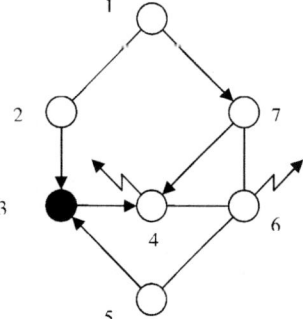

Fig. 2(a). An arbitrary initial state: nodes 1, 4, 5, and 7 are privileged to move

Fig. 2(b). The system state after one round; nodes 1, 3, 4, 6 and 7 privileged

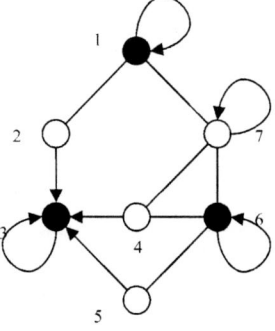

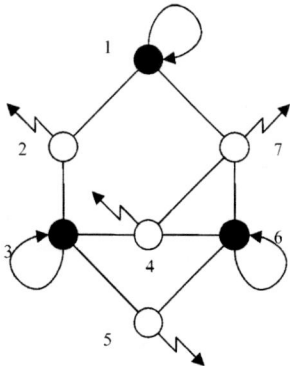

Fig. 2(c). System state after 2 rounds; nodes 1, 3, and 4 are privileged

Fig. 2(d). The system state after 3 rounds; protocol terminated

Proof: If x(i) changes from 0 to 1, then by R2, all nodes in the neighborhood N(i) should have x(j)=0. And by Lemma 1, they will stay at x(j)=0 after mode i moves. These neighbor nodes have at least one node i in the neighborhood that is in S (i.e., x(i)=1), so they won't go into S unless node i goes out. But by R4, i will never go out of S unless it is adjacent to some nodes in S.   ❐

Theorem 2: The protocol will terminate in at most 4n rounds starting from any arbitrary illegitimate state.

Proof: By lemma 2, each node will change its change x value at most twice. Therefore, there can be at most 2n changes of x values on all nodes in all the time. In our synchronous model, there can be at most 2n steps which contain changes of x value. Note that if there is no change in x value of any node in a round, then the move involves only changes in pointer values. Since the change in any pointer value is determined only by x values, there can be at most one step which does not contain changes of x value. Therefore, the upper bound of execution time in synchronized model is $4n + 1$ rounds.   ❐

Example: Consider a network graph of 7 nodes in Figure 2; each node is numbered from 1 through 7. We use shaded circle to represent node in S (i.e. x-value equals 1), and un-shaded circle to represent node not in S (i.e. x-value equals 0). The arrows on the edge represent the pointers from one node to another. If an arrow (i, j) is drawn, then P(i)=j. If a node i is pointing to it self, then P(i)=i. A zigzag arrow represents a null pointer.

6   Conclusion

We have proposed a self-stabilizing distributed algorithm for maintaining a minimal dominating set in a network graph. The algorithm (protocol) stabilizes in O(n) rounds in the synchronous model which can be used for ad hoc networks. The proposed protocol is general in the sense that the protocol can stabilize with every possible minimal dominating set of the graph. Consider an arbitrary minimal dominating set S of the graph G and then consider the following global system state: if i∈ S, then x(i)=1, if i∉ S, then x(i)=0, if (|S∩ N[i]| > 1) then P(i) = null, and if (|S∩ N[i]| = 1), then P(i)=j∈ S∩ N[i]. It is easy to see that the protocol is stable in this system state. The significance of this "completeness" of the protocol is that if the system is initialized to any minimal dominating set with the correct pointer settings, including minimal dominating sets that are not independent, then it will remain stable. While the protocol proposed in [17] can only stabilize with an independent set, the protocol proposed in this paper is capable of being stable with *any* minimal dominating set. The importance is that for some graphs no dominating set of smallest cardinality is independent. For example, consider the graph G formed by taking two stars $K_{\{1,n\}}$, and joining their centers by an edge. For this graph, the algorithm of [17] will stabilize with a set S having at least n+1 nodes, but the proposed algorithm can stabilize with a set S having the minimum cardinality of two (adjacent) nodes.

Clock Synchronization Algorithms and Scheduling Issues

Dhruba Basu[1,*] and Sasikumar Punnekkat[2,**]

[1] Department of Computer Science, Techno India Engineering College,
Salt Lake, Sector V, Kolkata, 700091, India
Dhruba_basu@hotmail.com
[2] Software Quality Assurance Division,Vikram Sarabhai Space Centre,
Thiruvanathapuram, 695022, India
sasi_punnekkat@hotmail.com

Abstract. Predictability is regarded as one of the key requirements of safety critical hard real-time systems. In cases where the functionality is achieved using distributed/multi processor systems, one widely studied topic is that of synchronization of drifting clocks for which several algorithms have been proposed in the literature. In this paper, we are mainly concerned about real-time systems in which the utilization of processors is very high. Our paper highlights some scheduling concerns, which arise while employing the well-known clock synchronization algorithms of Srikanth and Toueg[10]. We propose simple clock synchronization algorithms that can overcome such scheduling issues as a well as provide a means to solve the time discontinuity problem.

1 Introduction

Synchronization of individual clocks is an essential requirement for any distributed/multiprocessor systems. This becomes all the more important in case of hard real-time systems, where predictable performance is the foremost concern and one needs to preserve a total logical/temporal ordering of the tasks in the system. One application area is the onboard computer of aerospace and avionics systems, which typically contain multiple processors, in order to share the workload as well as to achieve separation of concerns related to different logical task groups. Normally these processors are replicated for fault tolerance, and they communicate through serial communication links. Typically processor utilization is very high, since time redundancy techniques are employed to overcome transient faults [1][7].

Each processor (alias node) in a distributed system has its own hardware clock, which normally drift due to temperature changes etc. In a distributed real-time system, the sensor data acquisition, which is carried out in one processor must always maintain a fixed timing relationship with the sensor data processing algorithms and typically

---

* This work was carried out while Dhruba Basu was with the Vikram Sarabhai Space Centre.
** An unabridged version of the paper can be obtained by e-mail from the authors.

[16]   S. Shukla, D. Rosenkrantz, and S. Ravi, "Developing self-stabilizing coloring algorithms via systematic randomization", In *Proceedings of Internátional. Workshop on Parallel Processing*, pages 668 – 673, 1994.

[17]   W. Goddard, S. T. Hedetniemi, D. P. Jacobs, and P. K. Srimani, "Self-Stabilizing Protocols for Maximal Matching and Maximal Independent Sets for Ad Hoc Networks", Proceedings of the Fifth IPDPS Workshop on Advances in Parallel and Distributed Computational Models, Nice, France, April 22-26, 2003.

Decomposable Algorithms
for Computing Minimum Spanning Tree

Ahmed Khedr and Raj Bhatnagar

Univ. of Cincinnati
Cincinnati OH 45221, USA
akhedr@ececs.uc.edu
Raj.Bhatnagar@uc.edu

Abstract. In the emerging networked environments computational
tasks are encountering situations in which the datasets relevant for
a computation exist in a number of geographically distributed databases,
connected by wide-area communication networks. A common constraint
in such situations of distributed data is that the databases cannot be
moved to other network sites due to security, size, privacy or data-
ownership considerations. For these situations we need algorithms that
can decompose themselves at run-time to suit the distribution of data.
In this paper we present two such self-decomposing algorithms for computing minimum spanning tree for a graph whose components are stored
across a number of geographically distributed databases. The algorithms
presented here range from low granularity to high granularity decompositions of the algorithms.

1   Introduction

Minimum Spanning Tree The minimum weight spanning tree (MST) problem is one of the most typical and well-known problems of combinatorial optimization. Minimum spanning tree algorithms for single and parallel processor
architectures have been studied and analyzed extensively [3, 5, 6, 8]. In the
last two decades efforts have been concentrated on developing faster algorithms
based either on more efficient data structures or multiple processors. Karger et
al. in [7] propose a linear expected-time algorithm. In many MST algorithms designed for closely coupled processor systems, a number of processors are assigned
to work on a shared dataset and they work together under the specification of
a parallel algorithm [1, 2, 9]. Gallager *et al* [4] present a distributed algorithm
that constructs the minimum weight spanning tree in a connected undirected
graph with distinct edge weights. A processor is assigned to each vertex of the
graph, knowing initially only the weights of edges incident on this vertex. The
total number of messages to be exchanged among the processors for a graph of N
vertices and E edges is at most $5N\ logN + 2E$ where each message contains at
most one edge weight and $log\ 8N$ bits.

Widely Distributed Knowledge In environments of geographically distributed data and knowledge, subgraphs may reside on different sites of a commu-

S. R. Das, S. K. Das (Eds.): IWDC 2003, LNCS 2918, pp. 33–44, 2003.

nication network. Communication time across the sites of a wide-area network is orders of magnitude larger than that for the processors of a set of closely-coupled processors. The closely coupled processors' model is, therefore, not applicable in this situation. The desired model is one in which the processor at each site performs large amount of computation with its own subgraph and periodically exchanges minimal amount of necessary information with processors at other sites to construct a correct global solution.

Algorithms for processing such distributed graph structures have received very little attention. The emerging networked knowledge infrastructure requires algorithms for such distributed data situations. The focus of research described in this paper is on such situations of widely distributed knowledge.

The situation addressed by us is very different from the one addressed by algorithms using Distributed Shared Memory (DSM). Two major differences from the DSM environments are: (i) In DSM model the frequent messages from one processor to the other are for reading and retrieving fine-grained data whereas in our model results of fairly large local computations are exchanged very infrequently among the processors. This is because the communication cost is overwhelming in the situations being modeled by us. (ii) The algorithms using DSM seek to minimize the number of participating processors whereas in our algorithms the number of participating databases, and hence the number of processors, are determined by the global problem to be solved and we seek to minimize the number of messages that need to be exchanged among them.

An Example Scenario We briefly describe here a real-life situation in which the decomposable algorithm would be very useful. Consider a number, say n, of airlines each of which has its own database about flight segments it operates. All the flight segments operated by an airline may be represented as a graph, and one such graph exists in the computer system operated by each airline. Some cities are served by more than one airline and some may be served by only one of them. Therefore, any two subgraphs may have some shared vertices.

Consider the situation in which these airlines decide to collaborate. Such collaborations do not mean a complete merger of their routes and operations but only sharing each other's flight segments to complete flight paths for customers. The *costs* along each edge of a subgraph may reflect price, seat availability, travel time etc. Therefore, it is not possible for an airline to share a fixed subgraph (along with edge costs) with the other airlines. It is required that the databases consult each other to determine the least cost paths whenever such paths are needed. The dynamic nature of costs along the edges of subgraphs requires frequent computation of least cost paths across the global graph formed by all the component subgraphs. An algorithm that can dynamically consult all the individual databases to infer the minimum cost paths is desirable, and our algorithms seek solutions for these and similar problems.

2   Algorithm Decomposition

The abstraction for the problem addressed by us in this paper can be described as follows. We assume that each of a number of networked data sites contains a subgraph. For each site, an implicit global graph exists and it is formed by combining the local subgraphs at all those data sites with which it can communicate. We present algorithms for computing minimum spanning tree in these implicit global graphs, without having to explicitly construct the global graph, and also with minimum communication among the data sites so as to preserve communication resources and data security at individual sites.

Our problem involves design of algorithms for computing minimum spanning tree in a graph that is stored as overlapping component subgraphs across various sites of a network. We operate under the constraint that data cannot be transferred between sites. The mathematical formulation of our problem can be described as follows:

Let us say a result R is obtained by applying a function F to a dataset $\mathcal{D}$ that is: $R = F(\mathcal{D})$

In the case of algorithms for distributed databases, $\mathcal{D}$ (the global graph) cannot be made explicit and is known only implicitly in terms of the explicit components $D_1, D_2, \ldots, D_n$. The implementation of F in the last equation can be redesigned by an equivalent formulation: $R = G(g_1(D_1), g_2(D_2), \ldots, g_n(D_n))$ That is, a local computation $g_i(D_i)$ is performed at $Site_i$ using the database D_i. The results of these local computations are aggregated using the operation G.

Granularity of Function Decomposition It may not be mathematically possible to decompose every function F into an equivalent set of G and g_i functions. In those cases, we consider a sequential algorithm for evaluating F which is a step-sequence composed of simpler functions as in the sequential steps of an algorithm to evaluate F. Thus, we represent F as: $F = f_1; f_2; \ldots; f_n$. Then, the functional decomposition can be found for each f_i step and the entire sequence of f_is is evaluated in a decomposed manner, one f_i at a time, to evaluate F. When the function F is directly decomposable into a set of G and g_i operators, we refer to it as a higher granularity decomposition. The more primitive the level of f_is employed to achieve the decomposition, the lower is the granularity of decomposition. Evaluation of each f_i may require exchange of some messages among the sites. Therefore, it is highly desirable to perform the decomposition of F at as high a level of granularity as possible. The algorithms we present in this paper range from low granularity to higher granularity. It is expected that the higher granularity decompositions would be performed with lower communication costs.

Distributed Representation The representation of a graph for the cases of closely coupled processors and the widely distributed sites can be compared as follows. In the first case a graph $\mathcal{G}$ with its vertices V and edges E is stored either on a single computer in the form of an adjacency table or in the shared memory addressable by each of the closely coupled processors. In a widely distributed

environment overlapping components of a graph may be stored on different sites
of a network. Figure 1 below shows an example graph that may be stored on
a single site.

Second part of Figure 1 shows two graphs that may be stored on two different
sites of a network. The two subgraphs together constitute the same graph as in
Figure 1 and share some of the vertices. This situation can be easily generalized
to n network sites, each containing a part of the complete graph.

One major assumption made by the *closely coupled processors* paradigm is
that the data can be efficiently moved around among the processors and can
be made available very quickly to a processor site that needs it. This assumption does not hold good in the knowledge environment where potentially huge
databases exist at geographically dispersed sites, connected by wide-area networks. From the perspective of elapsed time, it is very expensive to move data
around such networks. The alternative for us is to design algorithms that can
decompose themselves according to the distribution of data across the sites. Partial computations resulting from such decompositions can then be sent to the
sites where data resides and the results obtained then combined to complete the
desired computations. Since a different decomposition of an algorithm may be required for each type of distribution of data across sites, we would like to develop
algorithms that can decompose themselves into relevant partial computations to
suit the distribution of data. The algorithms should also minimize the communication cost among the sites. For example, we may want to determine the edges
that should be retained in a minimum cost spanning tree of the complete graph
by the network sites having to exchange minimum number of messages among
themselves. The vertices and edges of the two subgraphs of Figure 1(b) above
can be represented by the following two tables at their respective sites (Table 1
and Table 2).

Complexity Cost Models We choose the following three cost models for analyzing the complexity of our algorithms.

Cost Model #1 Communication Cost Only: In this cost model we count
the number of messages, N_m, that must be exchanged among all the participating

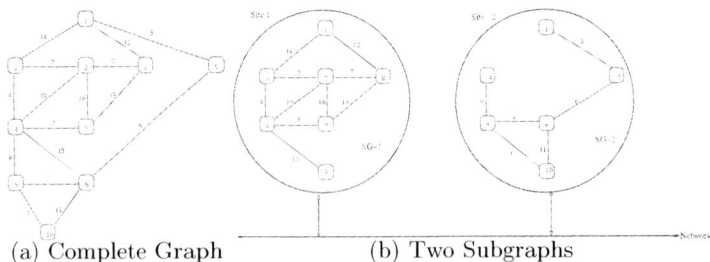

(a) Complete Graph   (b) Two Subgraphs

Fig. 1. An implicit complete graph is formed by multiple component graphs
stored at different network sites

Table 1. Database at Local Site 1

$vertex_1$	$vertex_2$	$weight$
1	8	12
4	3	4
4	6	15
1	3	14
3	2	5
2	8	7
2	7	16
7	4	8
4	2	10
8	7	13

Table 2. Database at Local Site 2

$vertex_1$	$vertex_2$	$weight$
9	1	3
9	6	6
6	5	2
5	4	9
5	10	1
10	6	11

sites in order to complete the execution of the algorithm. One message exchange includes one message sent by a site requesting some computation from another site and the reply message sent by the responding site. This cost model is relevant in situations where: (i) We need to analyze the number of messages exchanged because some critical resource, such as the battery power of sensors, is exhausted by the sending of messages; and (ii) Communication cost of messages is orders of magnitude larger than the cost of computations at local sites.

Cost Model #2 Communication + Computation Cost: In this model we examine a weighted sum of the number of messages exchanged and the number of local operations performed. If N_m messages are exchanged among the sites and a total of N_c computational units are performed at all the sites combined then the algorithm's cost is given by: $a * N_m + b * N_c$ where a and b are the weights representing the relative costs of exchanging a message and performing a local computational unit. This cost model is useful when the local computation time within a site is not negligible and must be included within the cost model. When the databases stored at the sites are huge, as in many scientific and data-mining applications, the time to execute a local computation may be comparable to the time taken for exchanging a message across a wide-area network.

Cost Model #3 Elapsed Time Cost: In this model we examine a weighted sum of the number of messages exchanged and the number of local operations performed, while accounting for parallel transmission of messages and simultaneous execution of local computations at the participating sites. If N_m messages are exchanged among the sites and a total of N_c computational units are performed at all the sites combined then the algorithm's cost is given by: $(a*N_m + b*N_c)/p$ where a and b are the weights representing the relative costs of exchanging a message and performing a local operation, and p is the average number of messages that can be exchanged in parallel. This cost model is useful when our criterion is the total elapsed time for executing the algorithm.

3   Low Granularity Decomposition

Here we start by adapting Prim's algorithm and decomposing each of its steps. Prim's algorithm works by building a tree which starts from an arbitrary root vertex r and grows until the tree spans all the vertices in V (V is the graph vertex set). At each step, an edge connecting a vertex in V_1 to a vertex in $V - V_1$ is add to the tree. This strategy is *greedy* since the tree is augmented at each step with an edge that contributes the minimum amount possible to the tree's weight. For a decomposable version of Prim's algorithm we implement each small step of the above algorithm in its decomposable form. Therefore, this is the lowest granularity decomposition. In this version we do not accumulate any work for a single node to perform and exchange results only after a bigger local result or summary has been generated and needs to be exchanged with neighboring subgraphs.

3.1   Algorithm Outline

Input: a connected graph with a vertex set V ($|V| = m$) and edge set E divided into n parts, $SG_1(V_1, E_1)$,$SG_2(V_2, E_2)$ ... $SG_n(V_n, E_n)$, each part residing at a different site and any two subgraphs may share some vertices, r The tree root (a vertex of V), W weight functions on $E_1, E_2, \ldots E_n$, $InTheTree[1 : m]$ global array initialized to F, and $Nearst[1 : m]$ global array initialized to ∞ except for, $Nearst[r]$ which is set to 0.

Output: $Parent[1 : m]$ global array of a minimum spanning tree.
        $MessageCounter = 0$
    $Parent[r] = 0$
    for step $= 1$ to $m - 1$ do
        select a vertex u that minimizes $Nearest[u]$ over
        all u such that $InTheTree[u] = F$
        set $InTheTree[u] = T$
        send message to every $site_i$
        select all vertices ($vertex_2$), $weight$ from SG_i where
        $vertex_1$=u, and there is an edge
        between $vertex_1$ and $vertex_2$
        $MessageCounter = MessageCounter + n$
        Results $= vertex_2$, $weight$ from SG_i
        while(R = NextElement(Result))
        where R is an object from local sites contains $vertex_2$
        and weight, and NextElement is a function take
        element by element from the local site results
            v = R.$vertex_2$ and w = R.weight
            if InTheTree[v] = F then
                Update $Nearest[v]$ and $Parent[v]$ for all
                $v \in V$ that are adjacent to u
                if $(Nearest[v]) > w)$

$$Nearest[v] = w$$
$$Parent[v] = u$$
      end if
     end if
    end while
  end for

3.2   Complexity Analysis

According to the cost models we defined in section 2 we derive below an expression for the number of messages that need to be exchanged for our algorithm dealing with the implicit set of tuples. Let us say: (i) There are a total number of m vertices in the whole graph and (ii) There are n relations, $D_1 \ldots D_n$, residing at n different network sites.

Cost Model #1: In this cost model we count the number of messages, N_m, that must be exchanged among all the participating sites in order to complete the execution of the algorithm. In this case the complexity can be explained as the following. We maintain a boolean array *InTheTree* to keep track of the vertices in/not in the tree. We then select the next edge uv to be added to the tree by maintaining an array $Nearest[1 : m]$ at the algorithm initiating site where $Nearest[v]$ is the distance from the vertex u to v in the tree and $Nearest[v] = \infty$ if v is not yet in the tree. In the algorithm we have $m - 1$ stages, and for each stage the querying site needs to send only one message to every other site. Therefore, a total of $n * (m - 1)$ messages will have to be exchanged to complete this algorithm.

Cost Model # 2: In this model we examine a weighted sum of the number of messages exchanged and the number of local operations performed. For each exchanged message, this algorithm performs one SQL query at the responding site. Therefore, the total cost for the algorithm will be $a * (m - 1) * n + b * (m - 1) * n = (a + b)n(m - 1)$ where a is the time taken to exchange a message and b is the average time taken to perform a simple SQL query at a site.

Cost Model # 3: In this model we examine a weighted sum of the number of messages exchanged and the number of local operations performed, while discounting the effects of messages and operations that can be executed in parallel, simultaneously at different sites. Therefore, the total cost for the algorithm will be $(a * (m - 1) * n + b * (m - 1) * n)/n = (a + b)(m - 1)$ where a is the time taken to exchange a message and b is the average time taken to perform an SQL query at a site.

4   Higher Granularity Algorithm for MST

The previous section take an algorithm for a single graph and decompose its steps to achieve a decomposable version. In this section we present an algorithm designed specially for the distributed environments. This algorithm works by taking each shared vertex at each local site as a fragment. We grow these fragments by adding the minimum outgoing edges this is described in **Local Computation** section below. Then we combine these fragments by the global computations described in **Global Computation** section below.

Fragment Definitions and Properties Before we present the second algorithm we introduce some definitions like fragment and outgoing edge.

Definition 1 A fragment f of an MST is a subtree of the MST.

Definition 2 An outgoing edge e of a fragment f is an edge that has one of its vertices in f and the other vertex out of f.

Property 1 Given a fragment of an MST, let e be the minimum weight outgoing edge of the fragment. Then joining e and its adjacent non-fragment vertex to the fragment yields another fragment of an MST. This result in the context of fragments has been shown in [4] and we briefly outline below their proof for this property.

Suppose the added edge e is not in the MST containing the original fragment. Then there is a cycle formed by e and some subset of the MST edges. At least one edge $x \neq e$ of this cycle is also an outgoing edge of the fragment, so that $W(x) \geq W(e)$ where $W(x)$ is the weight associated with the edge x. Thus, deleting x from the MST and adding e forms a new spanning tree which must be minimal if the original tree was minimal. Therefore, the original fragment including the outgoing edge e is a fragment of the MST.

We now present a self-decomposing version of an MST building algorithm for a graph that exists in parts across a network. The assumptions describing the situation are as follows.

- The complete global connected graph $\mathcal{G}$ is represented by n distinct but possibly overlapping subgraphs. Each subgraph is embedded in a database D_i residing at a different computer system (site) in the network. Each subgraph consists of vertices and edges such as: $SG_1(V_1, E_1), SG_2(V_2, E_2) \ldots SG_n(V_n, E_n)$ and $\mathcal{G} = \bigcup_i SG_i$.
- Each subgraph SG_i may share some vertices with other subgraphs. The set of vertices shared by SG_i and SG_j are: $V_{sh}(i,j) = V_i \bigcap V_j$.
- The set of all vertices that are shared by at least two subgraphs is: *Shared* $= \bigcup_{i,j} V_{sh}(i,j)$.
- Each edge e of a subgraph SG_i has a weight $W(e)$ associated with it.
- The weight of a tree in the graph is defined as the sum of the weights of the edges in the tree and our objective is to find the global tree of minimal weight, that is, the global MST.

4.1   Algorithm Outline

Data Structure A table called *Links* is maintained at the coordinator site and it stores information about candidate edges for linking various fragments to each other in order to complete the MST. This table has one row and one column for each shared vertex in the global graph. Each shared vertex name in the table is a representative of all the fragments, on all the different sites, that contain this shared vertex. Each entry in the table represents the site number and the weight of a potential edge that can link the row-fragment to the column-fragment. How these values are updated is given in the *Global Computations* paragraph below.

Local Computations The following steps are executed at each site on its locally resident subgraph. The goal is to create fragments within each subgraph at its local site.

- If the site has r shared vertices then each shared vertex is initialized as a separate fragment. That is, $f_i = v_i, i = 1, 2, \ldots, r$ and v_i is the $i^t h$ shared vertex.
- Expand fragments f_i's as follows
  - For each fragment f_i find its minimum weight outgoing edge e_i such that e_i does not lead to a vertex already included in some other fragment.
  - From among all the outgoing edges select the one with least weight. That is, $e_{min} = argmin_i W(e_i)$.
  - Add e_{min} to the fragment that selected this edge as its minimum outgoing edge.

Global Computations Each site, after completing its local computations, sends a message to the coordinator site informing it that the site is ready for global level coordination. After all the participating sites are ready the coordinator site performs the following steps to generate the global MST. In the final state each site knows the edges in its local subgraph that are included in the global MST. A complete MST is not generated at any single site.

- At each site, find the minimum-weight outgoing edge from each fragment to every other locally resident fragment. Each fragment is identified by the unique shared vertex contained in it. This step generates a tuple $<site-number, from-fragment, to-fragment, weight>$ for each fragment at the site. All these tuples are sent to the coordinator site.
- Update the cell *Links(row,column)* table with the above tuples received from all the sites as follows.
  - Include all the received tuples from the local sites in a set S where
    row = *from-fragment*; and
    column = to-fragment
  - From the set S select the tuple t with the minimum weight value.
  - Assign this tuple to *Links(row,column)*.
- Repeat the following steps till the *Links* table contains only one row and one column:

- Select in *Links* the cell with minimum value. This edge links its row-fragment (f_{row}) to its column-fragment (f_{col}) and is now selected to be in the global MST. This step also means that the fragments f_{row} and f_{col} are now merged into a new fragment, called $f_{row-col}$.
- Update the *Links* tuple as follows:
  * Create a new row and a new column in *Links* for the fragment $f_{row-col}$ and delete the rows and columns for the original two fragments f_{row} and f_{col}. The value for the cells in the new row and column can be determined as:
    · For every shared vertex d in the table if the minimum weight of outgoing edge from d to f_{row} is v_1 and to f_{col} is v_2 and $v_1 < v_2$ then the minimum weight of outgoing edge from d $f_{row-col}$ will be v_1.

End Algorithm.

4.2   Algorithm Correctness

An algorithm that grows and combines fragments until an MST for the graph is formed is given by Gallager et al. in [4]. Their algorithm is designed for a parallel processing environment in which one processor can be assigned to work at each vertex of the graph. Since our situation is different, we show below that our algorithm for growing and combining fragments will yield the global MST.

The set $Shared - Vertices$ contains all those vertices of the subgraphs that exist in more than one subgraph. We grow fragments starting from each member of $Shared - Vertices$ that exists at any site. That is, if k shared vertices occur in a subgraph at a site then the site will locally generate three distinct fragments each containing one and only one shared vertex. Then we combine these fragments by the global computations described in section 4.1 above.

We now show that (i) locally generated fragments are parts of the minimum spanning tree and (ii) the edges selected to combine local fragments are also part of the global MST.

Assertion: Each locally generated fragment is a part of the global MST.

Proof: The algorithm ensures that each fragment contains one and only one vertex that is shared with subgraphs at other sites. All the other vertices included in a fragment are completely local to the subgraph and the site on which the fragment resides. Therefore, in the global graph there does not exist an edge that connects a vertex in a local fragment to a vertex in another fragment that resides at some other site. This implies that edges between vertices of a fragment cannot be replaced by edges existing in some other subgraph residing at some other site.

Therefore, as long as a fragment is an MST of the local subgraph, it will be a fragment of the global MST.

Now we show that a fragment is part of the MST of its local subgraph i.e. each added minimum outgoing edge e is part of the MST.

Suppose the added edge e is not in the MST containing the original fragment. Then there is a cycle formed by e and some subset of the MST edges. At least one edge $x \neq e$ of this cycle is also an outgoing edge of the fragment, so that $W(x) \geq W(e)$ where $W(x)$ is the weight associated with the edge x. Thus, deleting x from the MST and adding e forms a new spanning tree which must be minimal if the original tree was minimal. Therefore, the original fragment including the outgoing edge e is a fragment of the MST.

Assertion: Edges selected to combine local fragments are part of the global MST.

Proof: Each fragment contains one and only one shared vertex. Therefore, joining two fragments by an edge is equivalent to bringing two shared vertices into one connected component of the MST. An edge that connects two shared vertices (and their respective fragments) may exist on more than one site. Our algorithm examines all possible sites on which the same two shared vertices (and their fragments) may be connected and selects the minimum-weight edge from among all the candidate edges on all the sites.

Now, if this does not result in the MST then there must exist an edge on a site that links two fragments such that it results in a lower weight MST. Suppose e_{ij} which is least weight minimum outgoing edge from fragment f_i to fragment f_j chosen from *Links* table is not the correct minimum outgoing edge from from fragment f_i to fragment f_j. Then there is another path from f_i to f_j with minimum weight less than the e_{ij} weight. Part of this path is an outgoing edge y from f_i with weight is less than e_{ij} which contradict with the assumption e_{ij} is the minimum outgoing edge from fragment f_i.

4.3   Complexity Analysis

We analyze below the MST finding algorithm for its complexity from the perspective of the three scenarios mentioned in section 2.

Cost Model # 1: In this cost model we count the number of messages, N_m, that must be exchanged among all the participating sites in order to complete the execution of the algorithm. In this case the complexity will be: (i) n exchanged messages to perform local fragments; and (ii) n exchanged messages to combine the local fragments. Then the total number of exchanged messages will be $2a * n$. where a is the time taken to exchange a message

Cost Model # 2: In this model we examine a weighted sum of the number of messages exchanged and the number of local operations performed. For each exchanged message, this algorithm performs one SQL query at the responding site. Therefore, the total cost for the algorithm will be $2(a + b) * n$ where a and b are the weights representing the relative costs of exchanging a message and performing a local operation.

Cost Model # 3: In this model we examine a weighted sum of the number of messages exchanged and the number of local operations performed, while discounting the effects of messages and operations that can be executed in parallel, simultaneously at different sites. Therefore, the total cost for the algorithm will be $2(a + b) * n/n = 2(a + b)$ where a and b are the weights representing the relative costs of exchanging a message and performing a local operation.

5 Conclusion

We have demonstrated that it is possible to execute graph operations for databases stored across wide-area networks. These algorithms are decomposable at run time depending on the set of shared vertices among the graph components stored at different network sites. We examined the complexity of our algorithms from the perspective of cost models that take into account the communication cost across the sites of a wide-area network. It turns out that these graph algorithms can be computed without too much communication overhead. These algorithms can perform more efficiently than having to bring all the data at one site while still preserving the privacy and ownership value of individual databases.

References

[1] Abdel-Wahab, H.; Stoica, I.; Sultan, F.; Wilson, K. *A Simple Algorithm for Computing Minimum Spanning Trees in the Internet* Information Sciences, Volume: 101, Issue: 1-2, September, 1997, pp. 47-69.

[2] Kenneth A. Berman and Jerome L. Paul *Fundamentals of Sequential and Parallel Algorithms* PWS Publishing Company 1997.

[3] Mcdiarmid, Colin; Johnson, Theodore; Stone, Harold S. *On finding a minimum spanning tree in a network with random weights* Random Structures and Algorithms, Volume: 10, Issue: 1-2, January - March 1997, pp. 187 - 204.

[4] Gallager R. G. and et.al A Distributed Algorithm for Minimum- Weight Spanning Trees ACM Transaction on programming and Systems, 5 66-77.

[5] Graham, R. L. and P. Hell.*ON the History of the Minimum Spanning Tree Problem* Annals Of the History of Computing,7 1985:43-57.

[6] toica, Ion; Sultan, Florin; *Keyes A Hyperbolic Model for Communication in Layered Parallel Processing Environments* journal of Parallel and Distributed Computing, Volume: 39.Issue: 1, November 25,1996, pp.29-45.

[7] Karger DR, Klein PN, Targan RE. *A randomized linear -time algorithm to find minimum spanning tree.* Journal of the Association for Computing Machinery ; 42/2:321-8.

[8] Nancy A. Lynch *Distributed Algorithms* Morgan Kaufman Publishers,Inc. San Francisco, California 1996.

[9] King, Valerie; Poon, Chung Keung; Ramachandran,Vijaya; Sinha, Santanu *An optimal EREW PRAM algorithm for minimum spanning tree verification* Information Processing Letters, Volume: 62, Issue: 3, May 14, 1997, pp. 153-159.

Clock Synchronization Algorithms and Scheduling Issues

Dhruba Basu[1,*] and Sasikumar Punnekkat[2,**]

[1] Department of Computer Science, Techno India Engineering College,
Salt Lake, Sector V, Kolkata, 700091, India
Dhruba_basu@hotmail.com
[2] Software Quality Assurance Division,Vikram Sarabhai Space Centre,
Thiruvanathapuram, 695022, India
sasi_punnekkat@hotmail.com

Abstract. Predictability is regarded as one of the key requirements of safety critical hard real-time systems. In cases where the functionality is achieved using distributed/multi processor systems, one widely studied topic is that of synchronization of drifting clocks for which several algorithms have been proposed in the literature. In this paper, we are mainly concerned about real-time systems in which the utilization of processors is very high. Our paper highlights some scheduling concerns, which arise while employing the well-known clock synchronization algorithms of Srikanth and Toueg[10]. We propose simple clock synchronization algorithms that can overcome such scheduling issues as a well as provide a means to solve the time discontinuity problem.

1   Introduction

Synchronization of individual clocks is an essential requirement for any distributed/multiprocessor systems. This becomes all the more important in case of hard real-time systems, where predictable performance is the foremost concern and one needs to preserve a total logical/temporal ordering of the tasks in the system. One application area is the onboard computer of aerospace and avionics systems, which typically contain multiple processors, in order to share the workload as well as to achieve separation of concerns related to different logical task groups. Normally these processors are replicated for fault tolerance, and they communicate through serial communication links. Typically processor utilization is very high, since time redundancy techniques are employed to overcome transient faults [1][7].

Each processor (alias node) in a distributed system has its own hardware clock, which normally drift due to temperature changes etc. In a distributed real-time system, the sensor data acquisition, which is carried out in one processor must always maintain a fixed timing relationship with the sensor data processing algorithms and typically

---

*   This work was carried out while Dhruba Basu was with the Vikram Sarabhai Space Centre.
**   An unabridged version of the paper can be obtained by e-mail from the authors.

S. R. Das, S. K. Das (Eds.): IWDC 2003, LNCS 2918, pp. 45-55, 2003.

performed in another processor. Similar requirements exist for processors, which command control system actuators. All these can be achieved only if all the clocks of the participating processors are synchronized. Similarly, the popular pre-run time scheduling algorithms [11] use the logical clocks of individual processors, and hence good clock synchronization is essential to guarantee precedence relations.

Clock synchronization is achieved normally by external and internal synchronizations. External synchronization tries to maintain processor clock within a specified deviation from an external time reference, using phased lock oscillators [9]. One needs to execute one pass of a synchronization algorithm initially or whenever a new processor joins in. Cesium or Rubidium based atomic clocks are stable but expensive. More over, in a distributed system, it may well be near impossible to use external clocks, as the cables interconnecting these clocks introduce distortions, which could be higher than the inherent stability of the external clock. Internal clock synchronization addresses these problems by using software algorithms, which ensure that, the logical clocks used by the processors in different nodes are consistent within limits, irrespective of the drift in the physical clocks. Our approach in this paper is from the viewpoint of a designer of distributed hard real-time systems (used for launch vehicle avionics applications), who does pre run-time scheduling [11] of tasks and has the responsibility to show a priori, that no deadlines are violated.

Furthermore, discrete clock synchronization algorithms can lead to the well-known time discontinuity problem in distributed real-time systems. Several methods have been suggested to solve this problem including hardware solutions using PLLs, continuous clock synchronisation [5] etc. Recently Ryu et al [8] have proposed dynamic remapping of timing constraints to achieve equi-continuity. Our work provides a simple, cheaper and computationally less demanding method, which can also solve the time discontinuity problem.

2   Previous Research

Lamport and Melliar-Smith [5] have presented *Interactive Convergence Algorithm* and *Interactive Consistency algorithm* and variants of them, which achieve synchronization even under Byzantine faults. Koptez and Ochsenreiter [4] have proposed Fault-Tolerant Averaging (FTA) algorithm in which the highest as well as lowest f clock values are discarded and the remaining clock values are averaged to resynchronize. Christian [2] has suggested a probabilistic clock synchronization algorithm, which measures the round trip delay by sending a clock read request to a remote node. The remote clock value is then estimated based on this measured delay. Srikanth and Toueg [10] have presented a clock synchronization algorithm, which can tolerate different types of failures including arbitrary ones. Synchronization is achieved by periodically adjusting the logical clocks forward. Here the values of the logical clocks can jump to a higher value and no logical clock value is moved backwards, thus avoiding the repetition of same clock values. The algorithm is optimal in the sense of accuracy as well as the number of faulty processors that can be tolerated.

3   Computational Model

We consider a distributed system having n processors (*nodes*). Each node is assumed to have its own hardware (physical) clock and a logical clock (essentially a software counter). Each node has a *logical clock process* (a highest priority task), which maintains its logical clock and synchronizes it with the other logical clocks periodically. All other processes executing on a particular node use the time obtained from the logical clock process of that node for scheduling, synchronization etc. In the context of this paper, we use the terms, *node* and *logical clock process* interchangeably. We assume the tasks are periodic, where each task is characterized by its period p_i, release time r_i, and deadline d_i [6][11].

We represent the real time with lower case letters such as t, and the logical time with upper case letters such as $C_i(t)$ which represents the value of the ith logical clock ($i=1,..,n$) at real time t. We assume the ability to initially synchronize the clocks. The logical clocks are resynchronized periodically, and the *resynchronization interval* is represented by P. Let $C_i^{K-1}(t)$ represent the logical clock value of the ith node at real-time t, after performing the $(K-1)^{th}$ round of resynchronization. When $C_i^{K-1}(t)$ reaches the value KP, then the Kth round of resynchronization commences.

3.1   Assumptions Regarding Clocks

We make the following assumptions, similar to those in [10], regarding clocks:

1. We assume a constant $D_{max} \geq 0$, which represents the maximum permitted deviation of any correct logical clock from real time during a resynchronization interval. In other words, $|C_i^{K-1}(t)-t| \leq D_{max}$ which means that, the difference between logical clock values of the fastest and slowest correct clocks can be at most $2D_{max}$
2. The drift (i.e., frequency difference) of correct or non-faulty clocks from real time is bounded by a known constant, $\rho \geq 0$ which in turn determines D_{max}

3.2   Assumptions on Faults and Fault Tolerance

For simplicity, we assume that each individual processor and communication medium is fault free. This could be a realistic assumption, if one builds each processor node with triple modular redundancy and includes sufficient redundancy in the communication links. Following [3], we also assume the processor nodes to be fail-silent. We initially assume that all the clocks are fault-free. Later on we relax this assumption and provide an algorithm, which takes into account faulty clocks.

A *faulty clock* is defined to be one, which drifts by more than D_{max} time units from real-time during a resynchronization interval. D_{max} is a known constraint value (derived from the allowed physical clock drift of ρ and the synchronization period, P [10]. By the term '*f* faults' we mean that, *f* among the clocks can be faulty.

It is significant to note that, D_{max} as defined in our paper is the deviation that a non-faulty clock can have during a resynchronization interval, from the real time. Since physical clocks have drift specifications such as '±x'ppm, this deviation can be positive as well as negative for different clocks. Thus the deviation between two good clocks can be $2D_{max}$ during a resynchronization interval.

3.3   Assumption on Communication Links

We assume that there exists a known bound on the maximum communication delay between processors, denoted by t_{del}. This comprises the end-to-end time required for the message preparation, transmission and reception. We further assume that the transmission delay t_{del} is equal for communication between all pairs of processors. At the outset, this may sound unreasonable. However, in many distributed hard real-time systems, the processors used are homogeneous and the main factor that causes variations in the t_{del} value is the variations in lengths of cables used to link different nodes. This problem can be overcome by measuring the length of the longest cable used to connect the nodes, and replace all the links with cables having this length.

3.4   Definitions

Ready Point: Similar to Srikanth and Toueg [10] each logical clock process broadcasts a '[READY K]' message whenever its logical clock value reaches an integral (say, K^{th}) multiple of the resynchronization interval, P.   These time points are referred to as *Ready Points* and they are $\{KP|\ \forall K=1,2,...\}$. The real time corresponding to the Ready Point of the ith process during the Kth round is denoted by $ready_i^K$.

Send_Sync Point: When based on a particular algorithm, a logical clock process knows that it is time to synchronize, it broadcasts a [SYNC K] message. This time point is referred to as a *Send_Sync Point* and is denoted by $send\_sync_i^K$, for the ith clock process during the Kth round.

Sync Point: The real time at which a logical clock process resynchronizes is referred to as a *Sync Point* and is denoted by $sync_i^K$, for the ith clock process in the Kth round.

For a typical logical clock process, these time points satisfy the property $ready_i^K \leq send\_sync_i^K \leq sync_i^K$. Note that, in our context the term broadcast is equivalent to sending a message to each of the other nodes.

4   Scheduling Problems
of Some Clock Synchronization Algorithms

We now look at the clock synchronization algorithms by Srikanth and Toueg[10]. The basic principle of these algorithms is that no logical clock is ever adjusted backwards, to prevent the occurrence of the same clock value twice. The main problem here is that logical clocks of slower processors are forced to jump forward. This could result in unfinished or unscheduled tasks in processors with high utilization, leading to unpredictability, which we have formalized as Lemmas 1 and 2. Lamport and Melliar-Smith [5] have briefly mentioned these problems in an earlier work.

Lemma 1. No process i scheduled with a release time $r_i \in (KP-D_{max}, KP+D_{max})$ can be guaranteed to execute.

Proof: By constructing a counter example (as given in full version of paper).   □

There can be an argument, that this problem can be circumvented by suitable tailoring of the scheduler to look for tasks with release points 'greater than or equal' to the current time. However this type of scheduling is not desirable in many hard real-time systems, where the release points need to be defined and adhered to very precisely.

Lemma 2. A process i with its deadline $d_i \in (KP-D_{max}, KP+D_{max})$ and is incomplete at time point $KP-D_{max}$, cannot be guaranteed.

Proof: It can be seen that, the logical time values can jump from the smallest value $KP-D_{max}$ to a value of $KP+D_{max}$. Any task with a deadline in between $(KP-D_{max}, KP+D_{max})$, cannot be assured to complete, by any scheduling algorithm.   □

Srikanth and Toueg[10] have also mentioned two shortcomings of their algorithms viz., a) ambiguity regarding which clock value to be supplied to an external process and b) the discontinuity in the logical clock. The remedy proposed in[10], based on [5], is to spread out each resynchronization adjustment over the next resynchronization period. This approach is cumbersome and introduces extra overheads for the clock maintenance process, which may not be admissible in hard real-time systems.

5   New Synchronization Algorithms

We now present two simple clock synchronization algorithms, whose fundamental concept is to synchronize the clocks with respect to the slowest among the correct clocks. Theoretically it is possible to have more than one clock with exactly the same drift, in which case, one should consider the term *slowest clock* as either referring to the group of slowest clocks or to any one among them. The major argument against synchronizing with the slowest clock could be that, it necessitates logical clocks of processors with faster physical clocks to jump backwards, thus repeating the same logical time values again. This problem can be solved, by modifying the processes, which maintain the logical clocks. We now present algorithm-A, which can overcome the above problem under the assumption of 'no faults'. This algorithm works by stopping the logical clock of a node (as if the processor is in a wait state), as soon as it reaches its Ready Point.

5.1   Algorithm under Assumption of 'No Faults'

Let nR_i^K represent the number of [READY K] messages (including its own) received by the logical clock process of ith node during K^{th} resynchronization round. There exist two separate processes, one for incrementing the logical clock value and another, which receives [READY K] messages and increments the counter nR_i^K. The following algorithm needs to be executed by every clock process i, during K^{th} round:

Algorithm-A:
begin
if $C_i^{K-1}(t)= KP$ **then**   // K^{th} Ready Point has reached
         Suspend logical clock incrementing process   // put in 'wait' mode
         Broadcast message [READY K]
             if $nR_i^K<n$ **then wait till** $(nR_i^K=n$)   // wait till getting all Ready msgs
             Broadcast message marked [SYNC K] // K^{th} Send_Sync Point has reached
             Load value t_{del} in watchdog counter W
             while $((W >0)$ **and** (no [SYNC K] message received)) **do**
                down-count W   // W could typically be a hardware down-counter
             enddo
             $C_i^K(t)= KP$ and restart the logical clock incrementing process
                                // K^{th} Sync Point has reached
         Ignore any [SYNC K] message received afterwards   // for the slowest node
         endif
end

Algorithm-A can be easily implemented and does not impose any constraints such as those given in Lemma-1 and Lemma-2, on the scheduler.

The observations given below follow directly from definitions and are noteworthy:

1. The node with the slowest clock (referred to as the *slowest node*) is the last one to reach its Ready Point during K^{th} resynchronization.
2. The Ready Point and Send_Sync Point can be the same for the slowest node, only if all the other nodes have already reached their Ready Points at least t_{del} time units before, so that the node having the slowest clock has already received all the other (*n-1*) [READY K] messages. It may be noted that, since t_{del} is very small compared to P, this simultaneous occurrence is more likely than not.
3. The simultaneous occurrence of Ready Point and Send_Sync Point can happen *only* for the slowest node.
4. The slowest node is the first to reach Sync Point, but only after *all* nodes reach their Send_Sync Points.

Theorem 1. The node with the slowest clock is the first to reach Send_Sync Point.

Proof: From definitions it can be seen that the slowest node is the last one in real-time to reach its Ready Point. By that time, all the other nodes would have already reached their Ready Points and would have sent their [READY K] messages.

Now, since slowest node is the last one to reach Ready Point,

$$ready_i^K < ready_n^K , \forall\ i \neq n$$

Adding $t_{del}>0$ to both sides preserves the inequality

$$\Rightarrow ready_i^K +t_{del} < ready_n^K +t_{del} , \forall\ i \neq n$$

Since it is true $\forall\ i \neq n$, it is true for the maximum among them as well.

$$\Rightarrow Max_{i \neq n} (ready_i^K +t_{del})< ready_n^K +t_{del} \qquad (1)$$

The LHS of the inequality (1) represents the time point at which all nodes except the slowest have sent their [READY K] messages, which have been received by the other nodes. Here two cases can arise:

Case 1. The slowest node reaches its Ready Point after $Max_{i\neq n}(ready_i^K +t_{del})$. In this case, as per algorithm-A, the Ready Point and the Send_Sync Point are the same time instant for the slowest node.

$$\Rightarrow send\_sync_n^K =ready_n^K$$

$$\Rightarrow send\_sync_n^K < ready_n^K +t_{del} \text{ (adding } t_{del}>0 \text{ to RHS)} \qquad (2)$$

RHS of the above inequality (2) represents the time point at which all the other nodes receive the [READY K] message sent by the slowest node. This time point is the Send_Sync Point for all the other nodes except the slowest node (i.e., $send\_sync_i^K =ready_n^K +t_{del} \forall i\neq n$).

$$\Rightarrow send\_sync_n^K < send\_sync_i^K , \forall i\neq n$$

Case 2. Slowest node reaches the Ready Point before $Max_{i\neq n} (ready_i^K +t_{del})$

In this case, the LHS of inequality (1) defines the time point, which is the Send_Sync Point of the slowest node.

$$\Rightarrow send\_sync_n^K < ready_n^K +t_{del} \qquad (3)$$

Using same argument as in case-1, we get

$$send\_sync_n^K < send\_sync_i^K, \forall i\neq n. \qquad \square$$

Lemma 3. The slowest node will not receive any [SYNC K] message during its t_{del} waiting period.

Proof: As per Theorem-1, the slowest node is the first one to reach send_sync point, and then sends [SYNC K] messages to all other nodes.

$$\text{i.e., } send\_sync_n^K < send\_sync_i^K , \forall i\neq n$$

So, [SYNC K] messages from any node other than the slowest can originate only at a later time point (in real-time) than $send\_sync_n^K$. This message can reach the slowest node only t_{del} time units after it is sent, since we have defined the communication delay t_{del} to be the same for all the links.

$$\text{i.e., } send\_sync_n^K +t_{del} < send\_sync_i^K +t_{del} , \forall i\neq n$$

This implies that the slowest node will not receive any [SYNC K] message during its waiting period ($send\_sync_n^K$, $send\_sync_n^K +t_{del}$). $\qquad \square$

Theorem 2. All the nodes reach the Sync Point at the same time, equal to t_{del} time units after Send_Sync Point of the slowest node. i.e.,

Proof: All the (n-1) nodes other than the slowest node, would have already reached their Ready Points earlier than the Ready Point of the slowest node and would be waiting for [READY K] message from the slowest node. By Theorem-1, the slowest node reaches its Send_Sync Point first and broadcasts the [SYNC K] message and all the waiting nodes receives this [SYNC K] message simultaneously at time point send_sync$_n^K$+t$_{del}$. The slowest node also reaches its Sync Point at the same time point, since it waits for t$_{del}$ time units after its Send_Sync Point during which it will not receive any [SYNC K] message as per Lemma-3. Hence all the nodes reach the Sync Points at the same time point i.e., sync$_i^K$= send_sync$_n^K$+t$_{del}$, $\forall i =1,..,n$.   □

In algorithm-A, the slowest node always reaches its Sync Point at the end of the time out t$_{del}$ and exits the **while** loop when W becomes zero, thus making the condition (W>0) false. Nodes, which reach their Ready Points at least t$_{del}$ before the slowest node reaches its Ready Point, reach their Sync Points at the entry of **while** loop itself, since the condition (no [SYNC K] message received) evaluates to false. Nodes that reach their Ready Points after ready$_{slowest}$- t$_{del}$, reach their Sync Points while executing the **while** loop, whenever the node receives a [SYNC K] message.

Algorithm-A satisfies all properties of a good synchronization algorithm and is an ideal one in the case of a 'fault-free' system, since the jump experienced in logical clock values is zero.

5.2   Algorithm under the Assumption of 'f faults'

We now discuss a few important results and then present our second algorithm, i.e., algorithm-B, which works in the presence of faults.

Theorem 3. If three clocks are arranged in ascending order of their Ready Points on the time axis, the Ready Point of a faulty clock cannot lie in between the Ready Points of two good clocks.

Proof: By contradiction, assuming that i+1 is a faulty clock interspersed between two good clocks i and i+2 and using basic definitions of good clocks.   □

Corollary 1. In an ensemble of *n* clocks arranged in ascending order of their Ready Points on time axis, i.e., ready$_{i+1}$≥ ready$_i$, faulty clocks lie either at the left end of time axis (indicating faster clocks), or at the right end (indicating slower clocks), or at both the ends, but can never lie interspersed between good clocks.

Proof: By induction on Theorem-3.   □

Lemma 4. If $n≥2f+1$, then the $(n-f)^{th}$ clock is guaranteed to be a correct clock.

Proof: Faulty clocks can be faster or slower as compared to the correct clocks.

   n≥2f+1 ⇒ n-f≥(2f+1)-f

   i.e., $n-f≥f+1$ ⇒ ready$_{(n-f)}$≥ ready$_{(f+1)}$

Considering cases where a) all the *f* faulty clocks are faster or b) all the *f* faulty clocks are slower, compared to the correct clocks, it can be proved that when $n>2f+1$, clocks with Ready Points, ready$_{f+1}$,..,ready$_{n-f}$ are all correct clocks, out of which the one with Ready Point $ready_{n-f}$ is the slowest among the set of 'obviously correct clocks'.

Lemma 5. The Ready Point of the slowest correct clock is within $2D_{max}$ from the Ready Point of the $(n-f)^{th}$ clock.

Proof: By Lemma-4, we have $(n-f)^{th}$ clock is a correct clock. In section 3.1, assumption a) states that logical clock values of any two correct clocks can be a most $2D_{max}$ apart in a resynchronization interval. So, the Ready Point of slowest correct clock will be within $2D_{max}$ from the Ready Point of $(n-f)^{th}$ clock.   □

Consider the extreme case, where f faulty clocks are all faster, as compared to the correct clocks. Here, when a node receives *(n-f)* [READY K] messages, it includes f messages from the faster incorrect clocks as well. However, at this time point, there are f slower correct clocks, which are yet to reach their Ready Points. If we synchronize based on the $(n-f)^{th}$ clock, though we are synchronizing with a correct clock, the f slower correct clocks are forced to jump forward, a situation which we want to avoid. Therefore, the system should be made to wait till all the correct clocks reach their Ready Points. Since the correct clocks can differ by at most $2D_{max}$ only, we can surely say that all the correct clocks will reach their Ready Points by $ready_{n-f} + 2D_{max}$.

Suppose, some of the f faulty clocks, say f_1 are faster, and the remaining $f - f_1$ are slower, as compared to the correct clocks. In this case, the $(n-f)^{th}$ clock is faster than the slowest correct clock, viz., $(n- f_1)^{th}$ clock, but still all correct clocks reach their Ready Points by $ready_{n-f} + 2D_{max}$.

A generalized algorithm which synchronizes with respect to the $(n-f)^{th}$ slowest clock, in order to tolerate up to f faults, is given below:

Algorithm-B:
  begin
        if $C_i^{K-1}(t)=KP$ **then**   // K^{th} Ready Point has reached
        Suspend logical clock incrementing process // put in 'wait' mode
            Broadcast message [READY K]
            if $nR_i^K < n-f$ **then wait till** $(nR_i^K = n-f)$ // wait till getting (n-f) Ready messages
            if $nR_i^K = n-f$ **then**
                wait for $2D_{max}$ time units
                broadcast message marked [SYNC K] // K^{th} Send_Sync Point reached
                Load value t_{del} in watchdog counter W
                while $((W > 0)$ **and** (no [SYNC K] message received)) **do**
                    down-count W
                enddo
            else
                // for slower correct nodes, which gets more than (n-f) [READY K]
                // messages by the time they reach Ready Points, i.e. $nR_i^K > n-f$
                wait till [SYNC K] message received
            endif
            $C_i^K(t)=KP$ and restart the logical clock incrementing process
                // ignore any [SYNC K] msg received after restarting logical clock
        endif
    end

The significant observations with respect to Algorithm-B are as follow:

a)   The nodes on reaching their Ready Points check to see if their respective nR_i^K is equal to $(n\text{-}f)$. This is because the $(n\text{-}f)^{th}$ clock on the time line is guaranteed to be a correct clock. Note that the $(n\text{-}f)^{th}$ clock is generally not the slowest clock.

b)   The Send_Sync Point is delayed by $2D_{max}$ time units from the point where the $(n\text{-}f)^{th}$ node receives $(n\text{-}f)$ [READY K] messages.

c)   The delay of $2D_{max}$ time units is to ensure that synchronization is with respect to the slowest correct clock in all situations.

d)   When a clock reaches its Ready Point after ready$_{n\text{-}f}$, but within the $2D_{max}$ waiting time of the $(n\text{-}f)^{th}$ slowest clock, there are 2 possible scenarios:

   1)   If this Ready Point happens to be within t_{del} after the Ready Point of the $(n\text{-}f)^{th}$ clock, i.e., if nR_i^K =n-f, then this clock sends a [SYNC K] message. However, this [SYNC K] message will be only after the [SYNC K] message of the $(n\text{-}f)^{th}$ clock and will therefore be ignored.

   2)   If this Ready Point happens to be after t_{del} time units from the Ready Point of the $(n\text{-}f)^{th}$ clock, i.e., if $nR_i^K>(n\text{-}f)$, then this clock waits for the first [SYNC K] message to resynchronize. This is enforced in algorithm-B by **wait** in the last **else** condition.

Note that, we are concentrating on faults in the clocks only, but not on the faults in the processors. If we want to deal with both, then we require $3f+1$ nodes to tolerate f faults in either processors (which are assumed to be fail-silent) or clocks.

6   Conclusions

We have highlighted a few practical problems of well-known clock synchronization algorithms, which could be against the very basic requirements such as predictability and preserving precedence constraints among tasks, of safety critical hard real-time systems. We have proposed new algorithms for clock synchronization to overcome these problems by synchronizing with respect to the slowest correct clock. These algorithms can also be used as a cost effective solution to the time discontinuity problem in distributed systems. A minor criticism against our algorithms could be that the overall utilization of the system decreases. Clearly there is a tradeoff between predictability and optimality of utilization/accuracy, but in the class of systems we are dealing with, the thrust is on predictable solutions rather than optimal ones, thus favoring our new algorithms.

References

[1]   T. Anderson and J. C. Knight, "A Framework for Software Fault Tolerance in Real-Time Systems", IEEE Transactions on Software Engineering, Vol. SE-9, No.3, 355-364, 1983.

[2]   F. Cristian, "Probabilistic Clock Synchronization", Distributed Computing, No.3, pp 146-158, Springer-Verlag, 1989.

[3]   H. Kopetz and G. Grünsteidl, "TTP- A Protocol for Fault-Tolerant Real-Time Systems", IEEE Computer, January 1994.

[4]   H. Kopetz and W. Ochsenreiter, "Clock Synchronization in Distributed Real-Time Systems", IEEE Transactions on Computers, Vol. 36, No. 8, pp 933-940, August 1987.

[5]   L. Lamport and P.M. Melliar-Smith, "Synchronizing Clocks in the Presence of Faults", Journal of the ACM, Vol. 32, No. 1, pp 52-78, January 1985.

[6]   C.L. Liu and J. Layland, "Scheduling Algorithms for Multiprogramming in a Hard Real-Time Environment", JACM Vol. 20, No. 1, pp46-61, 1973.

[7]   B. Randell, "System Structure for Software Fault Tolerance", IEEE Transactions on Software Engineering, Vol. SE-1, No.2, 220-232, 1975.

[8]   M.Ryu, J. Park and S.Hong, "Timing Constraint Remapping to Achieve Time Equi-Continuity in Distributed Real-Time Systems", IEEE Transactions on Computers, Vol 50, No. 12, pp 1310-1320, December 2001.

[9]   K. G. Shin and P. Ramanathan, "Transmission Delays in Hardware Clocks Synchronization", IEEE Transactions on Computers, Vol 37, No.11, pp 1465-1467, November 1988

[10] T.K. Srikanth and S. Toueg, "Optimal Clock Synchronization", Journal of the Association for Computing Machinery, Vol. 34, No. 3, pp 626-645, July 1987.

[11] J. Xu and D. L. Parnas, "On Satisfying Timing Constraints in Hard-Real-Time Systems", IEEE Transactions on Software Engineering, Vol. 19, No.1, pp 70-84, January 1993.

Estimating Checkpointing, Rollback and Recovery Overheads

Partha Sarathi Mandal and Krishnendu Mukhopadhyaya

Advanced Computing and Microelectronics Unit, Indian Statistical Institute
203, B T Road, Kolkata 700108, India
{partha_r,krishnendu}@isical.ac.in

Abstract. There are several schemes for checkpointing and rollback recovery. In this paper, we analyze some such schemes under a stochastic model. We have found expressions for average cost of checkpointing, rollback recovery, message logging and piggybacking with application messages in synchronous as well as asynchronous checkpointing. For quasi-synchronous checkpointing we show that in a system with n processes, the upper bound and lower bound of selective message logging are $O(n^2)$ and $O(n)$ respectively.

1 Introduction

Checkpointing and rollback recovery is a well known method to achieve fault tolerance in distributed computing systems. In case of a fault, the system can rollback to a consistent global state, and resume computation without requiring additional efforts from the programmer. A *checkpoint* is a snapshot of the current state of a process. It saves enough information in non-volatile stable storage such that, if the contents of the volatile storage are lost due to process failure, one can reconstruct the process state from the saved information. If the processes communicate with each other through messages, rolling back a process may cause some inconsistency. In the time since its last checkpoint, a process may have sent some messages. If it is rolled back and restarted from the point of its last checkpoint, it may create *orphan messages*, i.e., messages whose receive events are recorded in the states of the destination processes but the send events are lost. Similarly, messages received during the rolled back period, may also cause problem. Their sending processes will have no idea that these messages are to be sent again. Such messages, whose send events are recorded in the state of the sender process but the receive events are lost, are called *missing messages*.

A set of checkpoints, with one checkpoint for every process, is said to be *Consistent Global checkpointing State (CGS)*, if it does not contain any orphan message or missing message. However, generation of missing messages may be acceptable, if messages are logged by sender.

Checkpointing algorithms may be classified into three broad categories: (a) *Synchronous*, (b) *Asynchronous* and (c) *Quasi-synchronous* [5]. In asynchronous checkpointing [7, 22] each process takes checkpoints independently. In case of

S. R. Das, S. K. Das (Eds.): IWDC 2003, LNCS 2918, pp. 56–65, 2003.

a failure, after recovery, a CGS is found among the existing checkpoints and the system restarts from there. Here, finding a CGS is quite tricky. The choice of checkpoints for the different processes is influenced by their mutual causal dependencies. The common approach is to use *rollback-dependent graph* or *checkpoint graph* [6, 8, 11].

If all the processes take checkpoints at the same time instant, the set of checkpoints would be consistent. But since globally synchronized clocks are very difficult to implement, processes may take checkpoints within an interval. In synchronous checkpointing [1, 5, 7, 8, 9, 10, 11, 13, 18, 19] processes synchronize through system messages before taking checkpoints. These synchronization messages contribute to extra overhead. On the other hand, in asynchronous checkpointing some of the checkpoints taken may not lie on any CGS. Such checkpoints are called *useless checkpoints*. Useless checkpoints degrade system performance. Unlike asynchronous checkpointing synchronous checkpointing does not generate useless checkpoints.

To overcome the above tradeoff of *synchronous* and *asynchronous* checkpointing, *quasi-synchronous* checkpointing algorithms were proposed by Manivannan and Singhal [5]. Processes take checkpoints asynchronously. So there is no overhead for synchronization. Generation of useless checkpoints is reduced by forcing processes to take additional checkpoints at appropriate times.

Several works on performance evaluation of checkpointing and rollback recovery algorithms have been reported in the literature. Plank and Thomason [26] calculated the average availability of parallel checkpointing systems and used them in selecting runtime parameters like the number of processors and the checkpointing interval that can minimize the expected execution time of a long-running program in the presence of failures. Vaidya [24] proposed a two level distributed recovery scheme and analyzed it to show that it achieves better performance than the traditional recovery schemes. The same algorithm was also analyzed by Panda and Das [25] taking the probability of task completion on a system with limited repairs as the performance metric.

Section 2 describes the stochastic model used for the analysis. In sections 3, 4 and 5 we find expressions for checkpointing and recovery cost for synchronous, asynchronous and quasi-synchronous checkpointing respectively. In section 6, we compare the different schemes analyzed.

2   Stochastic Model

For the purpose of analysis, we consider the following stochastic model:

1. Time is assumed to be discrete.
2. The system consists of a loosely-coupled system with message passing. Each processor has exactly one process.
3. Message sending, checkpointing and faults occur independent of each other.
4. At any point of time, a process may generate a message with probability λ_m. The destination of the message is uniform over the rest of the $n - 1$ processes.
5. At any point of time, a process may start checkpointing with probability λ_c.
6. At any point of time, a process may fail with probability λ_f.

3   Synchronous Checkpointing

In synchronous checkpointing algorithms, processes communicate through system messages and make sure that they take a CGS. In the schemes proposed by Prakash and Singhal [11], and Cao and Singhal [8] the checkpointing initiator forces the dependent process to take checkpoints. The dependency relations are maintained by attaching an n-bit vector with every application message. Every message sent makes the receiver dependent on the sender. These two algorithms can attain in their worst case when checkpointing initiator directly or transitively depends on remaining $n-1$ processes. In that case all processes take checkpoints for the checkpointing initiator. We consider the algorithms [9, 10, 18, 19] where the checkpointing initiator forces all processes in the system to take checkpoints. The results of our analysis gives an upper bound for the overhead in the other algorithms.

3.1   Checkpointing Overhead

At any point of time, a process initiates checkpoint with probability λ_c. It takes checkpoint if at least one of the process initiates checkpointing. Probability that at least one process initiates checkpointing is $(1-(1-\lambda_c)^n)$. Expected intercheckpoint gap $= \frac{1}{1-(1-\lambda_c)^n}$

Assuming that it takes t_c units of time to take a checkpoint.

$$\text{Checkpointing overhead for a process} = \frac{(1 - (1 - \lambda_c)^n)t_c}{1 + (1 - (1 - \lambda_c)^n)t_c}$$

3.2   Rollback Recovery Overhead

If failures are rare, we can safely assume that when a failure occurs, it can be at any point between two successive checkpoints, with equal probabilities. Let C_{reco} denote the recovery cost of per unit time.

Thus the average rollback recovery overhead for a process is

$$= \frac{1}{2} \text{ (average intercheckpoint gap) } C_{reco} = \frac{C_{reco}}{2(1 - (1 - \lambda_c)^n)}$$

4   Asynchronous Checkpointing with Message Logging

In checkpointing and message logging protocols, each process typically records both the content and the receive sequence number of all the messages it has processed in a location that will survive the failure of the process. In case the process has to rollback, the logged messages are replayed from the stable storage; they need not be retransmitted by the sender. The messages which are not logged will have to be resent, and may force the sender to rollback too. A process may also periodically create checkpoints of its local state, thereby allowing message logs to be removed. The periodic checkpointing of a process state is only needed

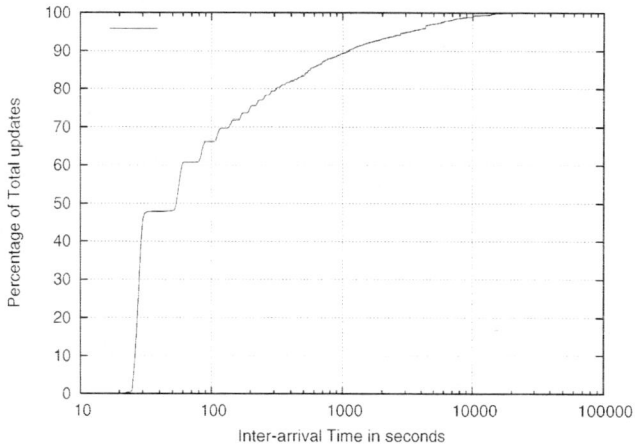

Fig. 4. Cumulative Distribution of Inter-arrival time for AS 18296 for Jan 25th 2003 from AS 7018

Table 1 shows the first few updates in this sequence. The penalty is calculated using Cisco default parameters.

We assume that the first withdraw creates a consistent routing state (i.e., no one in Fig. 6(a) has a path to AS 18296) and start our analysis from this point, with the penalty value set to 0. The second update announces a path via AS 701, which can only happen if AS 701 has sent such an announcement to AS 7018. Since it is a re-announcement following a withdraw, the damping penalty is zero. The third update shows that AS 7018 changes its next hop to AS 3549, and the fourth update withdraws the path again. The explanation of these two updates depends on AS 7018's routing policy. If AS 7018 prefers AS 701 over AS 3549, it could be that AS 701 withdraws its path first, causing AS 7018 to switch to AS 3549 and until AS 3549 withdraws its path as well. If AS 3549 is more preferred, the third update could be caused by an announcement from AS 3549, followed withdraws from both AS 701 and AS 3549 that cause the fourth update. In our analysis, instead of estimating an AS's routing policy, we simply

Damping Parameter	Cisco	Juniper
Withdraw penalty	1000	1000
Re-announcement penalty	0	1000
Attributes change penalty	500	500
Suppression threshold	2000	3000
Half time (min)	15	15
Reuse threshold	750	750
Maximum suppress time (min)	60	60

Fig. 5. Default parameters for route flap damping

logged sometime in the future so as to minimize logging overhead. This may be achieved by grouping several messages or logging during idle time of the system.

The receiver, P_i, of a message m_1 depends on the state of the sender, P_j. Suppose P_j received a message m_2 from P_k, before sending m_1. If the P_j fails without logging m_2, P_i will be orphan. Thus rollback of P_j may cause a rollback of P_i too.

This problem can be solved by optimistic recovery. Each process keeps track of its dependency on the states of other processes with which it communicates. The receiver, P_i will rollback to a state (checkpoint) before the dependency was created. If this rollback creates orphan messages of some other processes, they also roll back during recovery. Finally recovery based on optimistic message logging finds the most recent checkpoints of dependent processes so that no orphan message is created.

Let C'_{mlog} denote the average cost for logging a message in this scheme. Note that $C'_{mlog} < C_{mlog}$.

Total cost (checkpointing & message logging) per unit time ($\mathbf{E_{opt\_ckpt\_msg}}$) is

$$\mathbf{E_{opt\_ckpt\_msg}} = \frac{t_c}{T_p + t_c} + \lambda_m (C_{snr} + C'_{mlog})$$

Let the average gap between two loggings be $\frac{1}{\lambda_l}$. The average rollback recovery overhead ($\mathbf{E_{opt\_reco}}$) for a process is the sum of the recovery cost, the messages replaying cost and messages re-send cost.

$$\mathbf{E_{opt\_reco}} = \left[\frac{1}{2\lambda_c} - \frac{1}{2\lambda_l}\right] C_{reco} + \left[\frac{1}{2\lambda_c} - \frac{1}{2\lambda_l}\right] \lambda_m C_r + \frac{1}{2\lambda_l} \lambda_m C_{snr}$$

$$= \frac{1}{2\lambda_c} [C_{reco} + \lambda_m C_r] + \frac{1}{2\lambda_l} [(C_{snr} - C_r)\lambda_m - C_{reco}]$$

4.3   Causal Message Logging

Causal message logging protocols [2], [3] and [4] neither create orphans when there are failures nor do they ever block a process when there are no failures. Dependency information is piggybacked on application messages. In order to make the system f fault tolerant $f+1$ processes log the dependency information to their volatile storage.

Consider a message being sent by process P_i to process P_j. At any point of time, the probability of P_i sending a message to P_j is $\frac{\lambda_m}{n-1}$. Suppose the current time is τ. Probability that the last checkpoint before τ was taken at time $(\tau - t)$ is $(1 - \lambda_c)^{t-1} \lambda_c$.

$\mathbf{P}$(last message sent to P_j at $\tau - i$ | last checkpoint was taken at $\tau - t$)
$= \left(1 - \frac{\lambda_m}{n-1}\right)^{i-1} \frac{\lambda_m}{n-1} = q_i^t$ (say) for $i = 1, 2, 3, \cdots, t-1$.

$\mathbf{P}$(there was no message to P_j since the last checkpoint | last checkpoint was

taken at $\tau - t) = 1 - \sum_{i=1}^{t-1} q_i^t = \left(1 - \frac{\lambda_m}{n-1}\right) = r_i^t$ (say) for $i = 1, 2, \cdots, t-1$.
$\mathbf{E}($time gap since the last message to P_j **or** since the last checkpoint, if there
was no message $\mid$ last checkpoint was taken at $\tau - t) = \sum_{i=1}^{t-1} [iq_i^t] + tr_i^t$
$= \frac{1}{p^2} [2 + p + tp^2 - tp^3 - tp(1-p)^{t-1}] = s^t$ (say) where $p = \frac{\lambda_m}{n-1}$
Therefore, $\mathbf{E}($time gap since the last message to P_j **or** since the last checkpoint,
if there was no messages$) = \sum_{t=1}^{\infty} s^t (1 - \lambda_c)^{t-1} \lambda_c$

Let C_{pgb} be the cost for one piggybacking information. Let $\mathbf{E_{pgb}}$ be the
expected cost of piggybacking information. Therefore,

$$\mathbf{E_{pgb}} = C_{pgb} \lambda_m \sum_{t=1}^{\infty} s^t (1 - \lambda_c)^{t-1} \lambda_c$$

The average rollback recovery overhead ($\mathbf{E_{causal\_reco}}$) for a process is the sum
of recovery cost, messages and determinants [3] collection and message replaying
cost from the logs of the other processes. Let C_r' denotes the cost of replaying
a logged message from other processes.

$$\mathbf{E_{causal\_reco}} = \frac{C_{reco}}{2\lambda_c} + \frac{\lambda_m C_r' C_{snr}}{2\lambda_c} = \frac{1}{2\lambda_c}(C_{reco} + \lambda_m C_r' C_{onr})$$

5 Quasi-synchronous Checkpointing with Message Logging

5.1 Selective Message Logging

A *recovery line* (a globally consistent set of checkpoints) divides the set of all
events of the computation into two disjoint parts. When a process rolls back,
all those application messages whose send events lie to the left and the corresponding receive events lie to the right of the current recovery line are *lost
messages*. All such messages should be replayed. To cope with messages lost due
to a rollback, all such messages should be logged into stable storage. Manivannan
and Singhal [23] proposed *Selective message logging protocol* that logs only such
messages instead of all messages.

In a distributed computing system processors are connected through communication links. Let a single process run in a processor. The topology of the
system may be represented by a graph. A node represents a process and an
edge represents a communication link between a pairs of nodes. We assume that
one hop message passing time is constant (t_{hop}) for all edges. Edges are bidirectional. The *distance* ($d(i,j)$) between P_i and P_j is the length of the shortest
path between them.

Definition 1. *Let $G = (V, E)$ be any connected graph. For every node $v \in V$, we
define the pathsum of v, $pathsum(v) \overset{def}{=} \sum_{u \in V} d(u, v)$. The maximum pathsum
of G is defined as $MPS(G) \overset{def}{=} \max_{u \in V} \{pathsum(u)\}$.*

Lemma 1. *Let $T = (V, E)$ be a tree. If $MPS(T) = pathsum(v)$ for some $v \in V$, then v is a leaf node of T.*

Proof. If possible, let v be a non-leaf node such that $MPS(T) = pathsum(v)$. Let the nodes adjacent to v be $u_1, u_2, \cdots, u_k$ for some $k \geq 2$. The removal of v splits T into k different trees, with $u_1, u_2, \cdots, u_k$ in different trees.

Let the number of nodes in the tree having u_i be n_i for $i = 1, 2, \cdots, k$. Without loss of generality, let $n_1 \leq n_2 \leq \cdots \leq n_k$. Let $|V| = n$. $\sum_{i=1}^{k} n_i = n - 1$. Then $n - n_1 = ((n - 1) - n_1) + 1 = \sum_{i=2}^{k} n_i + 1 \geq n_2 + 1 > n_1$.

$pathsum(u_1) = pathsum(v) - n_1 + (n - n_1) > pathsum(v)$, which is a contradiction as $MPS(T) = pathsum(v) \geq pathsum(u)$ for any $u \in V$. $\square$

Lemma 2. *For a path graph P_n with n nodes, $MPS(P_n) = \frac{n(n-1)}{2}$.*

Proof. By *Lemma 1*, $MPS(P_n) = pathsum(v)$ where v is a leaf node of the path. For a leaf node v,

$pathsum(v) = 1 + 2 + 3 + \cdots + (n - 1) = \frac{n(n-1)}{2}$. $\square$

Lemma 3. *Let T_n be a tree with n nodes. Then $MPS(T_n) \leq MPS(P_n) = \frac{n(n-1)}{2}$.*

Proof. The result is true for $n = 1$. Suppose the result is true for $n = m$. Let v be any leaf node in T_{m+1}. Let $T_m = T_{m+1} - \{v\}$. $MPS(T_{m+1}) = MPS(T_m) + m \leq \frac{m(m-1)}{2} + m$ (induction hypothesis) $= \frac{m(m+1)}{2}$. $\square$

Definition 2. *For a connected graph G, we define the total pathsum of G to be $TPS(G) \stackrel{def}{=} \frac{1}{2} \sum_{v \in V(G)} pathsum(v)$.*

Lemma 4. *For a path graph P_n, $TPS(P_n) = \frac{1}{6}(n^3 - n)$.* $\square$

Theorem 1. *Let T_n be a tree with n nodes. Then $TPS(T_n) \leq \frac{1}{6}(n^3 - n)$.*

Proof. The result is true for $n = 1$. Suppose the result is true for $n = m$. Let v be a leaf node in T_{m+1}. $T_m = T_{m+1} - \{v\}$ is also a tree.

$TPS(T_{m+1}) = TPS(T_m) + pathsum(v)$
$\leq \frac{1}{6}(m^3 - m) + MPS(T_{m+1})$ (induction hypothesis)
$\leq \frac{1}{6}(m^3 - m) + \frac{m(m+1)}{2}$ (*Lemma* 3)
$= \frac{1}{6}((m + 1)^3 - (m + 1))$. $\square$

Theorem 2. *Suppose P_i and P_j are two processes which take checkpoints at t_i and t_j ($t_i \leq t_j$) respectively. Let $d(i, j)$ denote the distance between them. If $d(i, j) > t_j - t_i$, P_j will log the messages, sent by the P_i during the interval $[t_j - d(i, j), \quad t_i)$; otherwise P_j will not log any message sent by P_i. P_j will log the messages, sent by the P_i during the interval $[t_i - d(i, j), \quad t_j)$.*

Proof. Since a message takes $d(i, j)$ time to reach P_j from P_i, the messages sent in the interval $[t_i - (d(i, j) - t_j + t_i), \quad t_i)$ are the only ones which are sent before t_i (the checkpoint time for P_i) and reach after t_j (the checkpoint time for P_j). Hence, these are the only messages that are logged.

Similarly for the messages sent by P_j to P_i are logged if and only if they are sent in the given interval. $\square$

Table 1. Table comparing the checkpointing and recovery costs of different schemes

	Synchronous Checkpointing	Quasi-synchronous Checkpointing	Asynchronous Checkpointing		
		Selective Logging	Pessimistic Logging	Optimistic Logging	Causal Logging
Checkpointing cost	$\frac{(1-(1-\lambda_c)^n)t_c}{1+(1-(1-\lambda_c)^n)t_c}$	$\leq \frac{(1-(1-\lambda_c)^n)t_c}{1+(1-(1-\lambda_c)^n)t_c}$	$\frac{t_c}{T_p+t_c}$	$\frac{t_c}{T_p+t_c}$	$\frac{t_c}{T_p+t_c}$
Message logging cost (C)	0	$t_{hop}\lambda_m n \leq C \leq$ $\frac{2}{3}t_{hop}\lambda_m n(n+1)$	$\lambda_m(C_{snr}$ $+C_{mlog})$	$\lambda_m(C_{snr}+C'_{mlog})$	0
Recovery cost	$\frac{C_{reco}}{2(1-(1-\lambda_c)^n)}$	$\frac{C_{reco}}{2(1-(1-\lambda_c)^n)}$	$\frac{1}{2\lambda_c}(C_{reco}$ $+\lambda_m C_r)$	$\frac{1}{2\lambda_c}[C_{reco}+\lambda_m C_r]$ $+\frac{1}{2\lambda_l}[(C_{snr}-C_r)\times$ $\lambda_m - C_{reco}]$	$\frac{1}{2\lambda_c}(C_{reco}+$ $\lambda_m C'_r C_{snr})$
Piggybacking cost	0	$Const.$	0	$C_{pgb}n$	$C_{pgb}\lambda_m \sum_{t=1}^{\infty} s^t$ $\times(1-\lambda_c)^{t-1}\lambda_c$

Let us consider a distributed system with underlying topology $G = (V, E)$. Suppose process P_0, initiates checkpointing at t_0. Without loss of generality, let $d(0,1) \leq d(0,2) \leq \cdots \leq d(0,n-1)$. We assume that all messages take a shortest path to the destination and each hop takes t_{hop} units of time with no congestion delay. For the checkpoint initiated by P_0, a process P_i ($1 \leq i \leq n-1$) receives checkpoint request and takes checkpoint at $t_i = t_0 + d(i,0)t_{hop}$. We also assume that there is no other new request for checkpointing. Let E_{logged} be the expected number of messages logged by all processes.

Theorem 3. $t_{hop}\lambda_m n \leq E_{logged} \leq \frac{2}{3}t_{hop}\lambda_m n(n+1)$

Proof. Applying theorem 2, we see that P_i will log a message sent by P_j if and only if $i < j$. P_0 will log messages sent by P_i during $[t_i - d(i,0)t_{hop}, t_i + d(i,0)t_{hop})$, $1 \leq i \leq n-1$. So, the expected number of messages to be logged by P_0 is $\frac{1}{n-1}2t_{hop}\lambda_m \sum_{i=1}^{n-1} d(i,0)$.

Similarly, process P_i is expected to log $\frac{1}{n-1}2t_{hop}\lambda_m \sum_{k=i+1}^{n-1} d(k,i)$ messages from processes $P_{i+1}, P_{i+2}, \cdots, P_{n-1}$.

$E_{logged} = \frac{1}{n-1}2t_{hop}\lambda_m \left(\sum_{i=1}^{n-1} d(i,0) + \sum_{i=2}^{n-1} d(i,1) + \cdots + \sum_{i=n-2}^{n-1} d(i,n-3) + d(n-1,n-2) \right) = \frac{1}{n-1}4t_{hop}\lambda_m TPS(G)$ $(Def\ 2) \leq \frac{2}{3}t_{hop}\lambda_m n(n+1)$ $(Th\ 1)$

It is easy to see that in a complete graph the lowest number of messages would be logged. Checkpointing message reaches all other processes in the very next moment. A message would be logged only if the message is sent during the time when the message travels. So, $E_{logged} \geq t_{hop}\lambda_m n$. $\square$

6   Conclusion

In this work, we have calculated expected costs of different types of checkpointing algorithms such as synchronous, asynchronous and quasi-synchronous and their roll back recovery algorithms with message logging and without message logging. Table 1 shows the values of different components of overheads in some checkpointing schemes.

Acknowledgement

The first author is thankful to *Council of Scientific and Industrial Research (CSIR), India,* for financial support during this work.

References

[1] K. M. Chandy, L. Lamport: Distributed snapshots: Determining global states of distributed systems, *ACM Trans. Comput. Syst.,* vol. 3, no. 1, pp. 63-75, Feb. 1985. 57

[2] L. Alvisi, K. Bhatia, K. Marzullo: Nonblocking and orphanfree message logging protocols, in *Proc. 23rd Fault-Tolerant Computing Symposium,,* pp. 145–154, June 1993. 60

[3] L. Alvisi, B. Hoppe, K. Marzullo: Causality tracking in causal message-logging protocols, *Distributed Computing,* vol 15, pp. 1-15, 2002. 60, 61

[4] E. N. Elnozahy, W. Zwaenepoel: Manetho: Transparent rollbackrecovery with low overhead, limited rollback and fast output commit., *IEEE Transactions on Computers,* vol. 41, no. 5, pp. 526–531, May 1992. 60

[5] D. Manivannan, M. Singhal: Quasi-synchronous checkpointing: Models, characterization, and classification, *IEEE Trans. Parallel and Distributed Syst.,* vol. 10, no. 7, pp. 703-713, July, 1999. 56, 57

[6] Y. M. Wang: Consistent global checkpoints that contain a given set of local checkpoints, *IEEE Trans. Computers,* vol. 46 no. 4, pp. 456-468, Apr. 1997. 57

[7] K. Z. Meth, W. G. Tuel: Parallel checkpoint/restart without message logging, in *Proc. IEEE 28th International Conference on Parallel Processing* (ICPP '00) pp. 253-258, Aug. 2000. 56, 57

[8] G. Cao, M. Singhal: On coordinated checkpointing in distributed systems, *IEEE Trans. Parallel and Distributed Syst.,* vol. 9, no. 12, pp. 1213-1225, Dec. 1998. 57, 58

[9] P. S. Mandal, K. Mukhopadhyaya: Mobile agent based checkpointing and recovery algorithms on a distributed system, *in Proc. of 6th Int. Conf./ Exhibition on High Performance Computing in Asia Pacific Region,* Bangalore, India, vol. 2, pp. 492-499, Dec. 2002. 57, 58

[10] P. S. Mandal, K. Mukhopadhyaya: Concurrent checkpoint initiation and recovery algorithms on an asynchronous unidirectional ring network, *in Proc. 9th Int. Conf. on Advanced Computing and Communications,* Bhubaneswar, India, pp. 21-28, Dec. 2001 57, 58

[11] R. Prakash, M. Singhal: Low-cost checkpointing and failure recovery in mobile computing systems, *IEEE Trans. Parallel and Distributed Syst.,* vol. 7, no. 10, pp. 1035-1048, Oct. 1996. 57, 58

[12] J. L. Kin, T. Park: An efficient protocol for checkpointing recovery in distributed system, *IEEE Trans. Parallel and Distributed Syst.,* vol. 5, no. 8 pp. 955-960, Aug. 1998.

[13] R. Koo, S. Toueg: Checkpointing and rollback-recovery for distributed system, *IEEE Trans. Software Eng.*, vol. 13, no. 1, pp. 23-31, Jan. 1987. 57

[14] S. Venkatesan, Tony T-Y. Juang: Efficient algorithms for optimistic crash recovery, *Distributed Computing*, vol. 8, no. 2, pp. 105-114, June. 1994. 59

[15] D. Johnson, W. Zwaenepoel: Recovery in distributed systems using optimistic message logging and checkpointing, *Journal of Algorithms*, vol. 3, no. 11, pp. 462-491, 1990. 59

[16] D. Johnson, W. Zwaenepoel: Sender-based message logging and checkpointing, in *Proc. of 17th Annual International Symposium on Fault-Tolerant Computing, IEEE Computer Society*, pp. 14-19, June, 1987. 59

[17] A. P. Sistla, J. Welch: Efficient distributed recovery using message logging, in *Proc. of the ACM Symp. on Principle of Distributed Computing*, pp. 223-238, 1989. 59

[18] E. N. Elnozalhy, D. B. Johnsone, and W. Zwaenepoel: The performance of consistent checkpointing, in *Proc. 11th Symp. Reliable Distributed Systems*, pp. 86-95, 1992. 57, 58

[19] L. M. Silva, J. G. Silva: Global checkpointing for distributed systems, in *Proc. 11th Symp. Reliable Distributed Systems*, pp. 155-162, 1992. 57, 58

[20] M. Spezialetti, P. Kearns: Efficient distributed snapshots, in *Proc. of the 6th ICDCS*, pp. 382-388, 1986

[21] R. E. Strom, D. F. Bacon, S. Yemini: Volatile logging in n-fault-tolerant distributed systems, in *Proc. of 18th Annual International Symposium on Fault-Tolerant Computing*, pp. 44-49, 1988. 59

[22] R. E. Strom, S. Yemini: Optimistic recovery in distributed systems, *ACM Trans. On Computer Syst.* Vol. 3, no. 3, pp. 204-226, Aug. 1985. 56, 59

[23] D. Manivannan, M. Singhal: Asynchronous recovery without using vector timestamps, *J. Parallel Distrib. Comput.* Vol. 62, pp. 1695-1728, 2002. 61

[24] N. H. Vaidya: A case for Two-level Recovery Schemes, *IEEE Trans. Computers*, Vol. 47, pp. 656-666, 1998. 57

[25] B. S. Panda, Sajal K. Das: Performance evaluation of a two level error recovery scheme for distributed systems, in *Proc. 5th International Workshop on Distributed Computing, Pub. Springer-Verlag*, pp. 88-97, Dec, 2002. 57

[26] James S. Plank, Michael G. Thomason: Processor allocation and checkpoint interval selection in cluster computing systems, *J. Parallel Distrib. Comput.* Vol. 61, pp. 1570-1590, 2001. 57

Analysis of BGP Update Surge during Slammer Worm Attack[*]

Mohit Lad[1], Xiaoliang Zhao[2], Beichuan Zhang[2],
Dan Massey[2], and Lixia Zhang[1]

[1] University of California, Los Angeles, CA 90025, USA
[2] USC Information Science Institute, Arlington, VA 22203, USA

Abstract. Although the Internet routing infrastructure was not a direct target of the January 2003 Slammer worm attack, the worm attack coincided in time with a large, globally observed increase in the number of BGP routing update messages. Our analysis shows that the current global routing protocol BGP allows local connectivity dynamics to propagate globally. As a result, any small number of edge networks can potentially cause wide-scale routing overload. For example, two small edges ASes, which announced less than 0.25% of BGP routing table entries, contributed over 6% of total update messages observed at monitoring points during the worm attack. Although BGP route flap damping has been proposed to eliminate such undesirable global consequences of edge instability, our analysis shows that damping has not been fully deployed even within the Internet core. Our simulation further reveals that partial deployment of BGP damping not only has limited effect, but may also worsen the routing performance under certain topological conditions. The results show that it remains a research challenge to design a routing protocol that can prevent local dynamics from triggering global messages in order to scale well in a large, dynamic environment.

1 Introduction

The SQL Slammer worm [1] was released on Jan 25th, 2003 and exploited a known bug in MS SQL servers. Infected machines sent heavy traffic loads towards seemingly random destinations. If a destination shared the same MS SQL vulnerability, it would become infected and in turn attempt to infect another set of random destinations. Slammer infected at least 75,000 hosts in just over 30 minutes and is reported to be the fastest spreading worm to date [2]. Although the SQL Slammer worm attack was *not* directly targeted at the Internet routing infrastructure, the Internet Health Report [3] reported that a number of

[*] This material is based upon work supported by the Defense Advanced Research Projects Agency (DARPA) under Contract No DABT63-00-C-1027 and by National Science Fundation(NSF) under Contract No ANI-0221453. Any opinions, findings and conclusions or recommendations expressed in this material are those of the authors and do not necessarily reflect the views of the DARPA or NSF.

S. R. Das, S. K. Das (Eds.): IWDC 2003, LNCS 2918, pp. 66–79, 2003.
© Springer-Verlag Berlin Heidelberg 2003

critical AS-AS peering links were operating above critical load thresholds during the Slammer attack period. Further, [2, 4, 5] noted that the Slammer worm attack coincided in time with a large, globally observed increase in the number of BGP[6] routing update messages. In fact, such coincidences between Internet worm attacks and the surges in BGP routing update messages have been also observed during previous worm attacks, such as the Code-Red [7] attack in July 2001 and the NIMDA [8] attack in September 2001.

In this paper, we analyze the BGP log data collected from various monitoring points to understand the causes of the high surge in BGP update messages during the SQL Slammer attack. Our analysis shows that the current BGP routing protocol allows *local* connectivity dynamics to propagate *globally*. As a result, any small number of edge networks, such as those networks whose connectivity to the Internet was severely impacted by Slammer, can potentially cause global routing overload. Figure 1 shows the number of BGP path announce messages as observed from routers in three different ASes. The results, typical of that seen by monitoring points located in other ASes, clearly show a surge in BGP activity that coincides with the worm attack. As we will show later in the paper, two small edges ASes, which announced less than 0.25% of BGP routing table entries, contributed over 6% of total update messages observed, during the worm attack.

BGP route flap damping[9] was introduced specifically to prevent edge instability from flooding update messages globally. Route damping is applied on a $\langle peer, prefix \rangle$ basis and each update received from peer N for prefix p increases a penalty value associated with $\langle N, p \rangle$. Once the penalty exceeds a threshold value, N's updates regarding p are suppressed and the router behaves as if N has no route to p. The penalty decreases (decays) exponentially using a configured half-life and routes from N are again accepted after the penalty falls below a re-use limit. The RFC suggests that all Internet core routers deploy damping in an effort to "provide a mechanism capable of reducing router processing load caused by instability and in doing so, prevent sustained route oscillations". However, we will show that BGP route flap damping has not been fully deployed within the Internet core. Furthermore, even if a full deployment of route flap damping could reduce the number of global updates, it also results in a much longer routing convergence time as shown by both [10] and our simulations. Our simulation further reveals that partial deployment of BGP damping not only has limited effect but may also *worsen* the routing performance under certain topological conditions.

The paper is organized as follows. Section 2 presents our study of BGP behavior during the Slammer worm and demonstrates how local changes can propagate to create global events. Section 3 examines the impact of BGP route flap damping on the observed data and shows evidence that route flap damping could have reduced some dynamics if it had been deployed. Section 4 uses simulation to explore the impact of route flap damping deployment on both update counts and convergence time. Section 5 reviews the related work and Section 6 concludes the paper.remains a research challenge to design a routing protocol that can prevent local dynamics from triggering global messages in order to scale well in a large, dynamic environment.

2 Edge AS Instabilities With Global Effects

[2, 4, 5] noted that the Slammer worm attack coincided in time with a large increase in BGP routing update messages. To understand the worm's effect on BGP, we delve deeper into the update behavior during the SQL Slammer worm attack, with the objective of investigating the origins of the update bursts and their time distribution. Our results show that a small number of edge AS prefixes contribute greatly to the global routing update volume.

2.1 Methodology

Our study uses BGP update data collected by Oregon RouteViews[11]. Monitoring points at RouteViews peer with routers in a wide variety of Autonomous Systems. Each AS router treats the monitoring point as a BGP peer router and sends routing updates to the monitoring point, where they are logged and made available to researchers. We examined BGP update logs from January 22, 2003 to January 29, 2003, a period of 8 days around the Slammer worm attack. We apply the techniques from [12] to remove monitoring artifacts from the logs.

The resulting data shows a dramatic increase in the number of BGP route announcements and matches the observations reported in [2, 4, 5]. Figure 1(a) shows the number of route announcements sent by monitored routers in three difference Autonomous Systems. Early on January 25, the number of update spikes dramatically and tails off by January 26. Similar large surges in BGP updates are seen at all monitored AS. Figure 1(b) further shows the presence of a spike in the number of withdraw messages observed during the same worm period. This suggests that there was also loss of connectivity to some portions of the Internet.

2.2 Identifying Edge Instability

[2] reported that the worm caused high traffic congestion on the edge ASs. To better understand the edge behavior, we ranked the Internet ASs based on the average number of update messages per prefix originated by the AS, as observed on January 25 . Figure 2(a) shows the top five ranked ASs. Figure 2(b) shows the corresponding total updates for prefixes originating from these AS, as observed from all the monitoring points. For comparison, Figures 2(c) and 2(d) show the corresponding graphs on the day prior to the worm attack; note the maximum value on y-axis decreases by two orders of magnitude.

Figure 2(a) shows that prefixes from AS 18296 (belonging to a unversity in South Korea), accounted for the highest number of updates per prefix, averaging over 4500 updates per prefix on the day of Slammer attack. In contrast, this same AS, averaged only 47 updates per prefix on the day prior to the Slammer attack.

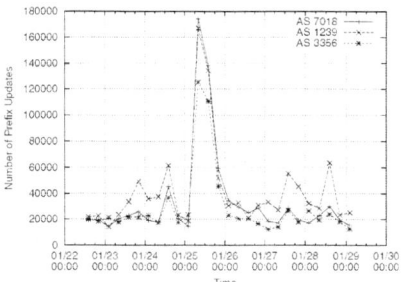
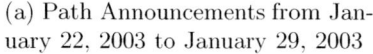
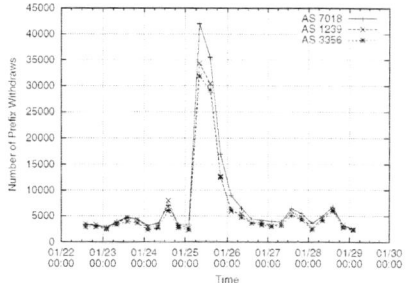

(a) Path Announcements from January 22, 2003 to January 29, 2003

(b) Withdrawals from three peers from January 22, 2003 to January 29, 2003

Fig. 1. Updates Messages During Slammer Attack

This massive change is not suprising, following reports in [13], suggesting that South Korea's connectivity was among the worst affected by the SQL Slammer worm. AS 18296 advertises only about 30 prefixes out of the roughly 120,000 prefixes in a typical global BGP routing table. Although AS 18296's 30 prefixes constitute less than 0.02% of the total prefix space, this AS generated about 1.7% of the total BGP updates observed on January 25. AS568, owned by the US Department of Defense, advertises a total of 238 prefixes and stands out in terms of total number of updates. Together AS 18296 and 5568 announce less than 0.25% of Internet prefixes, but contributed over 6% of total updates seen during the worm attack, as shown in Figure 3(a).

2.3   Analysis of AS 18296 and AS 568 Edge Instability

Figure 3(b) classifies the BGP update messages associated with the 30 prefixes in AS18296 as viewed from a particular ISP, AT&T (AS7018). The updates in Figure 3(b) are classified into five categories: *DPATH* updates indicate a change in the AS path; *New Announcement* updates announce a path to a previously unreachable prefix; *Withdrawal* updates remove the path to prefix and make the prefix unreachable; *Duplicates* convey no new information what so ever and are identical to the previous update for this prefix; finally, *SPATH* (Same AS Path) updates indicate no change the AS path, but do indicate a change in some other BGP attribute. On the worm attack day, DPATH messages (Different AS Path messages) were the dominant type of BGP update associated with the AS 18296 prefixes; the number of withdraw messages is also significant. Both DPATH and withdraw BGP updates convey real changes in the AS paths used to reach these prefixes. Similar results are obtained when the AS 18296 updates are analyzed from other ISPs and we note that these prefixes are seen as unstable from a wide

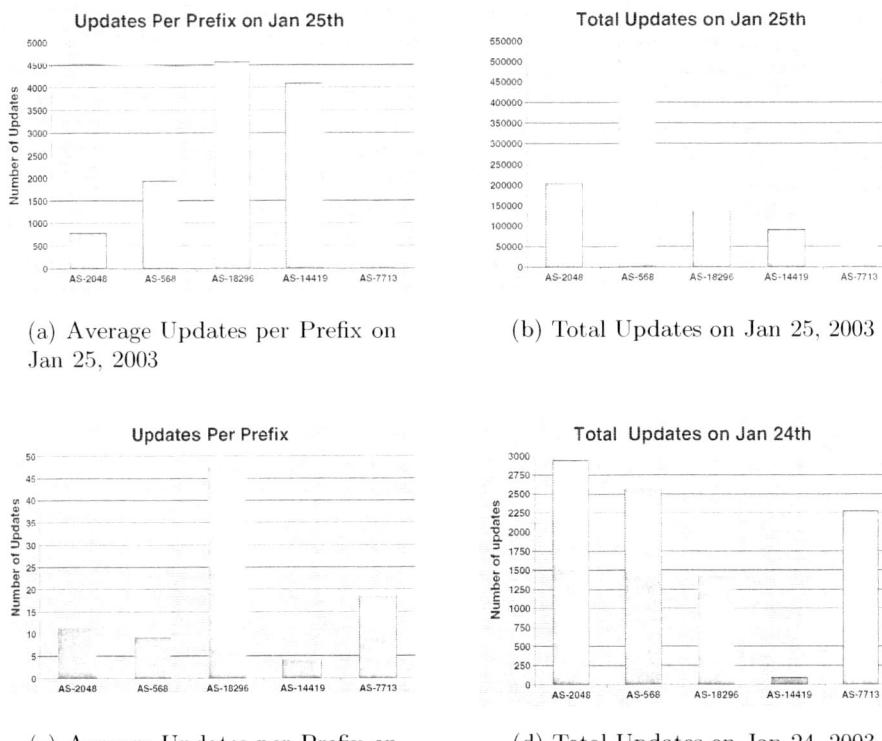

(a) Average Updates per Prefix on Jan 25, 2003

(b) Total Updates on Jan 25, 2003

(c) Average Updates per Prefix on Jan 24, 2003

(d) Total Updates on Jan 24, 2003

Fig. 2. Update counts during Slammer Worm attack

variety of locations, an indication that the cause of the instability is close to the origin AS.

AS 18296 primarily uses AS 9318 as a next hop to reach the Internet, although the data shows it is also multi-homed through AS 4766. Both prior to and after the Slammer attack, AS 18296 announced only a small number of BGP updates each day. A routing snapshot from Jan 22, 2003 showed that nearly all of the prefixes originating from AS 18296 had a next hop of AS 9318. We also observed that a total of 220 prefixes relied on AS 9318, thus AS 18296 originates roughly 14% (30/220) of the prefixes that rely on AS 9318. During the worm attack day, however, the AS 18296 prefixes accounted for 82% of BGP updates involving AS 9318. Every monitoring point observed high numbers of updates for the 30 prefixes originated by AS 18296, indicating a problem near the origin. Furthermore, the AS next to the origin, AS 9318, exhibited few changes in connectivity to destinations except for the prefixes originating from AS 18296. The above evidence strongly suggests that the problem is local to AS 18296, or the peering between AS18296 and AS9318. We can see from this example, that

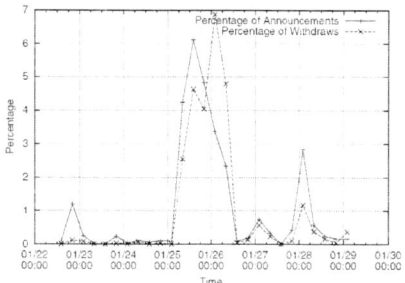

(a) Percentage of updates on pre-fixes belonging to AS18926 and AS568 from Jan 22nd to Jan28th

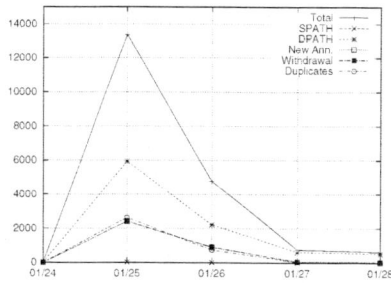

(b) Classification of Update Messages for AS 18296

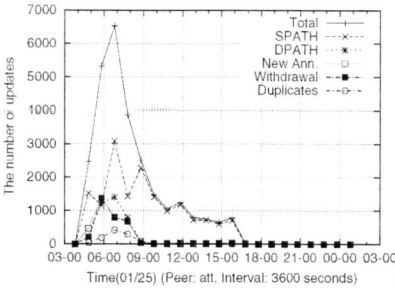

(c) Classification of Update Messages for AS 568

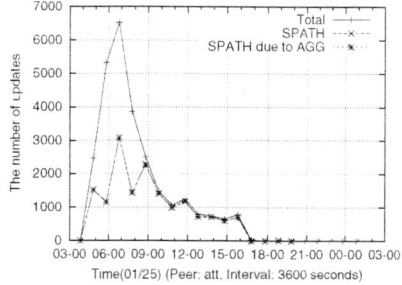

(d) Aggregator Attribute for AS 568 Updates

Fig. 3. Statistics for AS 568 and AS18296

the worm attack affected edge ASs, causing fluctuations in connectivity, and thus resulting in a burst of BGP updates.

Figure 3(c) shows the type of BGP update messages generated by AS 568's 281 prefixes. Our previous work in [14] demonstrated that congestion at AS 568 coincides with a local AGGREGATOR attribute changes and the BGP design propagates this local change globally. Figure 3(d) shows that nearly all of the SPATH updates reported a change in the AGGREGATOR attribute.

3   Worm Attack and BGP Damping

Section 2 showed that the worm caused a high number of updates and identified particular edge AS instability that was partly responsible for the update surge. This section investigates how BGP features designed to limit instability, fared during the worm attack. BGP includes a Minimum Route Advertisement Interval (MRAI) timer that places a lower limit on the time between two successive update messages The MRAI timer is intended to suppress routing instability during this hold time, and help BGP convergence by limiting AS path exploration [15]. Fig. 4 shows the per-prefix cumulative inter-arrival time between BGP update messages, as observed from AS 7018. We can see that about 50% of the update messages have an inter-arrival time per prefix of close to MRAI timer's default value, 30 seconds. While the update surge could have been worse without MRAI, the presence of MRAI timer alone was clearly not sufficient to prevent a sudden update burst. *BGP route flap damping* is also intended limit the impact of edge instability.

3.1   Route Flap Damping

While the MRAI timer works at the time scale of tens of seconds, *route flap damping* [9] was proposed to deal with route instability at a larger time scale. The objective of damping is to suppress the updates caused by an unstable link from propagating. Unlike MRAI timer operation that is implemented on the sender side, damping is implemented on the receiving end of any peering session. A BGP router maintains a penalty value for each prefix and peer combination it receives. Whenever a peer advertises a change in route to the prefix, its penalty is increased. When a route's penalty exceeds a *suppression threshold*, a BGP router will stop using this route, thus preventing future changes from propagating. The penalty decays exponentially over time and a suppressed route can be reused only after the penalty over time, drops below a *reuse threshold*. Figure 5 shows the default damping settings for Cisco routers and Juniper routers

[9] recommends that at least the Internet core routers enable damping to achieve reduced update messages. Damping has been said [16] to be widely deployed and is considered a major factor in keeping the Internet stable at BGP's early days, when faulty implementations caused lots of excessive updates.

3.2   Case Study on Damping

We used the monitoring point data to infer BGP activity on real operational links. Due to the limitations of BGP monitoring and data collection, some assumptions are needed to make the analysis feasible. As in most BGP research work, we model the Internet inter-AS topology as a graph, in which each AS is represented by a single node, and every AS link appearing in any update message is represented by a single link between the two AS nodes. In our analysis, we choose sequences of updates that exhibit simple route flap behavior, such as alternating between two paths, or up and down for a single path, so that we can safely infer remote BGP activity without making broad assumptions on routing policy. One such sequence of frequent updates was received from AS 7018 for the path to AS 18296, and we analyzed it to see if route flap damping should have been triggered. Fig. 6(a) shows the topology containing the relevant ASes and links. The link we are interested is the one connecting AS 7018 and AS 701.

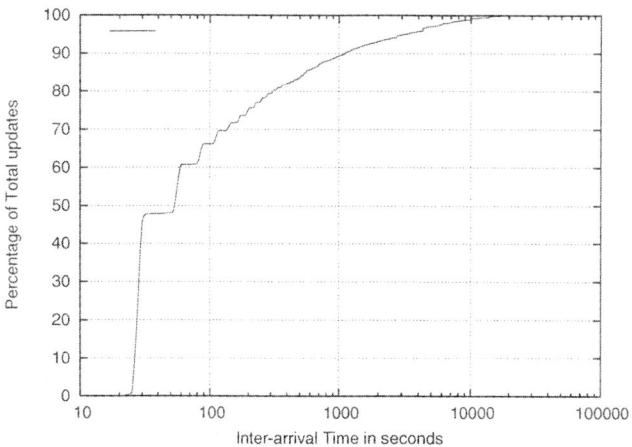

Fig. 4. Cumulative Distribution of Inter-arrival time for AS 18296 for Jan 25th 2003 from AS 7018

Table 1 shows the first few updates in this sequence. The penalty is calculated using Cisco default parameters.

We assume that the first withdraw creates a consistent routing state (i.e., no one in Fig. 6(a) has a path to AS 18296) and start our analysis from this point, with the penalty value set to 0. The second update announces a path via AS 701, which can only happen if AS 701 has sent such an announcement to AS 7018. Since it is a re-announcement following a withdraw, the damping penalty is zero. The third update shows that AS 7018 changes its next hop to AS 3549, and the fourth update withdraws the path again. The explanation of these two updates depends on AS 7018's routing policy. If AS 7018 prefers AS 701 over AS 3549, it could be that AS 701 withdraws its path first, causing AS 7018 to switch to AS 3549 and until AS 3549 withdraws its path as well. If AS 3549 is more preferred, the third update could be caused by an announcement from AS 3549, followed withdraws from both AS 701 and AS 3549 that cause the fourth update. In our analysis, instead of estimating an AS's routing policy, we simply

Damping Parameter	Cisco	Juniper
Withdraw penalty	1000	1000
Re-announcement penalty	0	1000
Attributes change penalty	500	500
Suppression threshold	2000	3000
Half time (min)	15	15
Reuse threshold	750	750
Maximum suppress time (min)	60	60

Fig. 5. Default parameters for route flap damping

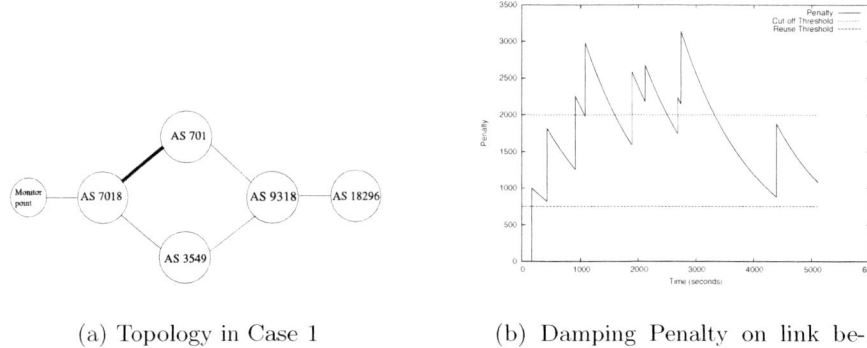

(a) Topology in Case 1

(b) Damping Penalty on link between AS 7018 and AS 701

Fig. 6. Case study of peering between AS7018 and AS701

calculate penalty for all possible cases and choose the one that gives the most conservative value.

Fig. 6(b) shows the damping penalty value on link between AS 7018 and AS 701 over a time period of about 80 minutes. The penalty was well above the suppression threshold, reaching more than 3000 while the suppression threshold is only 2000. However, we still see updates from AS 7018, inspite of calculations showing that the threshold was crossed. Had damping been turned on, update from AS 7018 should not have included AS 701 until the penalty dropped below the reuse threshold and this may have prevented the unstable path information from spreading to other parts of the Internet. Therefore, this case study suggests that, even in the Internet core, it is possible that AS 7018 did not implement route flap damping at its link to AS 701. Overall, our analysis suggests that the ideal core deployment has not been achieved in practice.

No.	Time (second)	Update observed	Possible update on interested link	Induced Penalty	Total Penalty
1	0	withdraw	withdraw	-	0
2	30	(7018 701 9318 18296)	re-announcement	0	0.0
3	141	(7018 3549 9318 18296)	-	0	0.0
4	167	withdraw	withdraw	1000	1000.0
5	281	(7018 701 9318 18296)	re-announcement	0	915.9

Table 1. Sample update sequence and analysis

4   Effectiveness of Route Damping

BGP route flap damping could have an impact on the number of updates, but the extent of damping is unclear based on data observed. In this section, we use simulations to evaluate the potential effectiveness of route flap damping in response to instability at an edge AS, such as that seen during the SQL worm attack.

4.1   Simulation Settings

We used the SSFNET BGP simulator [17] to simulate BGP behavior with different topologies. In our simulations, we set link delay to 2 milliseconds and the processing delay of each routing message to a random value between 0.1 and 0.5 second. The MRAI timer (discussed in the previous section) is set to the default value of 30 seconds with random jitter, the value used widely in real network operations. We used 110-node and 208-node topologies that are derived from real Internet routing tables [18].

For each simulation run, we randomly chose one node from the topology and attached an extra origin AS node to it. Throughout the rest of the paper we refer this extra AS node as the *flapping source*. To simulate edge instability, the flapping source withdraws its AS path every 100 seconds, and re-announces it 50 seconds later. We also simulated different rates, but the results were found to be similar. A pair of the withdraw and re-announcement is called a *pulse*. For each unstable source, we ran the simulation using one to 25 pulses. The route flapping starts after the network has been stable for a 1000 second period. Each simulation run ends when there are no more updates to send, *i.e.*, when BGP converges. We count the number of update messages during this time period in order to evaluate the effective damping has on reducing updates.

4.2   Results

Figure 7 shows the results for three different route damping deployment scenarios:

1. No damping deployed at any node
2. Full deployment of damping at every node
3. Damping deployment only at "core" nodes

In our simulation, a "core" node is distinguished by its node degree and nodes with degree larger than or equal to 13 as core nodes. 8 (7.3%) nodes were selected as core nodes. The X-axis represents the number of pulses generated by the instable edge AS (flapping source) and for each number of pulses, we run a set of simulations by randomizing the location of flapping source. The Y-axis represents the total number of updates in the network during the simulation. The results presented are the average values with 95% confidence interval.

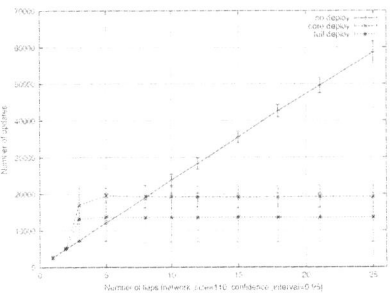

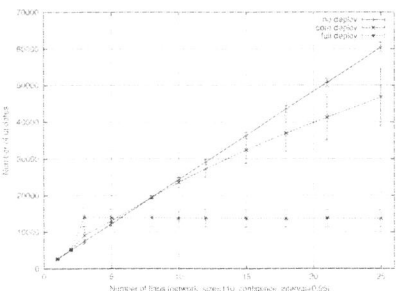

(a) Flapping source attached to core

(b) Flapping source attached to edge

Fig. 7. Number of Updates

The results show that full deployment of route flap damping dramatically reduces the total number of updates and the reduction in updates occurs regardless of the location of flapping source. In Figures 7(a) and 7(b), the number of updates sent with no damping deployed increases linearly as the number of flaps increases. However, with deployment of damping at every node, the number of updates initially grows and then remains nearly constant. The neighbors immediately adjacent to the flapping source will damp the prefix from the unstable edge AS and thus effectively ignore the updates caused by additional flaps, preventing these updates from propagating throughout the network. This can help reduce the routers' load when excessive updates were generated by highly intermittent connectivities, or stressful network events, such as worm attacks. In addition, our simulations (not shown in this paper) also confirm the findings in [10], that route flap damping can result in a much longer route convergence time.

However, even deployment of full damping raises issues that are not yet well understood. Note that for a small number of initial flaps (less than 8 for flapping source attached to the core and less than 5 for flapping source attached to an edge), the number of updates actually increases due to route damping. This behavior reflects the complexity of adding features to a complex distributed system. In this case, adding a route suppression feature actually *increases* the total update count for the system in the special case of a small number of origin AS flaps.

When damping is deployed only at core nodes as the BGP standard recommends, our simulation results show that damping has different effects depending on whether the flapping node is connected to a core node or an edge (non-core) node. As shown in Figure 7(a), when the flapping source was attached to core nodes, the damping effectively reduced the number of updates. In this case, the node directly next to the flapping source would damp future changes after a sufficient number of initial flaps. However, when flapping source was attached to a non-core node, the damping was not as effective in reducing the number of updates, as shown in Figure 7(a). In both the core and non-core cases, the deployment of damping will still incur a cost of delayed convergence.

In conclusion, while damping appears to be a useful technique to suppress updates due to unstable links, there is a lack of complete understanding of the effects of the deployment of damping. On one hand is the issue of increased convergence time, while on the other hand as showed in this paper, further work needs to be done on understanding the tradeoffs of partial deployment.

5   Background and Related Work

The Internet consists of a large number of Autonomous Systems (AS) that exchange routing information with each other to learn the best path to the destinations and BGP (Border Gateway Protocol) [6] is the de-facto inter-AS routing protocol. BGP is a path vector based routing protocol, designed to allow flexible routing policies and be able to adapt to routing dynamics such as link failures and topology changes. Each BGP router establishes a peering session with its neighbor BGP routers, and advertises the entire AS path information to destination prefixes. When a BGP session is set up between two peers, the complete routing tables are exchanged. After this initial exchange, routers only send update messages for additional changes.

Impact of worm attacks on Internet routing infrastructure has been studied before. Researchers have looked at BGP updates during stressful events such as the Code-Red worm attack. [19] first reported the surge of BGP updates coincided with the Code-Red and the NIMDA worm attack. According to [12], the worm attack had a big impact on some edge networks, and weaknesses in BGP's design and implementation substantially amplified the impact. They reached this conclusion by classifying BGP updates into several categories and examining the cause of each category. Another study [14] also showed that worm attack affected some edge networks like Department of Defense (DoD) networks.

Route flap damping [9] is designed to suppress unstable routes when they flaps frequently, so to prevent the local instability from spreading further. Damping is viewed by the network operation community as one of the major contributors to keep the Internet stable [16], but it has not been fully studied by the research community. In [10], the authors studied the effect of BGP slow convergence on route flap damping. They showed that due to the path exploration in BGP slow convergence, a single flap at one end of the network is able to trigger route damping at another place in the network. This undesired damping will in turn cause longer BGP convergence time.

6   Summary

This paper examined the surge in BGP updates that coincided with the January 2003 Slammer worm attack. Our analysis illustrates how two small edge

Autonomous Systems that announce fewer than 0.25% of BGP routing table entries, contributed over 6% of total update messages during the worm attack, as observed from the Oregon RouteViews monitoring points. We also showed that these two Autonomous Systems generated a large number of updates in different ways. The instability at edge ASes can trigger a large number of AS path changes (DPATH updates) in the global Internet, such as those triggered by AS 18296, and can also trigger a combination of different AS paths and changes in other attributes (SPATH updates), such as those triggered by AS 568. In both cases, BGP allows dynamics belonging to local networks to propagate globally. As a result, a small number of edge networks can potentially cause widespread routing overload to the entire Internet.

Route flap damping is the current BGP mechanism to defend against such undesirable global consequences caused by edge instability. Our analysis of BGP udpate data shows that damping has not been fully deployed even within the Internet core. But simple lack of deployment is not the only problem. Our simulation further reveals that partial deployment of damping not only has limited effect but may also worsen the routing performance under certain topological conditions. Even in the case of full deployment, our simulation results show that its effects can be mixed if the edge generates only a small amount of instability. The results presented here are the first step toward a better understanding of BGP instability and route damping. Overall, adding route flap damping or any feature to the global routing infrastructure results in complex and often unexpected behaviors. It remains a research challenge to design a routing protocol that can scale well in a large, dynamic network and insights obtained from understanding BGP instability under stressful events such as the worm attack can help make the Internet infrastructure more stable and scalable.

References

[1] CERT Advisory CA-2003-04, "SQL Slammer," http://www.cert.org/advisories/CA-2003-04.html.   66

[2] David Moore et. al., "The spread of the Sapphire/Slammer worm," http://www.cs.berkeley.edu/ nweaver/sapphire/.   66, 67, 68

[3] Internet Health Report, "Sapphire Worm Attack," http://www.digitaloffense.net/worms/mssql_udp_worm/internet_health.jpg.   66

[4] Tim Griffin, "BGP Impact of SQL Worm," http://www.research.att.com/ griffin/bgp_monitor/sql_worm.html.   67, 68

[5] Avi Freedman, "ISP Security Talk, Nanog 2003," http://www.cs.berkeley.edu/ nweaver/sapphire/.   67, 68

[6] Y. Rekhter and T. Li, "A border gateway protocol (BGP-4)," *Request for Comment (RFC): 1771*, Mar. 1995.   67, 77

[7] CERT Advisory CA-2001-19, ""Code Red" Worm Exploiting Buffer Overflow In IIS Indexing Service DLL," http://www.cert.org/advisories/CA-2001-19.html.   67

[8] CERT Advisory CA-2001-26, ""Nimda Worm"," http://www.cert.org/advisories/CA-2001-26.html.   67

[9] C. Villamizar, R. Chandra, and R. Govindan, "BGP route flap damping," *Request for Comment (RFC): 2439*, Nov. 1998. 67, 72, 77

[10] Z. Mao, R. Govindan, G. Varghese, and R. Katz, "Route flap damping exacerbates internet routing convergence," in *Proceedings of the ACM SIGCOMM*, Pittsburg,PA, Aug. 2002. 67, 76, 77

[11] Univeristy of Oregon, "The Route Views Project," http://www.antc.uoregon.edu/route-views/. 68

[12] L. Wang, X. Zhao, D. Pei, R. Bush, D. Massey, A. Mankin, S. Wu, and L. Zhang, "Observation and analysis of BGP behavior under stress," in *Proceedings of the ACM SIGCOMM Internet Measurement Workshop 2002*, Nov. 2002. 68, 77

[13] PC World, "Slammer worm slaps Net down but not out," http://www.pcworld.com/news/article/0,aid,108988,00.asp. 69

[14] X.Zhao, M. Lad, D. Pei, L. Wang, D. Massey, and L. Zhang, "Understanding BGP Behavior through a study of DoD Prefixes," in *DISCEX 2003*, Feb. 2003. 71, 77

[15] C. Labovitz, A. Ahuja, A. Bose, and F. Jahanian, "Delayed Internet routing convergence," in *Proceedings of the ACM SIGCOMM 2000*, August/September 2000. 72

[16] Geoff Huston, "Analyzing the Internet BGP Routing Table," *The Internet Protocol Journal*, March 2001. 72, 77

[17] ssfnet.org, "SSFNET modeling the global internet," http://www.ssfnet.org. 75

[18] B.Premore, "Multi-as topologies from bgp routing tables," http://www.ssfnet.org/Exchange/gallery/asgraph/index.html. 75

[19] J. Cowie, A. Ogielski, B. J. Premore, and Y. Yuan, "Global routing instabilities triggered by Code Red II and Nimda worm attacks," Tech. Rep., Renesys Corporation, Dec 2001. 77

TMS: A Scalable Transition Multicast Scheme

Lei Sun, Ke Xu, Weidong Liu, and Jianping Wu

Department of Computer Science,Tsinghua University
Beijing 100084, P. R. China
sunlei97@mails.tsinghua.edu.cn
{xuke,liuwd}@tsinghua.edu.cn
jianping@cernet.edu.cn

Abstract. In this paper, we propose a multicast scheme known as Transition Multicast Scheme (TMS) that is suitable for the network multicast services during the transition from IPv4 to IPv6. TMS has extended NATPT implemented in the edge router, which cooperates with Rendezvous Point (RP) of IPv6 domain for multicast communication between IPv4 and IPv6. In the process of communication, NATPT-router group acts as a joint to transfer multicast messages between IPv6 network layer members and IPv4 application layer members. We have simulated our scheme using the GT-ITM and the measurements show that TMS is effective in decreasing delay penalty, reducing link stress and balancing the load among extended-NATPT routers as well as serving for smooth transition for IPv6 multicast deployment.

1 Introduction

The introduction of IPv6 in the Internet requires a change in the network infrastructure. Since an instantaneous upgrade from IPv4 to IPv6 is neither possible nor desirable, viable transitioning mechanisms are essential to the evolution of the Internet. Two protocols will have to co-exist in the Internet for an interim period. Current efforts in the area of IP transitioning have largely considered unicast mechanisms such as NATPT[1].

While a number of challenges that are associated with the IP multicast transition make the problem distinct from the unicast scenario. As well known, the deployment of multicasting capabilities in IPv4 is still very far from being practical though it was formally added to IPv4 in 1988. So current feasible technique for multicast communication between IPv6 hosts and IPv4 hosts are due to application layer multicast implemented at the end-hosts instead of network routers. NATPT router located at the boundary of the network is the only traffic entry between IPv6 domain and IPv4 domain. If both sides run application layer multicast, all the traffic will pass through the NATPT router. If application layer multicast is just simply transplanted to IPv6, for its assumption only unicast support from underlying, it will aggravate the link stress-a kind of metrics to evaluate multicast scheme, more detail can be found in Sect. 4-especially at the boundary of the network. In the worst case, it even will make the edge router over-loaded.

S. R. Das, S. K. Das (Eds.): IWDC 2003, LNCS 2918, pp. 80–90, 2003.

As TMS we have presented, it has extended NATPT to handle multicast to and from IPv6 members between the other IPv4 members in the same application layer group. Thus all IPv6 members inside the isolated island are hidden from the NATPT router. IETF designers of IPv6 want to take advantage of the deployment of a new protocol to make sure that multicasting is available on all IPv6 nodes, and they make sure that all IPv6 routers could route multicast packets. Inside domain it is native IPv6, so we can use network layer multicast to do multicast communication. Outside it is IPv4, we use current solution-application layer multicast. In the process of communication, the extended-NATPT router serves as an agent for IPv6 members. It has been extended with some new functions for multicast services, such as translations between IPv4 datagram and IPv6 multicast datagram and vice versa, mapping IPv4 unicast address to IPv6 multicast group address and so on. While RP router runs a NATPT server inside IPv6 domain, gathers the information from extended-NATPT routers at intervals, redirects new NATPT requests to balance the load among them when necessary.

The rest of our paper is organized as follows. Sect. 2 presents related works. Sect. 3 describes our TMS architecture to support multicast in the transition network from IPv4 to IPv6. Sect. 4 presents the simulation of our scheme using GT-ITM,reports the performance measurement result of the experiment and Sect. 5 concludes the paper.

2   Related Work

For the sparse deployment of IPv4 multicast in current Internet, we have to resort to application layer multicast to implement multicast services. There are many application layer multicast implementations such as Scribe [2] and Host-Multicast [3]. From the statistic data of experiments, Scribe is the most scalable scheme which also has achieved best performance. So we choose it as the representative of application layer multicast scheme in TMS, also as a comparison in our experiment. In addition, the intention of Host-Multicast is similar to TMS. The differences are: Host-Multicast implements multicast functionality at end hosts for future fully IPv4 multicast services; TMS implements at NATPT edge router for future fully IPv6 multicast. For application layer multicast schemes assume unicast as underlying network, when deployed to the transition network, it will treat every IPv6 member as an ordinary IPv4 member even they locate at the same domain. This will exacerbate the link stress and lead to bad performance.

MTP [4] has proposed a mechanism which enables direct network layer multicast communication between IPv4 members and IPv6 members also based on NATPT. In MTP, NATPT is extended with a multicast translator and address mappers. It can only support one multicast group communication between small IPv6 isolated island and IPv4 group. And the IPv6 multicast addresses in use must be pre-defined, not dynamic allocated. For its limitations on topology of group members and poor scalability on multicast address management, the MTP draft was not adopted by IETF.

Above all, to solve the problem of transition multicast neither of the current mechanisms seems satisfied. So there comes our Transition Multicast Scheme.

3   TMS Architecture

3.1   Overview

In unicast scenario, NATPT router will maintain a mapping item composed by an IPv4 unicast address and an IPv6 unicast address for communication between IPv6 hosts and IPv4 hosts. During the process of communication, NATPT router will do translations according to the address mapping items and the types of datagram such as ICMP, TCP, and UDP etc.

First we will give some assumptions in TMS. Protocol Independent Multicast(PIM) [5] is a scalable, efficient and robust multicast routing protocol, capable of supporting thousands of groups, different types of multicast applications. The main applied scenario we discussed is early stage of the transition, we assume there has been an IPv4 application layer multicast group. IPv6 members scatter around the group, may join or leave. We choose PIM-Sparse Mode (PIM-SM) implemented as multicast support for IPv6 domain. Scribe is chosen as the application layer scheme running on NATPT router in TMS, for its scalability and good performance.

Then we will give the overview of our scheme-TMS. To support it, some extensions will be implemented in NATPT router that we called extended-NATPT router.

In TMS, we run a NATPT server at RP router inside IPv6 domain. So the RP router acts as not only the root of the multicast distribution tree inside IPv6 domain but also advertises NATPT-message periodically to inform the NATPT routers to join. When NATPT router is up or just decides to join, it will register at the RP and RP router will return it a natptID as an identifier. Thus NATPT-router group is formed. NATPT is responsible for multicast data communication. While the RP also aims at scheduling all the NATPT requests of the NATPT-router group, receiving and delivering the NATPT-messages between NATPT routers.

In NATPT router, multicast mapping items are introduced. The multicast mapping item is composed by an IPv4 unicast address and an IPv6 multicast address. Inside the IPv6 domain, the IPv6 multicast address is shared among the IPv6 hosts belong to the same group; Outside the domain, NATPT router will serve as an agent to communicate with the IPv4 application layer multicast group based on the IPv4 unicast address.

The Rendezvous Point (RP) that is defined in PIM-SM [5] is responsible for those specific IPv6 domain group-list concerned with application layer group. It will also help to form a multicast distribution tree.

Because Scribe is a large-scale and decentralized application layer multicast infrastructure, the source of the application layer multicast distribution tree is not limited, each member can be the speaker of the group. PIM-SM shares the same characteristic with Scribe on this point. So in TMS, every member can be either the sender or the receiver whatever it is IPv4 or IPv6.

3.2   Membership Management

When an IPv6 multicast receiver expresses its interest in receiving traffic destined for an application layer multicast group, it does this using MLD [6]. As well as PIM-SM, one of the receiver's local routers is elected as the Designated Router (DR) for that subnet. On receiving the receiver's expression of interest, the DR then sends a PIM Join message towards the RP for that multicast group. For RP route is extended with

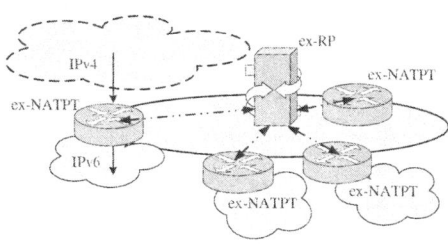

Fig. 3.1.   ex-NATPT and ex-RP in TMS

a NATPT server, it is also responsible for the application layer mapping groups. The distribution trees are also rooted at the RP.

As Fig. 3.2 described, in IPv6 domain, hosts will join the application layer multicast group with following sequence r1, r2, r3 (r1 and r2 belong to the same DR2, r3 belongs to DR1).When host r1 wants to join the group, DR2 will send PIM Join message to NATPT. On receiving the message, NATPT will forward the request to the RP to checks its group-list. RP finds the requested group item doesn't exist. Then the NATPT server running on RP will

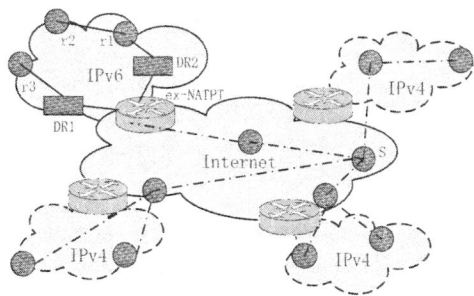

Fig. 3.2.   Network topology at early stage of transition

calculate its evaluation function to select out one NATPT router to handle the request.

The decision is made according to the combination of both routing metrics in application layer-multicast routing policy and current threshold of the NATPT, which can be described as following:

$$Ev(natptID, G) = R(natptID, G) + \frac{sysLoad_{current}(natptID)}{a * Threshold(natptId)} . \qquad (1)$$

The evaluation function is consist of two parts: one part is routing policy of application layer multicast: $R(natptID, G)$.The $natptID$ is the identifier of NATPT routers in the NATPT router group. While G refers to the specific group it wants to join. If the system load of the NATPT router is under some specific value, we will surely select the NATPT which is nearest to the group to handle the multi-cast session. If the nearest NATPT router has been working in a heavy system load, the second part: $\frac{sysLoad_{current}(natptID)}{a*Threshold(natptId)}$ will increase rapidly. a is a constant parameter in our simulation experiment, $Threshold(natptId)$ is the maximum processing ability of the NATPT router. The selection depends not only on the routing metrics between the NATPT router and the multicast group, but also on its current system load.

Once the NATPT has been selected, it will create a group item, associate with a newly allocated unicast IPv4 address, a newly allocated IPv6 group multicasts address. The NATPT router serves as an agent to join the application multicast group with the newly allocated IPv4 unicast address at IPv4 side.

When r2 wants to join, for r2 shares the same DR with r1, DR2 won't send out PIM Join message to NATPT, just return the IPv6 multicast address to r2.

When r3 wants to join, for r3 belongs to DR1, DR1 will also send PIM Join message to NATPT. This time RP found the requested group has existed, it will not send out the request to IPv4 side, just add DR1 to the multicast distribution tree.

When all receivers on a leaf-network leave the group, the DR will send a PIM (*,G) Prune message towards the RP for the multicast group. If all the DRs have left the group, the NATPT router will send the leave request to IPv4 application layer group. After successfully leaving the group, it will recycle the IPv4 unicast address and the IPv6 multicast address, the RP will also remove the group item.

3.3   Multicast Message Dissemination

Once the multicast distribution tree is setup inside IPv6 domain, all multicast communication will transfer along the distribution tree with the NATPT router acting as both the root and the source. It is different from the IPv4 application layer distribution tree; it is entirely IP layered and contains the shortest path.

As Fig. 3.3 described, when IPv6 hosts (r1, r2, r3) are receivers, all multicast messages received are from the NATPT router. Once receiving the datagram from other IPv4 member, the NATPT router will translate the datagram to DR1 and DR2, and the DRs will forward it to their hosts.When an IPv6 host e.g. r3 wants to send message to the whole group, the message will first transfer to DR2, then DR2 transfer the message to both DR1 and NATPT router. NATPT router will translate the datagram, then forward it to the near-est group node according the application mul-ticast routing rules to IPv4 side. DR1 will for-

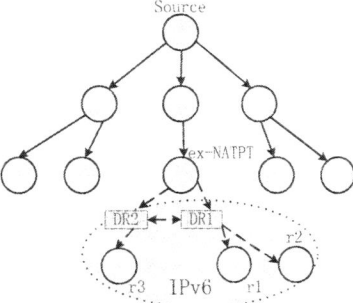

Fig. 3.3.   Application   layer multicast distribution tree

ward the message to r1 and r2. Thus all group members can hear the voice from r3.

3.4   Scalability

To guarantee the scalability, we extend the single NATPT router to NATPT-router group in IPv6 domain in TMS. As well known, there always exists one RP in IPv6 domain, while not only one NATPT router.

So we make RP also serve as a NATPT server, which is responsible for the message communication within the NATPT group and the schedule of NATPT

type	natptID	NATPT_PREFIX
threshold	current system load	data

Fig. 3.4. NATPT-message

session request and address allocation. The NATPT routers not only perform the multicast data communication, but also send NATPT messages to the NATPT server to report its system load and address mapping info. The message form is as Fig. 3.4 described.

Type field indicates the type of the NATPT-message, maybe REGISTER, UPDATE or URGENCY. On receiving the REGISTER messages, the NATPT server will return the client a natptID as an identifier. By UPDATE messages, NATPT server can make the decision whether redirects new NATPT request. When the NATPT reports its current system load is more than some value (depends on a),the server will re-calculate the evaluation function to select another NATPT to perform redirection by DNS-Application Layer Gateway [7].

Usually the new NATPT request will be redirected to a light-loaded NATPT. When NATPT sends URGENCY message to server, it means it doesn't want to provide NATPT services any longer or it will break down. The server will also re-calculate the evaluation function to select another NATPT to resume the multicast sessions. The NATPT server will assign the original PREFIX to the newly selected NATPT to resume the address-mapping. In this way, it can decrease the loss of communication to the least extend.

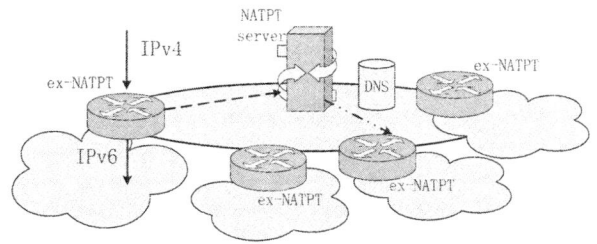

Fig. 3.5. DNS_ALG redirects DNS query when urgency

The new one can get the address-mapping information from data field of the URGENCY NATPT-messages.

TMS implementation of NATPT server on RP

```
1)  advertise_servermsg(msg)
2)  rcv_msg_from_client(msg)
3)  switch msg.type is
4)      REGISTER:  natptID=create_natpt_id()
```

```
5)      UPDATE:      update(natptID,msg)
6)      URGENCY:
7)                   newID=evaluate(G,newID.sysLoad)
8)                   reassign_prefix(newID,oldID)
9)                   reload_addr_mapper(newID.mapper,oldID.mapper)
10) rcv_natpt_request(request)
11) natptID=evaluate(G,natptID.sysLoad)
12) if(check_system_overload(natptID))
13)     newID=evaluate(G,newID.sysLoad)
14)     redirect_dns_query(newID ,oldID)
```

TMS implementation of NATPT router

```
1) if!(check_system_overload)
2)      send_update2server(serverID,msg)
3) else
4)      send_urgency2scrver(serverID,msg)
5)      send_sys_info(serverID)
```

3.5   Transition Adaptations

At the early stage of transition, the main part of the group is IPv4 application layer members. Main applied scene is IPv6 member to join IPv4 application group. With the transition of the network, the ratio of the IPv6 hosts to the whole group increases gradually. At the late stage of the transition, the main part of the group is IPv6 members, the IPv4 members locate sparsely in the group. And TMS will also adapt to the transition network, allow the sparse IPv4 application layer members

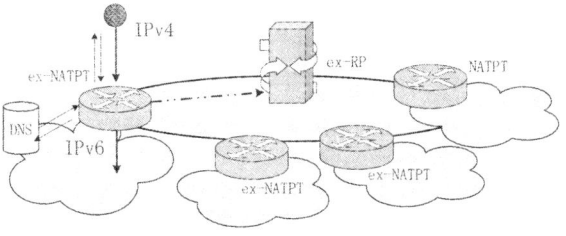

Fig. 3.6.   Single IPv4 host joins IPv6 multicast group

to join the IPv6 network layer multicast group based on the DNS-Application Layer Gateway implemented in NATPT router.

If a single IPv4 host wants to join an IPv6 network layer multicast group, it may send Scribe Join request based on the group domain name provided by the application program. As the NATPT server is viewed an IPv4 node running Scribe and also the traffic entry to the IPv6 group, it will get the DNS query message. The DNS-Application Layer Gateway will translate the query and forward it to the DNS server inside the IPv6 domain. If the DNS server replies an IPv6 DNS answer to the query, it shows that the domain name does exist inside the IPv6 domain. The selected NATPT router will create one multicast mapping items with the newly allocated IPv4 unicast address to the group. Thus the transition multicast structure has been setup, multicast communication can be carried out.

4   Experimental Evaluation

This section presents results of simulation experiments to evaluate the performance of a prototype TMS implementation. These experiments evaluate the performance along following three metrics:

1. The delay to deliver events to group members inside IPv6 domain when the percentage of IPv6 members increases.
2. Total network link stress inside IPv6 domain when the percentage of IPv6 members increases.
3. Load balance when NATPT requests increase, especially between many "single" NATPT routers and NATPT-router group defined in TMS. (The delay penalty is a measure of the increase in delay that applications perceive.

We refer to the number of identical copies of a packet carried by a physical link as the stress of a physical link).

We developed a simple packet-level, discrete event simulator to evaluate our scheme. The simulator models the propagation delay on the physical links. The simulations ran on a network topology with 5050 routers, which were generated by the Georgia Tech random graph generator using the transit-stub model. Scribe ran on 100,000 end nodes that were randomly assigned to routers in the core with uniform probability. Each end system was directly attached by a LAN link to its assigned router. The transit-stub model is hierarchical. There are 10 transit domains at the top level with an average of 5 routers in each. Each transit router has an average of 10 stub domains attached, and each stub has an average of 10 routers. We generated 10 different topologies using the same parameters but different random seeds. We ran all the experiments in all the topologies. The results we present are the average of the results obtained with each topology.

4.1   Transition Delay Penalty

The first set of experiments compares the delay to multicast messages using TMS inside IPv6 domain when the percentage of IPv6 members increases. To evaluate this delay, we measured the distribution of delays to deliver a message to each member of a group using both TMS and Scribe. We compute RAD as the metrics of delay penalty using these distributions: RAD is the ratio between the average delay using Scribe and the average delay using TMS. Fig 4.1 shows the cumulative distribution of the RAD metric when 20%, 30%, 60%, 80% IPv6 members in the whole multicast group.

In Fig 4.1,the y-value of a point represents the group numbers, the x-value represents the ratio between RAD of TMS and that of Scribe. Scribe surely increases the delay to deliver messages compare to TMS. What's more, it is conclude from Fig 4.1 that in TMS when IPv6 members increase the RAD will

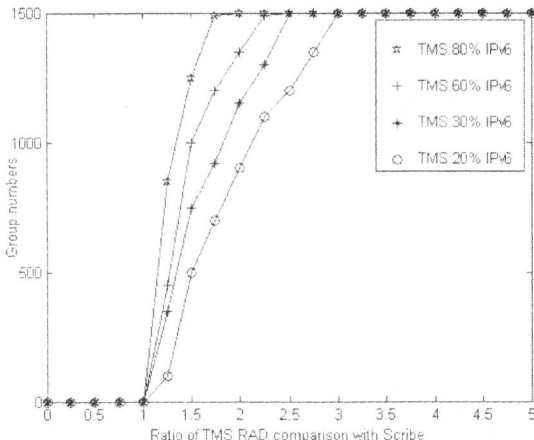

Fig. 4.1. Delay penalty decreases when IPv6 members increase

decrease and trends to 1 when the percentage reach 100%. It is easy to explain that inside IPv6 domain TMS using IPv6 network-layer multicast. When the percentage of IPv6 member reaches 100%, it is all network-layer multicast, so the RAD becomes 1.

4.2   Transition Link Stress

The second set of experiments compares the stress imposed by Scribe and TMS on directed link in the network topology with the transition to IPv6. We computed the stress by counting the number of packets that are sent over each link when a message is multicast to each of the 1,500 groups. Fig 4.2 shows the distribution of total link stress TMS compare to Scribe. The y-value of a point represents the ratio of total link stress in TMS to that in Scribe.

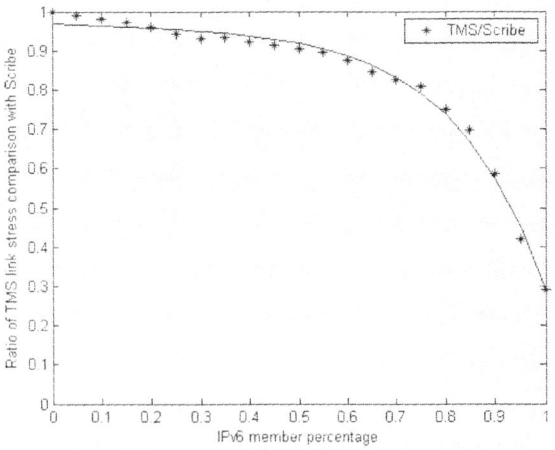

Fig. 4.2.   Total link stress decreases when IPv6 members increase

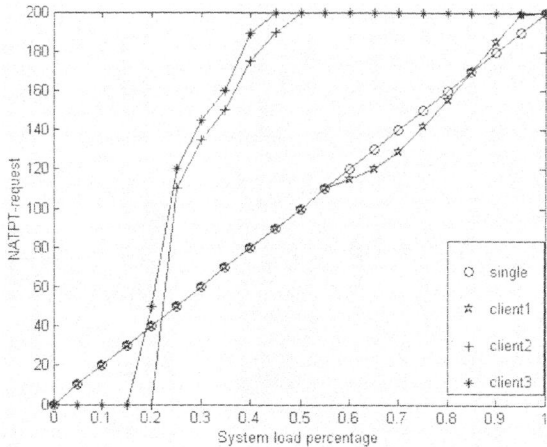

Fig. 4.3. Load balancing when NATPT requests increase

The x-value of a point represents the ratio of the IPv6 members to all members in the group.

Fig 4.2 shows when the percentage of the IPv6 members in the group increases to 100%, total link stress in TMS will decrease from 1 to about 1/3 compare to that in Scribe. Actually we will find current IPv4 application layer multicast and future IPv6 network layer multicast are the two extremes of our scheme. TMS is a general model for the deployment of multicast.

4.3   Load Balance when Scale

The final set of experiments compares the processing ability imposed by many "single" NATPT routers and NATPT-router group defined in TMS on directed link in the network topology with the transition to IPv6.Fig 4.3 shows the comparisons between many "single" NATPT router and NATPT-router group, mainly on the load balancing. The increase amplitude of incoming NATPT request is uniform.It shows when single NATPT router facing the increase of NATPT request, it has nothing to do. While in TMS, NATPT router client1 will share the load with client2 and client3 under the request-reschedule from NATPT server running on RP. Thus it can alleviate client1 when facing the increase of the NATPT requests.

5   Conclusions

We have presented Transition Multicast Scheme (TMS) for the network during the transition period from IPv4 to IPv6. It extends NATPT support in the edge router to handle multicast to and from IPv6 members among the other IPv4 members in the same application layer multicast group.TMS is implemented in network layer, so it is compatible with any application multicast schemes to supports large scale group multicast services. RP router is also extended with NATPT server performing load balance for NATPT-router group. So TMS possesses good scalability. At the early stage of transition, the main part of the group is IPv4 application layer ones. In the late of transition, the main part of the group is IPv6 members,IPv4 members locate sparsely around the group. TMS will also adapt to the transition network, allow the sparse IPv4 application layer members to join the IPv6 network layer multicast group based on the DNS-Application Layer Gateway implemented in NATPT.We have simulated our scheme and the measurements show its good performance in link stress and delay penalty compare with the application layer scheme.

Acknowledgements

We would like to thank the National Natural Science Foundation of China under Grant No. 90104002 and 69725003, and the National High Technology Development 863 Program of China under Grant No. 2001AA112132

References

[1] G. Tsirtsis, P. Srisuresh: Network Address Translation - Protocol Translation (NATPT), RFC 2766 (2000)
[2] Miguel Castro, Peter Druschel, Anne-Marie Kermarrec: SCRIBE: A Large-scale and Decentralized Application-level Multicast Infrastructure. IEEE Journal on Selected Areas in Communication, vol. 20, no. 8 (2002)
[3] Beichuan Zhang, Sugih Jamin, Lixia Zhang: Host-Multicast: A Framework for Delivering Multicast to End Users. In Proceeding of IEEE INFOCOM (2002)
[4] K. Tsuchiya, Hitachi: MTP http:www.ietf.org/internet-drafts/draft-ietf-ngtrans-mtp-03.txt
[5] S. Deering, D. Estrin, D. Farinacci, V. Jacobson, C. Liu, L. Wei: "The PIM Architecture for Wide-Area Multicast Routing," IEEE/ACM Transactions on Networking, vol. 4, no. 2 (1996)
[6] S. Deering, W.Fenner , B.Haberman: Multicast Listener Discovery(MLD) for IPv6, RFC2710 (1999)
[7] P. Srisuresh, G. Tsirtsis , P. Akkiraju: DNS extensions to Network Address Translators (DNS_ALG), RFC 2694 (1999)

User Relevancy Improvisation Protocol

K. Satya Sai Prakash and S. V. Raghavan

Network Systems Laboratory
Indian Institute of Technology Madras, Chennai, India
ssai@ieee.org
svr@cs.iitm.ernet.in

Abstract. In the present web search, mining the web repository and yielding the most relevant result set that is minimal in size is the most sought after requirement. In this paper we proposed a new protocol namely User Relevancy Improvisation Protocol (URIP) that employs "push" mechanism to obtain the most current and concise summary of the web sites. We strongly feel that context identification and relevancy characterization are best achieved through collaboration between web servers and search engines. Relevancy being user specific, we introduced two parameters, Knowledge Quotient (K_q) and Knowledge Factor (K_f) to rank the web pages. URIP clients at the respective web servers compute these parameters for each web page. Search Engine Server embodies the URIP server that gets the update notifications from all the web servers as and when the web sites are created or modified. User input contains query terms, his/her knowledge level ($K_{q'}$) and domain expertise ($K_{f'}$). Relevant document set is retrieved with closely matching $<K_q, K_{q'}>$ and $<K_f, K_{f'}>$. Apparent advantages of our proposal are, (1) Does not need constellation of thousands of systems to index the web data, (2) Search engine site need not maintain a snapshot of the web to respond to the users, (3) Eliminates spider activity and most importantly (4) Yields minimal and most relevant result set. We compare our results with the popular search engine Google. A Finite State Machine (FSM) based specification for the protocol is given and the protocol is analyzed using M/G/1 model.

1   Introduction

Web is a huge repository of information that is dynamic and all pervasive. SE is one of the most sought after tools to search the information on the WWW. In user's point of view SE need to fulfill two important specifications.

Recency: "It is the policy of the SE to render the most recent web document". Any document d_i rendered by SE with timestamp t_i should have the same timestamp at the original web site

S. R. Das, S. K. Das (Eds.): IWDC 2003, LNCS 2918, pp. 91–101, 2003.
© Springer-Verlag Berlin Heidelberg 2003

Relevancy: "It is the policy of the SE to render precise document the user is looking for". Any document d_i rendered for the input keyword (s) must be exact match to the user expectation.

Recency and *Relevancy* are the two sustaining core features of any SE. An *ideal* and *successful* SE must render absolutely recent and relevant information. A protocol based solution to maintain recency in real time is given in [1]. Unlike recency, relevancy is user specific feature. Relevancy depends on user's ability to grasp concept ($K_{q'}$) and user's awareness in the specific domain ($K_{f'}$). Current search engines adopt a static page ranking policy that fails to fit into the user's ability and knowledge potential. Hence the ranking methodology is revamped in this paper. Every document has specific (K_q) and certain (K_f) values with respect to their content. Computation of these parameters is a big task at hand. It is a semi-automatic process in a sense that the author is the best judge to assign the appropriate values to these parameters. Otherwise the parameters are computed according to specific formulae given in section 3. In the later stages, this information needs to be part of the SE index to yield the most relevant and minimal result set.

The best form of ensuring such communication between web servers that are spread across globe and SE is in the form of a *protocol*. Protocol ensures a formal and clear specification of the tasks and requirements. It enables the transactions to flow smoothly and regulates the communication process between the server and the clients.

In this paper we have given the formal description of the protocol and highlighted the need for such a protocol. We also show that this protocol has number of advantages like, a) Reduction of the large result set, b) Elimination of spider activity and c) Minimal infrastructure requirement.

This protocol enables the ubiquitous presence and distributes the page ranking computation by virtue of its deployment and definition, thus achieving the above stated advantages.

The limitation of this work is implementation of the protocol in real Internet scenario. We are hopeful that the success of collaborative technologies enables our protocol to come into practice.

The rest of the paper is organized as follows. In Section 2, we describe the motivation and related work. In section 3, we have given the relevancy characterization for users as well as documents. Section 4, gives a Finite State Machine (FSM) based formal specification and description of the proposed URI Protocol. We also gave an M/G/1 model based analysis for URIP in this section. Section 5, gives the cost analysis of this protocol-based technique and highlights its superiority. In section 6, we discuss the results and the future work and conclude in section 7.

2   Motivation and Related Work

Information retrieval is an important as well as integrated functionality of any SE. Well-known metrics in the literature to measure the relevancy of the result set for a given keyword(s) are, *precision* and *recall* [14].

Precision = relevant documents/retrieved documents

Recall = retrieved documents/relevant documents

But these metrics only serve as absolute reference. Actual way to improvise the relevance is by identifying all the contexts [3][10]. Various clustering algorithms like hierarchical clustering, k-means clustering, EM Clustering are in vogue. The client-server paradigm based technique proposed in [3] to yield highly relevant context sensitive results has strongly motivated our work. Wisenut that has the motto "search exactly" is yielding quite a small result-set. It is an added motivation to our work. Teoma and Vivisimo are also interesting search technologies to look at and observe.

In [2] personalized search is highlighted and described as an important issue to be addressed. [4] and [7] discuss two different techniques for person specific search on the Internet. Sine personalization and invasion of privacy have a thin break up line [8] addresses this issue and gave a technique for user specific search. We identified that relevancy and personalization can be achieved without transgressing the privacy by modeling user's knowledge quotient ($K_{q'}$) and knowledge factor ($K_{f'}$).

A hierarchical model based technique is presented in [5] to index the pervasive data. These meritorious works have helped us to formulate the protocol-based solution that carries out the distributed computation of K_q, K_f of pervasive web data at local web servers and uses client-server paradigm to communicate to the SE.

[11][12] and [13] gave an insight into protocol proposal, specification, description, testing, implementation and conformance. Thus web servers collaborate with SE to yield most recent and relevant information. In contrast to [9] we showed in section 5 that implementation of our technique reduces the storage and other infrastructure cost by about 90%.

3   Relevancy Characterization

Relevancy Characterization is done for both user and documents as well.

3.1   User Relevancy Characterization

User relevancy is characterized with the following parameters.

Knowledge Factor (K_f). Changes with Context. Every individual possesses varying knowledge levels in various fields. GM Vishwanath Anand is a *Chess* maestro. His K_f will be close to 1in the filed of Chess. But his capability tested over Nuclear Physics may result a value close to zero (0.1 or 0.2).

How to compute K_f? K_f is a normalized factor lies between 0 and 1. It is the ratio of individual knowledge to the World Knowledge Pool (WKP). WKP is a collection of all domains (D), where in each domain is a complete collection of facts (F). *Encyclopedia Britannica* would serve as one example for such a WKP.

$$K_f = \Sigma \ F_i/F \text{ for each domain D, i} = 1 \dots N$$

Knowledge Quotient ($K_{q'}$). Constant for a user. Like IQ every one has a definite ability to acquire knowledge. Let us term that as K_Q. It can be computed in the similar lines as IQ is measured. The weighted questions are collected across all disciplines

covering breadth and depth. Queries cover various learning processes such as
learning by example, learning by reading and practice etc.

3.2   Document Relevancy Characterization

Document relevancy is characterized with the following parameters.

Knowledge Factor (K_f). Each web document contains the information related to
some field. The depth at which the information is presented constitutes the K_f of a
document. For example, K_f of a Java tutorial page is given as 0.3, where as a page
describing the "multi threading concept with exception handling" in the same website
will have a value like 0.75.
   Computation of K_f is tricky. We take the standard document classification word
dictionary (ex: Tutorial, Illustration, Theorem Proving, Description etc.,) and look for
the presence of such words in the documents. Presence of these words associated with
positional weights yield a normalized value between 0 and 1.

Knowledge Quotient (K_q). very document that is created is meant for a class of users
with certain $K_{q'}$. This estimated ability is assigned as K_q of a document.
   For example, ACM, IEEE web sites are meant for a class of users who has serious
academic interest and pursue research, where as newspaper sites and sports sites are
for more general consumption.
   We need to add two more tags $<K_q></K_q>$ and $<K_f></K_f>$ to the existing HTML
tag set to enable the authors to rank their web documents. XML by default permits
such additional tag definitions. Authors of the web documents are best judges to rank
their documents. So they make use of the above tags to mark the targeted audience
and the depth of the content. K_q and $K_f \in [0,1]$.

Illustration:

```
<HTML>
<Head>
<Title>Welcome to the IEEE</Title>
. . .
<K >0.6</K >
 q          q
<K >0.2</k >
 f          f
</Head>
<Body>
. . .
</Body>
</HTML
```

4   User Relevancy Improvisation Protocol (URIP)

URI Protocol is an application layer protocol that enables the computation of the
relevancy parameters for the documents at their corresponding web sites and
communicates the update to SE site. Thus the centralized computation of present
search engines towards indexing and ranking is distributed to all the web servers

themselves. In this way URIP is distributed, pervasive computing protocol that instantaneously updates the SE when ever page is created or modified.

4.1   URIP Specification

We give Finite State Machine (FSM) based protocol specification for URIP. Initial handshake for connection establishment, connection closing and abort procedures are in accordance with the other application layer protocols. Hence no explicit mention is made in this paper. URIP client resides in the web servers and URIP server is embodied in the SE.

Web Server Functionality (Protocol Client):

- Invoke *Information Agent (IA)* whenever there is a creation or modification of web page
- Update IA with the URL Tag, key words, $K_{q'}$, $K_{f'}$, CSI and summary.

Search Server Functionality (Protocol Server):
Read the IA update - Update the document information - Read the content description - Get the change frequency of the URL - Rank the URL

URIP Server States:
Idle: Starting state in which the server waits for the client update.
Receive_Update: Upon receiving the client update, Server gets into this state. In this state it reads the packet header to check for the correctness. On success, it moves to next state (Process_Update) or *On_Error* it goes to Reject_Update state.
Process_Update: Parses the update packet and checks the correctness and validity of the update. Upon which it either moves to next state (Update_DB) or *On_Error* goes to Reject_Update state. It may invoke the spam guard functions incase of spurious updates in terms of content or frequency.

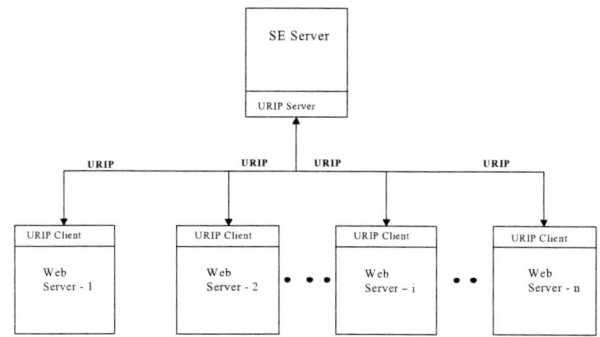

Fig. 1. User Relevancy Improvisation Protocol

Update_DB: In this state, server opens a connection with the database and updates the obtained information. Incase of any error, throws an appropriate error flag and moves to Idle state or Receive_Update state to process the next update request (if any)

Reject_Update: This state looks for pending requests in the queue and on finding them it moves to Receive_Update state otherwise settles in Idle state.

URIP Server Functions:
Read_Header (): A function to read the Update packet sent by the information agent from the client. If the header is not conforming to the protocol, an error flag is raised by this function.
Parse_Packet (): This function parses the Update packet and looks for the protocol conforming URL, and Description fields. Incase of incompatibility raises the appropriate erro flag.
Write_DB (): An important function that opens the connection with the local Database and writes the updated information.

Special Functions:
Spam_Guard(): This special function is invoked if the SE is flooded with spam packets (both in terms of spurious frequency and/or keywords).
Block_Sender(): This function is invoked as a remedy for spamming.

URIP Client States:
Idle: Start state in which the client is waiting for the OS to raise an Update interrupt.
Prepare_Update: In this state the client prepares the Update packet in the prescribed format.
Dispatch_Update: Sends the prepared Update packet with the help of the *information agent.*

URIP Client Functions:
Create_Update (): This is the function that enables the client (web server) to fill the packet with the correct information.
Send_Update (): Enables the client to send the Update packet to the server (SE).

4.2   FSM for URIP

Finite State Machine specification for URIP is given in this section.

URIP Server:
The FSM for URIP Server is a quintuple $<I, O, S, \delta, \lambda>$, where, $I = O = \{I_1, I_2, I_3, I_4, I_5, I_6, I_7\}$, $S = \{S_1, S_2, S_3, S_4, S_5\}$, $\delta = I \times S \rightarrow S$ and $\lambda = I \times S \rightarrow O$. Further
I_1 = On_Receive_Interrupt, I_2 = On_Receive_Update, I_3 = On_Process_Update,
I_4 = On_Update_DB, I_5 = On_Reject_Update, I_6 = On_Error_Update,
I_7 = On_Completion_Update.
And S_1 = Idle, S_2 = Receive_Update, S_3 = Process_Update, S_4 = Update_DB,
S_5 = Reject_Update

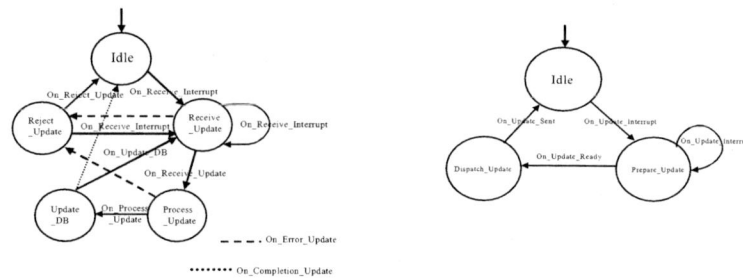

Fig. 2. FSM for URIP Server   **Fig. 3.** FSM for URIP Client

Table 1. State transition table for URIP server and Client

Inputs States	I_1	I_2	I_3	I_4	I_5	I_6	I_7	Inputs States	I_1	I_2	I_3
S_1	S_2	-	-		-	-	-	S_1	S_2	-	-
S_2	S_2	S_3	-	-	-	S_5	-	S_2	S_2	S_3	-
S_3	-	-	S_4	-	-	S_5	-	S_3	S_1	-	S_1
S_4	-	-	-	S_2	-	-	S_1				
S_5	S_2	-	-	-	S_1	-	-				

Problems related to the Database update are not highlighted in this protocol. Server takes the necessary measures for successful writing of the data to the database.

URIP Client:
The FSM for URIP Client is a quintuple <I, O, S, δ, λ>, where, I = O = {I_1, I_2, I_3}, S = {S_1, S_2, S_3}, δ = I x S → S and λ = I x S → O. Further
I_1 = On_Update_Interrupt, I_2 = On_Update_Ready, I_3 = On_Update_Sent
And S_1 = Idle, S_2 = Prepare_Update, S_3 = Dispatch_Update

URIP Errors:

URIP101 – Invalid Update Packet URIP201 – Invalid URL, URIP202 – Invalid Timestamp URIP203 – Invalid Data URIP301 – Blocked URL for Keyword Spam URIP302 – Blocked URL for Repeated Update URIP401 – Error Opening Database URIP402 – Error Reading/Writing Database

4.3   URIP Format

Typical URIP packet contains the URL information, calculated K_q, K_f , CSI and summary.

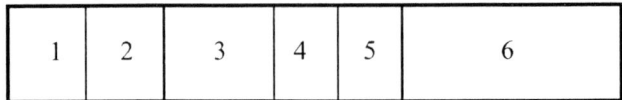

Fig. 4. URIP Packet Format

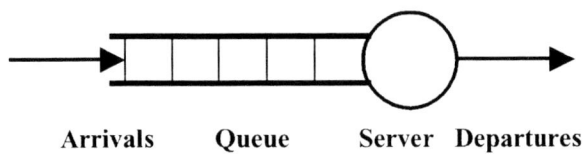

Arrivals   Queue   Server Departures

Fig. 5. M/G/1 queuing model

1 – Packet Header, 2 – URL, 3 – Keywords, 4 – K_f, 5 – K_q, 6 – Document Summary

4.4   URIP Analysis

M/G/1 Model. Information updates from URIP Clients to URIP Server are modeled as M/G/1 system in queuing parlance. As the information processing at each web server is an independent event the update arrivals at SE site are Poisson with arrival rate λ. Service rate at the SE is assumed to be general distribution with mean E(t) and variance var(t), as it depends on database read/write, user query processing and response time apart form processing the URIP Client updates. Let the service rate be $\mu = 1/E(t)$ and $\rho = \lambda E(t) < 1$ (Steady state condition). Then various measures such as server utilization (ρ), average response time (A), mean number of updates in the queue/buffer (W_q) can be given by the following expressions.

Server Utilization (ρ) = $\lambda E(t)$
Mean number of updates (L_s): $\lambda E(t) + [\lambda^2(E^2(t) + var(t))]/2(1-\lambda E(t))$
Average Response Time: $E(t)/(1-\lambda E(t))$
Mean number of updates waiting to be processed (W_q): $(L_s - \lambda E(t))/\lambda$

5   Cost-Benefit Analysis

SE establishment incurs variety of costs. Current SE maintains a snapshot of the web to yield the results. This requires enormous storage and computation power. They also employ spiders to download the web data that incur the communication cost. In this section we made a general estimate for computation cost, communication cost and storage cost.

Communication Cost. Google indexes about 3 billion pages in 28 days and also 3 million important pages every day [9]. Hence Spiders are reading the documents locally at web servers and downloading the data that is of above proportions (1.3 TB/day). But URIP need to send the updates about newly created or modified pages only. It is about $1/5^{th}$ or $1/6^{th}$ of the web that changes every day [5]. Hence about 60 million updates of size approximately 1KB are communicated. That amounts to 60GB/day, which corresponds to 95.38% reduction.

Computation Cost. About 3 billion pages need to be indexed, ranked and stored in databases. Adding to that computation for each query term, relevant result-set needs to be retrieved. On an average Google responds to 1000 requests/sec.

In contrast URIP server need to process the client updates according to the $<K_{q'},$ $K_{f}>$ and stores in the database. In place of complex indexing and ranking algorithms we have simple look-up and insertion procedures in place.

Storage Cost. Assuming that on an average each page is about 12KB, to store 3 billion pages, it needs about 36Tera Bytes. (Image and other formats that need more storage are not considered)

URIP on the other hand sends the update that is at most 1KB. (Exact packet size will be estimated in our future work). Hence the cost cut down on the total storage space is about 91.7%.

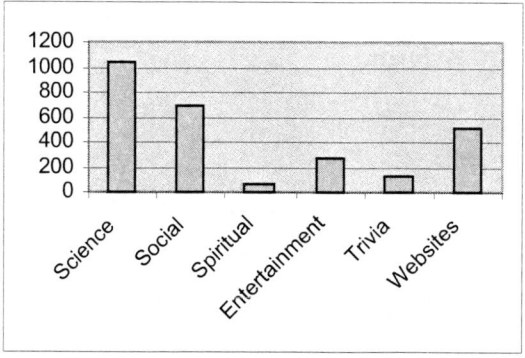

Fig. 6. Query Classification

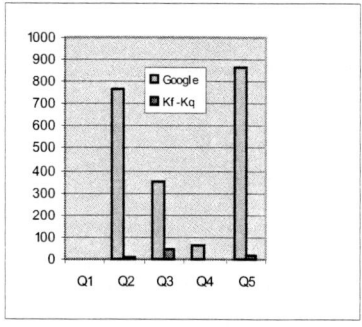

Fig. 7. Google vs. K_q-K_f retrieval

6   Results

This section contains the experiment description, limitations of our work and the future scope. Here we are showing our preliminary results derived from relatively small set of documents and keywords to show the validity and usefulness of concept introduced. Elaborate results based on exhaustive query terms on huge document set will be reported in future work. Typical questionnaire is given in Appendix – A, that shows how the $K_{q'}$ and $K_{f'}$ are obtained from users.

6.1   Experimental Setup and Description

Using TCPDUMP utility all the queries posted to Google [10] were traced and collected over a period of 10 days. The proxy server caters to the whole institute (IIT Madras). The traffic is generated across eleven departments and approximately 1000 students (under graduate, graduate and doctoral), faculty and other staff.

Following observations are made from the log analysis

Query Classification: All the query terms (2720) are classified into 6 general categories shown in figure 6.

Comparison of result sets from Google and K_q-K_f based retrieval for 5 types of queries: Queries specific to a website (http://iccc2002.ernet.in) resulted exact hit by both. But single key, multiple terms and phrases retrieved larger result sets by Google. They were sieved by applying K_q (0.2) and K_f (0.4) after which the result set is reduced by 95% to 99%.

6.2   Limitations & Future Work

The client specification of the protocol needs to be incorporated in all web servers to test the protocol in real Internet scenario. We believe that the success of collaborative technologies will help the limitation to be surpassed.

We are implementing the protocol in all web servers at our institute level. Workload characterization and performance analysis of the protocol will be done based on the collected traces. We are also planning to characterize the general distribution at the SE server by observing the database transactions and user query patterns.

7   Conclusion

In this paper we have characterized the relevancy, which strongly depends on the user's ability and domain expertise. Our conceptualization of relevancy naturally brings in the personalization without violating privacy. We proposed URI protocol to characterize the relevancy of a page and communicate the same to the SE, in order to cater to the pervasive web. We also showed that our technique has many advantages over the current search techniques such as 1) Minimal infrastructure for SE, 2) Elimination of Spider activity and 3) Minimal Result Set

References

[1]   K. Satya Sai Prakash and S. V. Raghavan, "Web Recency Maintenance Protocol", *LNCS 2571*, Springer, Proceedings of International Workshop on Distributed Computing, Kolkata, December 2002, pp. 35-44

[2]   James Pitcow et al., "Personalized Search", *Communications of ACM*, Vol.45, No. 9, September 2002, pp. 50 –55

[3]   Lev Finkelstein et al., "Placing Search in Context: The Concept Revisited", *ACM Transactions on Information Systems*, Vol.20, No.1, January 2002, pp. 116-131

[4]   Glover E. J. et.al, "Web Search Your Way", Communications of the ACM, 44(12), pp. 97 – 102, 2001

[5]   Paul Castro and Richard Muntz, "An Adaptive Approach to Indexing Pervasive Data", Second ACM international workshop on Data engineering for wireless and mobile access, California, May 2001, pp. 14-19

[6]   Brin E. Brewington and George Cybenko, "How Dynamic is the Web", *9th World Wide Web Conference*, Amsterdam, 15 – 19 May 2000, http://www9.org/w9cdrom/264/264.html

[7]   Pretschner, Susan Gauch, "Ontology based Personalized Search", Proceedings of the 11th IEEE International Conference on Tools with AI (ICTAI'99), Chicago, November 1999, pp. 391 – 398

[8]   Philip Chan, "A non-invasive Approach to Build Web User Profiles", Proceedings of ACM SIGKDD International Conference, SanDiego, August 1999, pp. 7 – 12

[9]   Sergy Brin and Lawrence Page, "The Anatomy of Large Scale Hyper textual Web Search Engine", *7th World Wide Web Conference*, Brisbane, Australia 14 – 18 April 1998. http: //www7.scu.edu.au/programme/fullpapers/1921/com1921.htm

[10]   Ellen Riloff and Wendy Lehnert, "Information Extraction as a basis for High Precision Text Classification", *ACM Transactions on Information Systems*, Vol.12, No.3, July 1994, pp. 296-333

[11]   Colin H. West, "Protocol Validation in complex Systems", Proceedings of ACM SIGCOMM'89 Symposium: Communications, Architecture and Protocols, Austin, September 19 – 20, 1989. Computer communications Review, Vol. 19, No. 4, pp. 303 – 312

[12]   Lin F., Chu P. and Liu M., "Protocol Verification and Using Reachability Analysis", Proceedings of ACM SIGCOMM'87, pp. 126 – 135

[13]   K. Tarny, "Protocol Specification and Testing", Plenum Press, ISBN: 0-306-43574-8

[14]   Robert E. Williamson, "Does Relevancy feed back Improve Document Retrieval Performance?", Proceedings of the 1st annual international ACM SIGIR conference on Information storage and retrieval, May 1978, pp. 151-170

[15]   http://www.searcenginehguide.com/wi/2002/0211-wi1.html

Characterizing Web Workloads –
A Transaction-Oriented View

Shubhashis Sengupta

Software Concept Laboratory, Infosys Technologies Limited
Electronics City, Hosur Road, Bangalore 561229, India
Tel: 91 80 852 0261   Fax: 91 80 852 0740
shubhashis_sengupta@infosys.com

Abstract. Workload characterization is a critical component in capacity planning for Web-based *n*-tier systems. Web workloads are characterized by different patterns and intensities of different transactions. In this paper, we present a technique for characterizing Web workloads in a transaction-oriented manner through offline analysis of Web server logs. The results can be used in performance evaluation of the current system, extrapolating the load for increased user base for capacity projection of an augmented system, and in preparing performance testing strategies for the augmented system.

1   Introduction

The performance of any computer system cannot be modeled reasonably accurately without knowing the workload, that is, the requests being processed. In case of a 3-tier Web architecture, consisting of a set of Web servers, application servers and database servers, the requests arrive at the Web (user interface) layer. The practitioners commonly use Web workload in terms of the metrics like concurrent requests per second, average number of active sessions, and the service rate (number of bytes transferred per second) at the ports etc. as inputs for sizing the infrastructure. While the number of concurrent users gives a fair approximation of load at the Web server, it is equally important to track the navigation of a user, through a series or sequential and related click streams, along a transaction path. A Web transaction is defined as a series or interactive activities between the user and the Web-site in which the user navigates through one or more pages or Universal Resource Locators (URLs) at the site to perform a logical business functionality in a browsing session. A request at any point of a transaction may trigger a set of demands at the resources in various tiers. In Figure 1 we show how a Web request for displaying ticket number for an employee results in visits to Web, Application and Database servers. In the figure, the transaction is started by an HTTP Get request to a page (URL). The Web server passes the query to a business object (B.O.) at the application layer. The B.O. may perform some business logic calculation and may throw up another page to the user as a response. The user may send in a second request against this page, which ultimately may get routed to the DB tier and may result in display of a third page showing the data. In this paper, we use the terms URL and page interchangeably.

S. R. Das, S. K. Das (Eds.): IWDC 2003, LNCS 2918, pp. 102-111, 2003.
© Springer-Verlag Berlin Heidelberg 2003

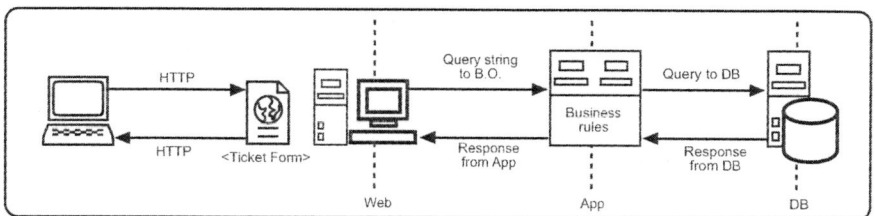

Fig. 1. Tier-wise decomposition of a Web request

For an existing Web-based system, main transaction paths are known beforehand. In such cases, we need to capture key characterization metrics such as ratio of completed transactions of different types, average time taken to complete a transaction, average visit counts for each of the URLs etc. to do an evaluation of the current workload. These metrics are often used to see how the current usage profile can affect the system performance when the user load is increased and / or new functionalities are added, and to devise an appropriate testing strategy for the augmented system. They are also used to derive important parameters like visit-to-buy ratio that are used in Web benchmarks like TPC-W [1].

Very few tools and techniques support transaction-oriented characterization of Web workloads. Tools like LogAnalyzer and WebLogMiner [2] analyze user navigational patterns. These tools have limited analyzing capabilities as they generate summary statistics like cumulative views for individual pages etc. Other techniques, such as those based on hypertext probabilistic grammar, are used to mine statistically significant navigational trails [3] from the Web logs and client usage information. These techniques are used for predicting most likely user surfing patterns for personalization purpose, and cannot be used for workload characterization.

In this paper we introduce a technique for workload characterization of Web-based OLTP systems through offline analysis of Web server access logs. The rest of the paper is as follows. In the next section we briefly review the previous works in workload characterization. In section 3 we introduce some useful concepts and an algorithm for transaction-oriented characterization of Web workloads. Section 4 presents computational experiences in real-life case studies. In section 5 we conclude by highlighting the drawback of the current techniques and proposing future work to overcome this.

2   Prior Work

A key aim of workload characterization is to derive a model to capture and reproduce the behavior of the workload quantitatively and isolate the most important features. An excellent survey of workload characterization of a wide variety of computing systems, including batch, interactive, database, network-based and parallel systems is presented in [4]. An impressive volume of work has been reported in determining the pattern of request arrivals to Web-based systems. However, although researchers have associated Web workloads to a sequence of pages or URLs denoting a transaction [5], not enough works have been reported to characterize Web workloads in transaction-oriented way.

One of the early works in modeling workload of interactive systems was the notion of the User Behavior Graphs [6]. The nodes of a UBG correspond to the commands, while the arcs, with their associated probabilities represent sequence of commands issued by the users. With the advent of graphical user interfaces and client / server environments, the hierarchical nature of the workload became apparent. A view of the system infrastructure components has been proposed in "network-based" workload characterization environment [7]. The methodology is based on a layered structure that corresponds to a logical subdivision of hardware components into three layers; i.e., the users, the processing nodes and the network. The UBG can then be broken down into a sequence of requests and their processing loads at each of the nodes and also on the network.

Krishnamurthy and Rolia [5] are among the first to classify the workload on the basis of the server-side URLs. The URLs are broadly classified into three categories – static HTML pages, pages generated due to CGI scripts, and forms for fixed database queries. The authors came up with different request workload classes by looking at how each workload (transaction) is composed from these URLs. The authors also proposed the advantages and the disadvantages of several workload abstractions for assessing the QoS offered by the server for estimating response times. The results were used to support call admission algorithms. At a different level, Arlitt et. al. have analyzed the data from Web servers from domains to characterize typical workload patterns [8].

The most significant effort of characterizing Web workloads is due to Menascé et. al.[9]. They introduced the concept of a state transition graph called Customer Behavior Modeling Graphs [CBMG] to describe the behavior of groups of customers who exhibits similar navigational patterns. The set of related and consecutive requests are called sessions, akin to the notion of transactions introduced here. The construction of a workload model relies on two algorithms proposed in the paper. The first one takes as input conventional HTTP logs and generates a session log. The second algorithm takes as input the session logs and performs a clustering analysis which results in a set of CBMGs that can be used as a compact representation of the e-commerce workload. The authors proposed an elegant model to profile the customers on the basis of the trace behavior for a session.

3   Towards Transaction-Oriented Workload Characterization

We start by defining path of a Web transaction. A transaction path (TP) is defined as a chain of URLs or pages that constitute a transaction. Figure 2 is a simple schematic description of an Internet auction site depicting the pages along the transactions. The figure shows two transactions – *Bid* and *Sell*. The transaction *Bid* has the path *login -> browse -> bid -> transact -> logout*. The path *login -> browse -> sell -> transact -> logout* represents transaction *Sell*. We define *Average Page Service Time* (APST) as the difference between the time at which a page is requested and the time by which the request is served at the server side. The *Average Session Length* (ASL) is defined as the average difference between time when the last page associated with a transaction associated with a transaction has served the request and time when the first page is requested. The ASL would include the network roundtrip time, the client think times at each of the pages and the server residence (processing + queue times) for the request at each of these pages. These values can be obtained from the log timestamps and from the service time fields in the logs. The pages or nodes in the TP can be forward, back or self-reference able. In the figure, the page *browse* (*b*) has a forward reference from the page *login* (*a*), back references from pages *bid* (*c*) and *sell* (*d*) along transaction paths for *Bid* and *Sell* respectively. The page *b* is also self-reference able as it can be invoked repetitively by the client. For example, the client can repetitively call the *browse* page with different query criteria to view the items put up for bidding in different categories. The TPs for a particular site can be clubbed together to form a Transaction Board (TB). A TB can be depicted as a directed graph where nodes are pages that occur along a path and the edges represent transition of one page to the other. A TB can be represented easily through a data structure similar to an adjacency matrix. The TB for the auction site is shown in Figure 3. The board depicts the linkages between various pages in the site. The value of cell(*i*,*j*) = 1 if there exists a path between pages *i* and *j* and 0 otherwise.

It is often seen that only some percentage of the transactions are concluded in the logical sense. In our example, only 25% of the requests for the *browse* page may eventually reach *bid* page. Such dropping off along a transaction path may happen due to business or performance related reasons. We define Transition Rate (TR) as probability of a request that getting forwarded from one page to the next along a transaction path. Suppose *i* and *j* are two pages along transaction path(s), the TR between pages *i* to *j* is given by $\rho(i\text{-}j) = T(i\text{-}j)/C_i$ where C_i is the total number of request arrivals to page *i* and $T(i\text{-}j)$ is the number of request that have reached *j* from *i*. The matrix in Figure 4 presents the illustrative transition rates between various pages in the auction site example, where *cell(i,j)* represents $\rho(i\text{-}j)$.

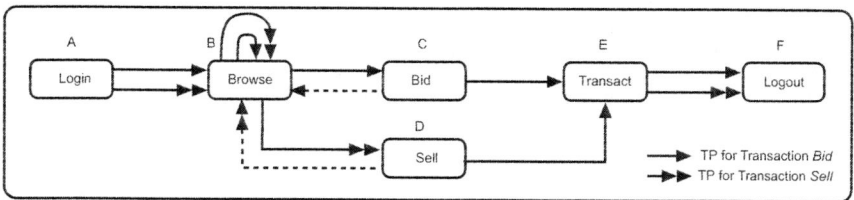

Fig. 2. Main transactions in a simplified auction site

	Login	browse	Bid	Sell	transact	logout
Login		1				
Browse		1	1	1		
Bid		1			1	
Sell		1			1	
Transact						1
Logout						

Fig. 3. Transaction board for the auction site

	Login	Browse	bid	Sell	transact	logout
Login		0.50				
Browse		0.30	0.25	0.15		
Bid		0.05			0.10	
Sell		0.12			0.10	
Transact						1.0
Logout						

Fig. 4. Transition rates between the URLs

In the above figure cell(i,j) gives the TR (ρ) from page i to j. Notice that the probability values at the third row (i.e., transitional rates out of page b) add up-to 0.7. This indicates that 30% of the clients drop out at the *browse* page. We term this as *exit* state. The transition rate can be utilized to calculate average number of visits or visit counts to a particular URL for a transaction. As discussed in [9], let V_m be the average number of visits to page m, then a set of linear equations can be written for the auction site transaction board as

$$V_a = 1.0 \qquad\qquad \dots\text{(i)}$$
$$V_b = \rho_{a\text{-}b}\,V_a + \rho_{b\text{-}b}\,V_b + \rho_{c\text{-}b}\,V_c + \rho_{d\text{-}b}\,V_d \qquad \dots\text{(ii)}$$
$$V_c = \rho_{b\text{-}c}\,V_b \qquad\qquad \dots\text{(iii)}$$
$$V_d = \rho_{b\text{-}d}\,V_b \qquad\qquad \dots\text{(iv)}$$
$$V_e = \rho_{c\text{-}e}\,V_c + \rho_{d\text{-}e}\,V_d \qquad \dots\text{(v)}$$
$$V_f = V_e \qquad\qquad \dots\text{(vi)}$$

In general, we can write VS = 1 and $V_j = \sum_{i=1, i \neq j}^{n} V_i$ where S is the starting page(s) of the TPs in the transaction board and j is any other page. For example, visit count for page browse in our auction site example is 0.75. A transaction is called active when a client makes a request for an URL along a TP. The variable *session_timeout* represents the time interval within which two consecutive requests from the same client along a transaction path must occur, otherwise the transaction is deemed as aborted. In our case, the TPs are taken as an ordered set of URLs and the Web log can be viewed as per-client ordered set. Each line from the access log contains information on a single request for a URL. The log entry for a normal request can be of the form

hostname- - -[dd/mmm/yyyy:hh:mm:ss tz)   request(URL)   status   bytes
For example,
10.177.93.8 - - [11/Mar/2003:02:16:43 +0000] "GET dbGRS/login.jsp HTTP/1.1" 200 8528

From each log entry it is possible to determine the IP address of the host (client) machine making the request, the time that the request was made, and the URL of the requested resource. The most popular formats of recording logs are those of Apache and Microsoft IIS Web servers. They, however, vary slightly in formats.

We now propose a Depth First Search (DFS) algorithm variation to find key workload metrics. Let S be the set of start URLs of all the TPs in a TB. In our example auction site $S = \{login\}$. Each of the TPs in a TB is associated with a *transaction_count* and each of the URLs is associated with a *page_count*. Each TP and page is also associated with ASL and APST respectively. From the log file we pick up the first entry that is not visited, i.e., that is not already in a closed or aborted transaction. If the URL is not in S we tag this URL as *visited* and search for the next non-visited entry. If the URL is in S, we try to explore each of the TPs the URL is part of. If we find a TP, we report that and update the associated parameters. If a node is created from a log entry and if the node is evaluated against a TP, we tag the entry as *visited*. We finish when the EOF is reached, i.e., when there are no unvisited entries remaining. We describe the algorithm in the form of the following pseudocode.

Algorithm FindMetrics

Step 1:   Create the TB for all TPs, initialize the set S for all TPs, set the *transaction_count* for all TPs and *page_count* for all URLs to zero, set the values for ASL (for each TP) and APST for each page as zero, set *session_timeout* equal to t seconds;

Step 2:   While there is an unvisited entry in the log file and the URL belongs to set S

    Step 2a: Select the first such entry, create node I with the following elements: let *client_id* = hostname, and *log_time* = current timestamp, *curr_URL* = current URL; initialize a global list called *curr_tran*, append *curr_URL* to this list;

    Step 2b: From the TB get all URLs that are connected to *curr_URL* through forward, backward or self references, initialize list *prob_tran* for the node I and append the URLs to this list;

    Step 2c: Insert node I in the tree;

    Step 2d: Call procedure *GetTP(I)*;

Step 3:   Tag the entry as visited;

Step 4:   Calculate transition rates between various URLs in the TB, calculate the visit counts using Gauss-Jordan elimination method;

Step 5:   Output workload metrics like transaction counts (number of completed transactions of a given type in a given time period), transaction mix (ratio of various transactions in a given time period), ASL, APST, transition rates, and visit counts.

Procedure *GetTP* (X) /* searches for possible TP the current path */

Step 1:   If the current node X is a leaf node, i.e., $X.prob\_tran$ = NULL
    Add $X.curr\_URL$ to   the   *curr_tran*   list,   increment   the *transaction_count* of the TP matching with *curr_tran;*

Increment page_count for X.curr_URL;   /* get the new
average values for ASL   and APST */
Update APST for *X.curr_URL* with the new timestamp;
Update ASL for the TP;
Tag entry as *visited*;
Return *success*;

Step 2:   While there is one unexplored URL in *X.prob_tran*

Let *U_Node* be the next URL in *X.prob_tran*;
Create node *Y* from the next unvisited entry in the log file such that
Y.client_id = *X.client_id* and *Y.curr_URL* = *U_node* and *Y.log_time*
– *X.log_time* ≤ *session_timeout*;

Increment page_count for Y.curr_URL;
Update APST for Y.*curr_URL*;
Insert node *Y* in the tree, tag entry as *visited*;
Append Y.curr_URL to list curr_tran;
Call proc. *GetTP(I)*
If the proc returns *success*
Return success;

Essentially, the bounding of this search is based on the *session_timeout* cut-off.
The timeout can be taken as an input from the experts. Normally a transaction is taken
as aborted when there is no user interaction within 5 minutes (300 secs). However, for
complicated transactions higher values are chosen. In an enhanced version of this
algorithm we have forced a gradually stricter bound by updating this value with the
maximum value of the APSTs obtained by partial analysis of the logs, if that value is
lower. We also constrict the search space by not searching immediate entries in the
log to get the successors to a particular node.

For systems which are already deployed and operational, the Web server access log
data is available and critical transaction paths are well known. There may be other
cases where the log data is available, but the TPs are not clearly known. This may
happen when the site is too big and complicated, or the navigational patterns are not
well understood. There can still be other cases where system is yet to be developed
and only the design artifacts like Use Case Maps etc are available. Clearly, for the
second case, we need to identify the transactions before we start characterizing the
workload. For the third case, we need a schematic way of translating the UCMs to
transactions and TPs and populating the same with synthetic workload trace.
Additionally, while analyzing the possible transaction paths it should be borne in
mind that some pages in this path can primarily be static information containers, i.e.,
these get populated from server or browser cache or have in-built verification schemes
that do not require any fresh visits to a resource. Other pages in the path are
dynamically generated. These require visits to resources and have, at least, one
element that cannot be cached statically. In today's advanced server-side
programming, even application objects or in-memory process data can be cached or
pooled to decrease the communication cost. The visit counts can be decomposed to
visit(s) to each of the tiers $1,2,...,n$. For example, the resource demands for the tiers

associated with page b in the TB can be visualized as $V_b * \sum_{i=1}^{n} d_i$, where d_i is demand for resource at tier i and can be a real number including 0. The resources can themselves be modeled further to lower levels of resources and solved through variations of hierarchical Queuing network theories like Layered Queuing Networks (LQN) also view computer systems in terms of layers [10]. The transaction mixes obtained from characterization can help in partitioning system resources accordingly. Deployed systems often have dedicated resources (for example, LDAP directories for authorization and authentication of users). Therefore, a proper estimation of the proportion of the load visiting those resources can avoid over-provisioning. Similarly, scripts for the test tools can be written in such a manner that only percentage of the total load can be made to flow through each page during stress testing of the augmented system.

4   Experimental Results

The algorithm is converted into a tool (called WCM) and was tested on a series of real-life data at Infosys. The tool is used in conjunction with a Log Extractor module. The Extractor takes huge log files as inputs and does the merging and formatting of the files to output the required fields for the WCM. The WCM, written in Visual C++, can take in the TPs either graphically or through text files.

We describe characterization of a Web-based car rental site as a real-life case study. The system provides functionalities like registration, rental inquiry, location and map services, new booking, booking updation etc. for customers. Figure 5 highlights the critical TPs in reservations for new and existing customers. For the purpose of system performance auditing, we analyzed two critical transactions which were facing performance bottlenecks in terms high response times and low throughputs. One is a relatively simple one for updating an existing reservation (*Update Booking*) with 5 URLs (A → B→ C →D → E) and the other is a complex one of creating new reservations (*Create New Reservations*) with 10 URLs (A → F→ G →H → I → J→ K →L → M → E). Several of these pages have self and backward references. Three different input files were created having 1, 10 and 45 days of log data and were fed into the WCM tool. The sizes of the input files were 10, 100 and 900 MBs respectively. To reduce the search space, the Log Extractor module filters out all other log entries than the union of these TP sets. We repeat the test runs with different *session_timeout* values of 300, 1000, and 3600 seconds respectively. Analyzing the site log data for 45 days with a *session_timeout* of 3600 seconds we find that while the *page_count* of the initial URL (A) is 256732 hits, that of the end point of the transactions (E) is 9528 hits only. This means that only 3.7% of the users have actually completed the reservation (or updated them). We also obtained page visit statistics of important pages like those for checking costs (G) – which is 62.3% and for amending reservations (D) – which is 4.2% of the total visits at the default page. The transaction count for *Update Booking* is 285 and that for *Create New Booking* is 619, thereby giving a transaction mix ratio of about 0.4. The ASLs are 578.9 secs and 1046.56 seconds respectively. The maximum value of APST is for the page that creates and validates reservation form (K) and is found to be 438 secs. This indicates that certain completed transactions were missed out with a *session_timeout* cut-off of 300 seconds. The lowest value of transition rate (0.096) is found between the pages payment details (J) and making reservations (K) indicating the highest drop out of the customers at that stage. It is noted that while there was no significance difference in the results obtained with a cutoff of 1000 and 3600 seconds, the results obtained with a cutoff of 300 seconds show significant deviations. Similarly, the results obtained from 900 MB data files are statistically more significant than those obtained from files having less data points. The algorithm performs satisfactorily. In a desktop running Windows XP OS, with 1.70 GHz Pentium processor, 256 MB RAM and 37.2 GB of Hard Disk capacity; WCM crunches 900 MB of data in approximately 584 seconds. The following figure shows the performance of the tool with two different transaction mixes as described above. The tests with 4 GB of log data and up-to 30 transactions are currently on a bigger workstation.

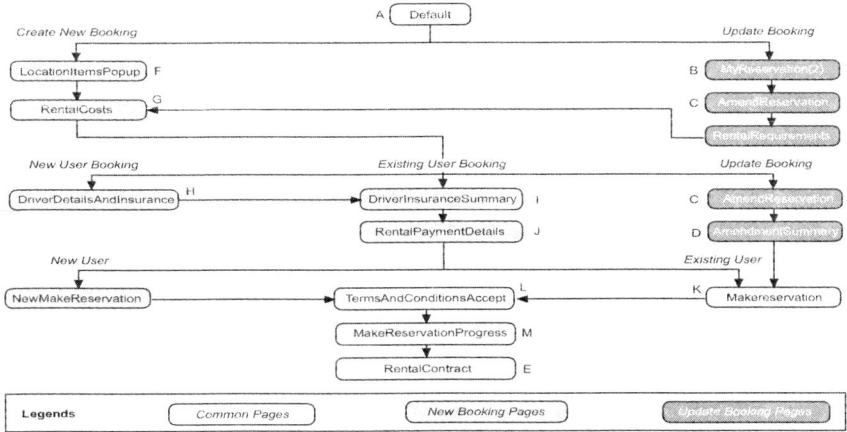

Fig. 5. Navigational patterns for main transactions in the car rental site

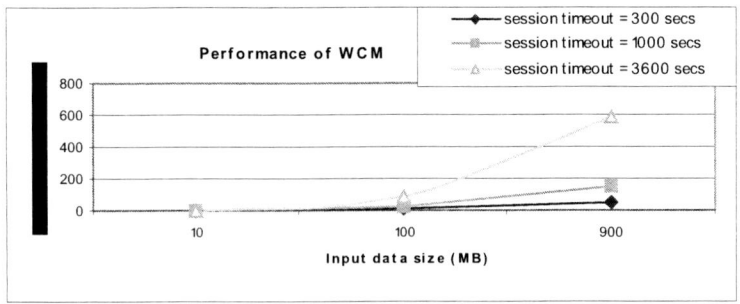

Fig. 6. Performance of the WCM on varying log data size and *session_timeout*

5   Conclusion and Future Work

In this paper we have introduced a simple yet effective algorithm of characterizing workloads for Web-based systems from a transaction-oriented manner. The method works by offline analysis of Web logs and come up with useful metrics like transaction intensity (counts and mixes), probability of a URL to be visited by a user (visit count), rate of propagation of transaction through various states (transition rates) etc. The metrics can be used to do the capacity provisioning of resources at various tiers of the system. The method currently works only on the systems where critical transaction paths and navigations are known beforehand. We propose to extend this method to identify statistically significant transactions (using clustering techniques like k-means or minimum spanning tree method) for those sites where the user navigation patterns are not well understood. Additionally, a model for resource usage at the backend corresponding to the workload through a Layered Queuing Network will be of great help for sizing the system infrastructure accurately.

References

[1]   D. A. Menascé, TPC-W: a benchmark for E-commerce, *IEEE Internet Computing*, May/June, 2002

[2]   Web LogMiner: a comprehensive tool to analysis of NCSA combined/extended web server logs, at
www.cet.nau.edu/Academic/Design/Path_to_Synthesis/EGR486/CSE/

[3]   J. Borges and M. Levene, A Fine Grained Heuristic to Capture Web Navigation Patterns, *SIGKDD Explorations*, Volume2, Issue1, 2000, pp. 40-50

[4]   M. Calzarossa and G. Serazzi, Workload Characterization: a Survey, in *Proceedings of the IEEE*, vol. 81, No. 8, August 1993

[5]   D. Krishnamurthy and J. Rolia, Predicting the performance of an E-commerce server: those mean percentiles, in *Proceedings of First Workshop on Internet Server Performance*, ACM SIGMETRICS, June 1998

[6]   D. Ferrari, On the foundations of Artificial Workload design, *in Proceedings of ACM SIGMETRICS Conf*, 1984, pp 8–14

[7]   M.K. Acharya, R.E. Newman-Wolfe and H.A. Latchman, Real-time hierarchical traffic characterization of a campus area network, in R. Pooley (eds.) *Computer Performance Evaluation'92 – modeling techniques and tools*, 1992

[8]   M.F. Arlitt and C.L. Williamson, Internet Web servers: Workload Characterization and Performance Implications, *IEEE/ACM Transactions on Networking*, vol 5, No. 5, October 1997

[9]   D.A. Menascé, V.A.F. Almeida, R. Fonseca and M.A. Mendes, A Methodology for Workload characterization of E-commerce sites, in *Proceedings of the 1st ACM conference on Electronic commerce*, 1999

[10]   P. Maly and C. Murray Woodside, Layered Modeling of Hardware and Software, with Application to a LAN Extension Router, *Computer Performance Evaluation / TOOLS*, 2000 pp:10-24

Internet Auctions: Some Issues and Problems

Amitava Bagchi and Atul Saroop

Indian Institute of Management Calcutta
Diamond Harbour Road, Joka, Kolkata 700104, India
bagchi@iimcal.ac.in
atul@email.iimcal.ac.in

Abstract. Recent advances in client-server, web-server and networking technology have made it possible to hold auctions over the Internet. The popularity of such auctions has grown very rapidly in the last few years, giving a new impetus to the study and analysis of auctions. Apart from technological problems, there are a variety of implementational, economic, and behavioral aspects that need to be studied. In this paper we describe a few representative issues, and outline promising quantitative and computer-based approaches that can lead to solutions. The three issues we examine are timed bids, forecasting of final bid values in incomplete auctions, and last-minute bidding. These and other related features of Internet auctions could have a major impact in the near future on the volume of business transactions over the Internet.

1 Introduction

One way to sell merchandise is by means of auctions. This method is preferred over others when the exact value of the item to the buyers is not known, so that the seller is uncertain what price to post on the item. The most popular type is the English auction, in which bidders publicly place their bids for the item on sale At any point in time, the highest bid placed up to that instant determines the going price. A bidder tries to outbid all others by quoting a price higher than the going price. As a result, the going price gets pushed upwards, until all bidders except one cease to place any more bids. The item is then allocated to the last remaining bidder at the price quoted by him. Conventional closed room auctions, such as tea auctions for example, are usually of short duration, lasting only a few hours at most. Such auctions have been extensively studied in the Economics literature (Refer [5] and [4] for details). In a historical survey, [3] has described the various prevailing types of auctions. [12] has shown that English auctions and Dutch auctions are equivalent under equilibrium conditions in terms of expected payoffs to the bidders and the auctioneer. This result is now known as the Revenue Equivalence Theorem. [7] and [2] have analyzed auctions with discrete bid increments.

The popularity of auctions has grown with the spread of the Internet. In recent years the reach of the Internet has extended to every corner of the globe, spurring the development of a variety of Internet-based consumer marketing

S. R. Das, S. K. Das (Eds.): IWDC 2003, LNCS 2918, pp. 112–119, 2003.

models. Some of the resulting e-commerce ventures have not fulfilled their original promise, but Internet auctions continue to grow in popularity. The participants in an Internet auction can be geographically dispersed, so the number of potential participants is essentially unlimited, but there are compensating disadvantages. If a physical item is to be sold, it can only be displayed in facsimile on the PC screen of a client machine, and there are logistic problems in delivering the item to the eventual winner. Unlike a conventional closed room auction, however, an Internet auction can extend for many days. There are four main types of such auctions:

1. Business-to-Business (B2B);
2. Business-to-Consumer (B2C);
3. Consumer-to-Consumer (C2C) and
4. Consumer-to-Business (C2B) or Reverse Auctions

Here we are mainly concerned with C2C auctions. Among C2C auctions sites, eBay is the most well known. In the second quarter of 2003, eBay has reported net revenue of $509.3 million (U.S.), which is 91% higher than that generated in the 2nd quarter of 2002. This large volume of commercial activity has focused research interest on the properties and characteristics of Internet auctions among both information technologists and economists.

There are a variety of issues of research interest in connection with Internet auctions. In this paper we generally assume that the auction is of the C2C type and is selling a single unit of a single item, but many of our remarks apply to other types of auctions as well. We choose three aspects for more detailed discussion:

1. Timed bids in Internet auctions;
2. Forecasting the value of the winning bid of an incomplete C2C auction;
3. Last-minute bidding in eBay auctions.

2   Timed Bids

A serious problem that arises when an Internet auction extends over a long period, say one week or ten days, is that bidders who have an urgent need of the item on auction are forced to wait until the very end of the auction to get the item. For example, if the item on auction is a raw material, its delivery at the end of the auction might be too late for the production schedule of a bidder. One way to resolve this difficulty is to allow timed bids A timed bid in an English auction held over the Internet has a bid value and a time of action. The bid value denotes the price the timed bidder is ready to pay to the auctioneer in case the item is allocated to him before the time of action. Typically, this is much higher than the current going price, and the time of action is much earlier than the auction closing time. If the auctioneer accepts the timed bid, the timed bidder gets the item at his bid value, and the auction ends if there is no other item on auction. If the auctioneer rejects the timed bid, or is unable to take any decision

by the time of action, the timed bid stands annulled. In this case we require the
timed bidder to leave the auction. This inhibits collusion with the auctioneer to
raise the going price to a high level by repeatedly placing timed bids.

Timed bids can be *public* or *private*. A public timed bid is announced by
the auctioneer to all bidders as soon as it is placed. Suppose, for example, that
coal is being auctioned by a coal mining company. A typical bidder would be
a representative of an organization that uses large quantities of coal as input
raw material, such as a steel plant or a thermal power plant. Such bidders would
have a clear valuation of the item in their minds, so there would be no harm
in broadcasting a timed bid to all bidders as soon as it is placed. The bidders
would want to procure the coal according to their production schedules. Some
might have received flash orders and would require an immediate supply, and
they might be willing to pay premium prices. Those who need the supplies
later would not want to augment their inventory right away and incur an early
expenditure of cash, so they would place regular bids.

A private timed bid is a secret contract between the auctioneer and the timed
bidder and is not announced to other bidders. Consider the sale of office space in
the downtown area of a metropolitan city. A startup company that needs office
space at a short notice might place a timed bid at a much higher price than the
going price. Since other bidders might not be very clear about their valuations,
broadcasting the timed bid to all bidders could cause apprehension among them
and reduce participation in the auction. It would therefore seem advisable for
the auctioneer to keep the timed bid private, and decide whether to accept it
on the basis of his forecast of the expected revenue if the auction were to run to
completion.

Timed bids are yet to be implemented in Internet auctions. The incorporation
of such bids, whether public or private, is likely to prove convenient for bidders
who have an urgent need of the item. As shown in [9], it is possible to show that if
certain broad conditions are satisfied, the use of timed bids tends to increase the
expected payoffs under equilibrium conditions of both the timed bidder and the
auctioneer. As an example of the kind of result that can be proved, we determine
the expected gain of a timed bidder who has placed a public timed bid, under
the following set of assumptions:

- The auction under consideration is a single-item single-unit auction. More
  than one bidder take part in the auction, and all of them participate in
  bidding from the instant the auction starts. (This implies the auctioneer
  allows the placement of timed bids in the auction only after all the bidders
  login to the auction.)
- Bidders have independent private valuations of the item on auction [5, p 3].
  The bid value of a bidder never exceeds his valuation.
- Bidders who participate in the auction are risk-neutral. (This means that
  they place bids that maximize their expected gains.)
- The valuations of the bidders are random draws from a uniform $U[0, 1]$ distribution. (In practice, the valuations are likely to be random draws from an

interval $[I, J]$, where I and J are real numbers, $I < J$. We map this interval onto the interval $[0, 1]$.)

– There is at most one bidder who has an urgent need for the item on auction. Under normal conditions, when he does not have an urgent need, his valuation of the item is V. But when he does have an urgent need of the item, his valuation rises to W, which is significantly greater than V. If he does decide to place a timed bid, then W is his valuation of the item in the immediate neighborhood of the time of action of his bid. Irrespective of whether he wins the item or not, the timed bidder leaves the auction at the time of action of his timed bid.

Let us suppose that there is sufficient difference in time between the moment the public timed bid is placed and its time of action. This permits other bidders to react to the timed bid and to place counter-bids if they want to do so. We determine the equilibrium conditions for the bidders and compute the gain realized by the timed bidder from the timed bid.

A bidder who does not have an urgent need of the item follows the same equilibrium strategy as a bidder in an English auction, i.e., he repeatedly places a bid with the smallest possible bid increment over the current going price until his valuation is reached or until no counter-bids are placed by other bidders To determine the equilibrium strategy for the timed bidder, we have to determine his expected gain. The timed bidder wins the item only when all other bidders have valuations smaller than his bid value b. So his expected gain is $E(\text{Gain}) = (W - b)b^{(N-1)}$. Maximizing the expected gain yields an equilibrium bid value $b^* = \frac{(N-1)}{N}W$ and an expected gain $E(\text{Gain}) = W^N \frac{(N-1)^{(N-1)}}{N^N}$. The bidder would place a timed bid only when his expected gain from placing the timed bid is higher than his expected gain in a normal English auction, which is V^N/N. So, the timed bidder would place a timed bid only when $\frac{W}{V} \geq \left(\frac{N}{N-1}\right)^{\left(\frac{N-1}{N}\right)}$.

3 Forecasting the Value of a Winning Bid

Consider methods that predict the final winning price in an incomplete Internet auction that follows an English auction format. The study of such prediction mechanisms can help in the identification of factors that favorably affect the revenue of the auctioneer and the gain of the winning bidder. It is well known that a large proportion of bidding in Internet auctions takes place in the closing stages. [6] found that in eBay auctions, 20% of the bidding took place during the last hour of the auction; they also found that in auctions that sold computers, 40% of the bids were registered in the last 5 minutes. These are very short intervals of time, since auctions at eBay usually last from 7 to 10 days. Thus the prediction of the winning price in an incomplete auction is a difficult task. [8] adopted a heuristic approach in predicting winning prices in Internet auctions, and suggested that the predictions could be put to use in a Decision Support System for auctions that allowed private timed bids. An alternative approach,

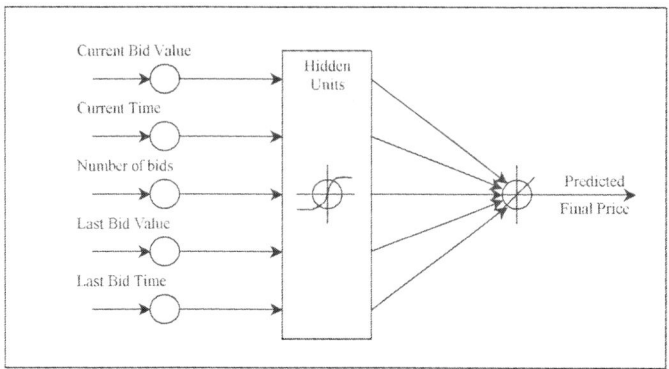

Fig. 1. ANN Architecture for final price prediction for one auction class

followed by [10], would be to try to predict winning prices with the help of Artificial Neural Networks (ANNs).

The model based on ANNs uses back-propagation networks as building blocks. Auctions are classified into different auction classes, and each class has its own neural network. An incomplete auction is put into its appropriate class, and the corresponding neural network predicts the winning price.

The ANN architecture is shown in Fig 1. It takes information about the current and last bids as inputs, and outputs the bid-value, normalized with respect to the final winning price. It has a single hidden layer, and all neurons have sigmoidal activation functions. In the system designed by us, the hidden layer had five neurons. The system was trained on data taken from a set of 1,000 completed eBay auctions. Validation was then carried out on data from another set of 1,000 completed auctions. Completed auctions that formed the training data were segregated into auction classes. For a particular auction, the error in the output of an ANN was computed as follows: The percentage deviation of the ANNs output from the actual bid value was determined at each instant of time when a bid was placed, and these values were then averaged. Since we had no prior knowledge of the constitution of auction classes, the first auction was put into a class of its own. Thenceforth, unclassified completed auctions were simulated on the ANNs of all the existing classes. If the smallest of the errors was above a threshold value, the unclassified auction was put in a new class. Otherwise it was put in the auction class corresponding to the ANN that generated the least error. Experiments were performed on three data sets, one each for camera, laptop and phone auctions on eBay. In phone auctions, the method predicted the final price with an error of 36% on average at a time when only 40% of the auction was complete. At such a time, an auction is at a very preliminary stage and a quite a few have not even registered a single bid. Similar results were obtained for camera and laptop auctions.

4   Last-Minute Bidding

C2C auctions on the Internet are typically of two types, hard close and automatic extension. A hard close auction ends at a pre-announced closing time, while an automatic extension (or "going, going, gone") auction is automatically extended by an agreed period of time (say ten minutes) as soon as a bid is placed. This implies that an auction of the latter type can get extended beyond its initially announced closing time. Unlike other auction sites such as Amazon that holds only automatic extension auctions, or Yahoo that holds auctions in both formats, eBay holds only hard close auctions. Earlier studies [6] have shown that in a hard close auction, a significant proportion of bidders feel an urge to place bids very late, such as in the last one minute prior to the closing time. This phenomenon, which has become very prevalent in the last few years, is commonly known as sniping. A snipe is a bid that is placed very late in a hard close auction, in the hope that other bidders would not have the time to place a counter-bid.

The economic analysis of auctions has until now concentrated mainly on iso- lated individual English auctions. It is well known that in such auctions, when bidders are risk-neutral and have independent private valuations, the final win- ning price attains a level that is marginally higher than the second highest valu- ation. C2C Auctions on eBay are also second-price auctions. But on eBay a very large number of auctions are in progress concurrently or in close succession, many of them selling similar types of items. A bidder would not be interested in participating in all the auctions, since he would want only one unit, or a small number of units, of an item. Moreover, the cost of participation and other prac- tical considerations would prevent him from engaging actively in a large number of auctions simultaneously. In such a situation, an experienced bidder might well decide to conserve time and energy by restricting his involvement to the last minute of an auction. Since many auctions are selling closely similar items, the bidder can hope to be lucky in his last-minute bidding (also called sniping) and win one of the auctions. Sniping has the added advantages that he does not have to face any price wars, and if the item has a common value component, as most items do, then his valuation would not get revealed to other bidders.

An analytical model has been recently proposed in [11] to explain the patterns observed in eBay auction data, with regard to both the number of snipes and the snipe timings. Further research is needed to assess the overall impact of sniping on hard close auctions, and to determine whether such auctions are likely to degenerate into just another variety of second-price sealed-bid auctions [1].

We now briefly describe an analytical model that helps to explain the pat- terns found in the data. It should be remembered that auctions are held one after another on eBay without respite. We assume that a steady state has been reached, and that the bidding data mirrors the steady state situation quite ac- curately. Let us try to allot S snipes independently to N auctions using a simple probabilistic binomial model. A snipe s gets allotted to a specific auction with probability $1/N$. The expected number E_i of auctions with exactly i snipes is given by the formula $E_i = {}^S C_i \left(\frac{1}{N}\right)^i \left(1 - \frac{1}{N}\right)^{S-i}, i > 0$. On examination, we find that this yields a value of E_1 that is larger than the observed number of auctions with exactly one snipe. On the other hand, the calculated values of E_2, E_3 and E_4 turn out to be smaller than the observed values, suggesting that there is a positive correlation between the snipes that hit an auction, i.e., an auction that has already been hit by a snipe tends to attract other snipers. In our improved model, we allot snipes to auctions a little differently. A snipe is now allotted with probability p_0 to an auction with no snipes, and with probability p_1 to an auction with one or more snipes, where $0 < p_1 \leq \frac{1}{N} < p_1 < 1$. Among the given set of N auctions, let n_0 be the number of auctions with no snipes and $n_1 = N - n_0$ the number of auctions with snipes. Since each snipe has to be allotted to an auction, we must have $p_0 n_0 + p_1 n_1 = 1$. The expected number of auctions with exactly i snipes now equals

$$E_i = {}^S C_i \left[n_0 p_0^i (1 - p_0)^{S-i} + n_1 p_1^i (1 - p_1)^{S-i} \right] \tag{1}$$

In the steady state we should have $n_0 = E_0$, so that $n_0 = n_0(1-p_0)^S + n_1(1-p_1)^S$. Let $\theta(p_0) = n_0(1 - p_0)^S + n_1 \left(1 - \frac{1 - n_0 p_0}{n_1} \right)^S - n_0$. We find that in the range of values of interest to us, the function $\theta(\cdot)$ has the following features:

1. $\theta(0) > 0$, i.e., the function has a positive value at the origin;
2. $\theta''(p_0)$ is always positive, so the function has an increasing slope;
3. $\theta'(\frac{1}{N}) = 0$, i.e., the function has a minimum at $p_0 = \frac{1}{N}$.

So the equation $\theta(p_0) = 0$ will have just one solution in the interval $0 < p_0 < \frac{1}{N}$ provided $\theta(\frac{1}{N})$ is negative. Once p_0 and p_1 are determined, we can compute the number of auctions with one, two, three or more snipes.

5   Conclusion

Internet auctions are beginning to play an important role in the world economy. Such auctions have become possible as a result of the rapid development of client- server, web-server and networking technology. The technology has now stabilized, and the interest is now beginning to focus on the business implications. The major issues now are implementational, economic and behavioral in nature. In this paper we have described three new problem areas that are of current interest to auction researchers. These and many other related topics deserve further investigation by both informational technologists and economists.

References

[1] Ravi Bapna. When snipers become predators: Can mechanism design save Online auctions? *Communications of the ACM*, 2003. forthcoming, downloaded from `http://www.sba.uconn.edu/users/rbapna/research.htm` in August, 2003.   117
[2] Ravi Bapna, Paulo Goes, and Alok Gupta. Analysis and design of business-to-consumer Online auctions. *Management Science*, 49(1):85–101, 2003.   112

[3] Ralph Cassady. *Auctions and Auctioneering*. University of California Press, 1967. 112

[4] Paul D Klemperer. Auction theory: A guide to literature. *Journal of Economic Surveys*, 13(3):227–286, 1999. 112

[5] Vijay Krishna. *Auction Theory*. Academic Press, San Diego, California, USA, 2002. 112, 114

[6] Alvin E Roth and Axel Ockenfels. Last minute bidding and the rules for ending Second-Price auctions: Evidence from eBay and Amazon auctions on the Internet. *American Economic Review*, 92(4):1093–1103, 2001. 115, 117

[7] Michael H Rothkopf and Ronald M Harstad. On the role of discrete bid levels in Oral auctions. *European Journal of Operations Research*, 74:572–581, b 1994. 112

[8] Atul Saroop and Amitava Bagchi. Decision support system for Timed bids in Internet auctions. In *Proc. WITS-2000, Tenth Annual Workshop on Information Technology and Systems*, page 229–234, Brisbane, Australia, 2000. 115

[9] Atul Saroop and Amitava Bagchi. Expected revenue in Internet auctions in the presence of Timed bids. In *Proc. WITS-2001, Eleventh Annual Workshop on Information Technology and Systems*, page 151–156, New Orleans, USA, 2001. 114

[10] Atul Saroop and Amitava Bagchi. Artificial Neural Networks for predicting final prices in eBay auctions. In *Proc. WITS-2002, Twelfth Annual Workshop on Information Technology and Systems*, Barcelona, Spain, 2002. 116

[11] Atul Saroop and Amitava Bagchi. Sniping in eBay auctions: A data driven analysis, 2003. In preparation. 117

[12] William Vickrey. Counterspeculation, auctions, and competitive sealed tenders. *Journal of Finance*, 41:8–37, 1961. 112

Transaction Management in Distributed Scheduling Environment for High Performance Database Applications

Sushant Goel[1], Hema Sharda[1], and David Taniar[2]

[1] School of Electrical and Computer Systems Engineering
Royal Melbourne Institute of Technology, Australia
s2013070@student.rmit.edu.au
hema.sharda@rmit.edu.au
[2] School of Business Systems, Monash University, Australia
david.taniar@infotech.monash.edu.au

Abstract. Synchronising the access of data has always been an issue in any data-centric application. The problem of synchronisation increases many folds, as the nature of application becomes distributed or volume of data approaches to terabyte sizes. Though high performance database systems like distributed and parallel database systems distribute data to different sites, most of the systems tend to nominate a single node to manage all relevant information about a resource and its lock. Thus transaction management becomes a daunting task for large databases in centralized scheduler environment. In this paper we propose a *distributed scheduling* strategy that uses a distributed lock table and compares the performance with centralized scheduler strategy. Performance evaluation clearly shows that multi-scheduler approach outperforms global lock table concept under heavy workload conditions.

1 Introduction

Continuously growing volume of data and expanding business needs have justified the needs of high performance database systems like *Distributed* and *Parallel Database Systems* (PDS) [2,8,10,11]. Distributed databases are the logically integrated systems that make the distribution of the data transparent to the user [11]. When the volume of data at a particular site grows large and the performance of the site becomes unacceptable, parallel processing is needed for data servers. PDS supports automatic fragmentation and replication of data over multiple nodes [2].

Single scheduler approach has been used as the correctness criterion for transaction management. With the increase in volume of data, message and locking overhead also increases in centralized locking approach. We will refer single scheduler transaction management approach and centralized locking interchangeably throughout the paper.

S. R. Das, S. K. Das (Eds.): IWDC 2003, LNCS 2918, pp. 120–130, 2003.

Most database management systems use single-scheduler transaction management approach [2,3,8,9,10,11]. But with increase in volume of data, managing transactions with single scheduler may become difficult and computationally expensive. Single scheduler strategy may not meet the requirements of high performance database systems as the database scales to terabyte sizes. Distributed lock managers have been proposed in the literature to distribute the load among multiple sites [9,10,11]. Unfortunately distributed lock manager nominates one node to manage all the information about the resource and its locks. The distributed lock manager still has to manage a global lock table [11], thus the architecture is not truly distributed and the number of messages in the network is also high due to internode communications. With increase in amount of the data stored, some of the single scheduler inherent weaknesses become prominent, e.g. big lock table, increased number of messages in the system, deciding the coordinator for the transaction etc.

This motivated us to distribute the scheduling responsibilities of the data items to their parent nodes where the data is residing and consider *distributed schedulers* for transaction management and to maintain the consistency of data items. Migrating from single-scheduler transaction management approach to distributed scheduler transaction management approach may pose a threat to the consistency of the data items. The underlying problem in direct migration is discussed later in the paper. We use multi-scheduler and distributed scheduler interchangeably in this paper.

We discuss the problem in migrating the single scheduler algorithms to multi-scheduler environment in the following sections. We will discuss the case for *distributed memory* also known as *Shared-Nothing* parallel database systems [2] in this paper but this concept can be modified to meet the requirements of other high performance data-centric applications.

We would like to highlight that mainly three types of transaction management strategies are used for distributed memory architecture in the literature [10]: 1) *centralized locking* 2) *primary-copy locking* 3) *distributed locking*. All three strategies either manage the locking table centrally or manage a portion of the *global lock table*. In this paper we focus to reduce the load of global lock table and remove the requirement of centralized scheduling.

Rest of the paper is organized as follows: Section 2 discusses the related work done in distributed database system and multidatabase systems, Section 3 explains our working environment and elaborates the proposed algorithm, section 4 shows and the performance comparison for single and multi-scheduler strategies. Finally, Section 5 concludes the paper and discusses the future extension of the work.

2   Related Work

In this section we discuss different high performance database systems like multidatabase, distributed database and parallel database systems. We also very briefly discuss the environment of transaction management required by these database systems.

2.1   Multidatabase Systems

Multidatabase system is an interconnected collection of autonomous databases. A multidatabase management system is the software that manages a collection of autonomous databases and provides transparent access to it [9]. Multidatabase management system provides Database Management System (DBMS) at two different levels – *i)* local *ii)* global. Transaction manager at local sites can guarantee correct schedules for local transactions only. A global transaction manager spanning all sites has to be designed to meet specific requirement. For detailed discussion please refer [9]. Multidatabase system (MDS) is a good option when individual databases have to be combined logically. If large volume of data has to be managed and data distribution is an important factor in performance statistics, then MDS may not be the preferred design option. MDS is best suited when autonomy is the key but certain transactions may span more than one database.

2.2   Distributed Database Systems

Data can be manually partitioned over geographically separated databases but can still appear as one unit to the user. The data is transparently distributed over nodes of distributed database. Distributed DBMS manages transactions by global transaction manager and global lock manager [10,11]. Distributed databases have high locking and message overhead due to the distributed nature of application. Lock table may become a performance bottleneck despite the distribution of data over multiple sites due to its global nature. Global lock table maintains the locking information about all data items in the distributed database thus the scheduling strategy act as a centralized scheduling scheme. Thus, despite most of the applications distribute data to multiple sites they still use a centralized scheduling scheme and maintain a global lock table.

2.3   Parallel Database Systems

The enormous amount and complexity of data and knowledge to be processed by the systems imposes the need for increased performance from the database system. The problems associated with the large volumes of data are mainly due to: 1) Sequential data processing and 2) The inevitable input/output bottleneck. Parallel database systems have emerged in order to avoid these bottlenecks. Different strategies are used for data and memory management in multiprocessor environment to meet specific requirements. In *shared memory architecture* [2], processors have direct access to the disks and have a global memory. In *shared disk architecture* [2], processors have direct access to all disks but have private memory per processor. *Shared-nothing architecture* [4] has individual memory and disk for each processor, called *processing element (PE)*. The main advantage of shared-nothing multiprocessors is that they can be scaled up to hundreds and probably thousands of PEs that do not interfere with one another, refer [1, 2, 4, 5, 6] for detailed discussion.

Though shared-nothing database systems have individual data partitions, to achieve correctness of schedule and consistency of data these database systems also rely on central transaction management schemes. Looking at the drawbacks of centralized transaction management schemes we distribute the transaction management responsibilities to individual PEs.

2.4   Basic Definitions and Problem Identification

We would like to briefly define the *transaction* and properties of transactions before we discuss the problem in direct migration from single scheduler to distributed scheduler environment. From the user's viewpoint, a transaction is the execution of operations that accesses shared data in the database; formally, a *transaction* T_i is a set of read (r_i), write (w_i), abort (a_i) and commit (c_i). T_i is a partial order with ordering relation $\prec_i$ [7].

A *history* or *schedule* indicates the order in which the operations of the transactions were executed relative to each other. Formally, let $T = \{T_1, T_2, \ldots T_n\}$ be a set of transactions. A complete history (or schedule) H over T is a partial order with ordering relation $\prec_H$ [7].

A history H is *Serializable (SR)* if its committed projection, $C(H)$, is equivalent to a serial execution H_s. A database history H_s is serial iff

$$(\exists p \in T_i, \exists q \in T_j \text{ such that } p \prec_{Hs} q) \text{ then } (\forall r \in T_i, \forall s \in T_j, r \prec_{Hs} s).$$

For detailed description of the definitions refer [7]. A history (H) is serializable iff serialization graph is acyclic. A transaction must possess ACID properties [7] and these properties must be enforced by the concurrency control and recovery management techniques implemented in the DBMS.

Problem identification: The algorithms developed for single scheduler might produce contradictory serialization order at different nodes in multi-scheduler environment, if two transactions access more than one processing elements simultaneously. We have already discussed the inherent weaknesses of single scheduler algorithms applied to high performance database systems [13]. We propose a new serializability, *Parallel Database Quasi-serializability,* and compare the performance of this distributed scheduling strategy with the centralized scheduling strategy in this paper.

3   Proposed Algorithm

Before discussing the algorithm we first present the working environment for our model in subsection 3.1, we then briefly discuss the correctness criterion for the multi-scheduler concurrency control algorithm namely *Parallel Database Quasi-serializability* (PDQ-serializability) in subsection 3.2. Finally, subsection 3.3 explains the proposed *Timestamp based Multi-scheduler Concurrency Control (TMCC)* algorithm. We have already proposed the algorithm in [13]. Here we give an overview of the algorithm, and we focus on the performance comparison between the two strategies.

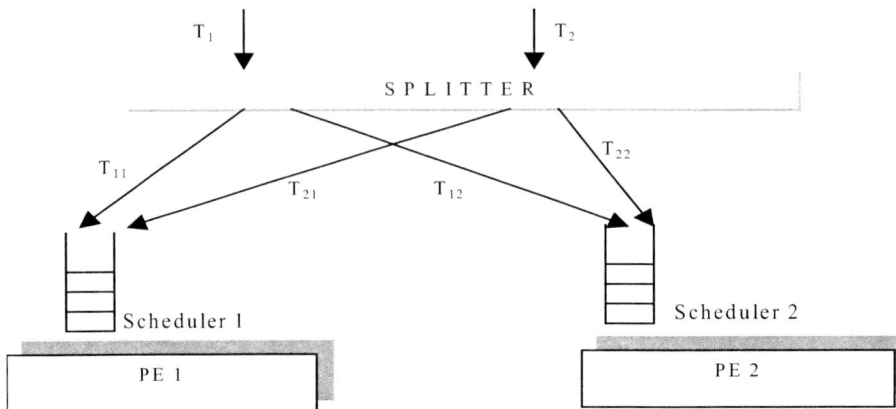

Fig. 1. Conceptual model of multi-scheduler concurrency control

3.1   Working Environment

The model presented in this paper has additional scheduling responsibility of transactions per processing element. The traditional model shown has centralized scheduling responsibility and a global lock table. We try to distribute the scheduling responsibility to respective PEs, where data is located, (see fig. 1) in order to take maximum advantage of the abundant computing power available.

We have already said that single scheduling strategy may produce incorrect serialization order in multi-scheduling environment [13]. For the sake of simplicity we consider only two processing elements to demonstrate our algorithm. We assume that local schedulers are capable of producing serializable schedule.

3.2   PDQ-Serializability

Scheduling responsibilities in PDS is distributed to respective PE according to the partitioned data. Problem in migrating from single-scheduler to multi-scheduler have been discussed in [13]. Thus, we conclude that operation level concurrency control cannot ensure a correct schedule to be produced in multi-scheduler environment. Even, subtransaction level granularity in itself is not sufficient to produce correct schedule. Some additional criterion has to be enforced to ensure multi-scheduler concurrency control in addition to subtransaction level granularity.

A new serializability criterion is proposed that separates two types of transactions – transactions having only one subtransaction and transactions having more than one subtransaction. The following definition states the correctness criterion for PDS.

Definition 1: A *Multi-Scheduler Serial (MS-Serial)* history is considered correct in parallel database system.

Definition 2: A history in multi-scheduler environment is PDQ-serializable iff it is equivalent to a *MS-Serial* history (definition of equivalence ($\equiv$) from [7], page 30).

The PDQ-serializability of the history is determined by analysing *Parallel Database Quasi-Serializability* (PDQ-serializabilty) graphs. Only the committed projection [7] of the history *C(H)* is considered in the definition. The following definition describes the PDQ-serializability graph.

Definition 3: At any given instance, histories of all the schedulers at each PE can be represented using a directed graph defined with the ordered three: ($\mathbf{T}^1$, $\mathbf{T}^n$, A). The graph would be referred as *Multi-scheduler Serializability Graph (MSG)*.

$\mathbf{T}^1$ and $\mathbf{T}^n$ are the set of labeled vertices representing transactions with one *subtransaction* and more than one *subtransaction* respectively. A is the set of arcs representing the ordering of transaction in each PE. In rest of the paper we would only consider transaction having more than one subtransaction and denote that as T, without the *superscript*. Transactions with single subtransaction are taken care by the individual PE and do not pose any threat to the concerned problem.

Based on the definition of MSG we next formalize the following theorem:

Theorem 4: A history in multi-scheduler environment is PDQ Serializable iff MSG is acyclic.

Proof: Please refer [13] for proof.

Our earlier work [13] discusses the above definitions and theorem in detail. In this paper we focus on the performance evaluation of the proposed algorithm and comparison with global lock table (single scheduler) approach.

3.3   Timestamp Based Multi-scheduler Concurrency Control Algorithm

In this section we propose a *Timestamp based Multi-scheduler Concurrency Control* (TMCC) algorithm that enforce *total order* in the schedule to ensure PDQ-serializability. *Total order* is required only for those conflicting transactions that accesses more than one PE being accessed by other active transactions. Functions that the algorithm uses are *Split_trans(T_i)*, *PE_accessed(T_i)*, *Active_trans(PE)*, *Cardinality()*, *Append_TS(Subtransaction)*. Function names are self-explanatory, definitions of the functions can be found in [13].

Working of the TMCC algorithm is explained below:

1. When transaction arrives at the splitter, *split_trans(T_i)* splits the transaction into multiple subtransactions, depending on allocation of data.
2. If there is only one subtransaction required by the transaction, the transaction can be submitted to the PE immediately without any delay.
3. All the transactions having more than one subtransaction are added to the **Active_Trans** set.
4. If multiple subtransactions are required by the transaction, the *splitter* appends a timestamp with every subtransaction.
5. If there are active transactions that access one or less than one PEs accessed by the transaction being scheduled then the subtransactions can be scheduled immediately. This condition does not pose any threat to the serializability.

6. If there is active transactions that access more than one of the PEs accessed by the transaction being scheduled then the subtransactions are submitted to the PE's *wait_queue*.
7. When all subtransactions of any transaction complete the execution at all the sites, the transaction commits and is removed from *Active_trans(PE)*. (*Active_trans(PE)*) takes the processing element as an argument and returns the set of transactions running at that PE)

4   Performance Evaluation

In this section, we discuss the results obtained by simulation for the two scheduling strategies namely single scheduler and multi-scheduler. We run our simulation for various sizes of the system, under various loading conditions (heavy and light) and also at different write probability of the transaction. We used CSIM [12], a process-oriented, discrete-event simulation package to obtain our results. The simulation code was written in C++. We discuss the performance metrics and parameter settings next, before the performance experiments and results.

4.1   Performance Metrics and Parameter Settings

Performance metrics for most of the high performance database systems are *response time* and *throughput*. But our perspective of this research is different and hence the performance metrics are also different. In this paper our focus is to evaluate the performance of single and multi-scheduler.

Table 1. Parameter settings for simulation

Description	Value
Number of Processors Write probability Minimum number of data items accessed Maximum number of data items accessed	2-8 15%-60% 4 12
CPU access time for data (*cpu_accesstime*) Disk access time for data Lock factor (*lock_factor*) Read time for a data object	1.5 millisecond 35 millisecond 10 % of disk access time ((cpu_access_time) + (lock_factor * lock_table_size))
Write time for a data object Maximum wait before the transaction aborts Number of pages per node (*Num_pages*) Data partitioning strategy	25 % more than read time. 4 second 125, 185 Range partitioning
Inter-arrival time between transactions (exponential distribution with mean *m*)	1 – 3.5 seconds.

Our first performance metric is the *abort-ratio*. Abort ratio is calculated by dividing the number of aborting transactions by the number of submitted transactions. We measure the abort ratio against different loading conditions (light and heavy). Depending on type of transaction, transactions can have different *write-probability*. We consider different write-probability to measure the blocking of transactions. Write-probability of the transaction can be thought as: how much percentage of read data item is being written. Finally, we increase the number of processors and measure the *utilization* of the processors for both scheduling strategies.

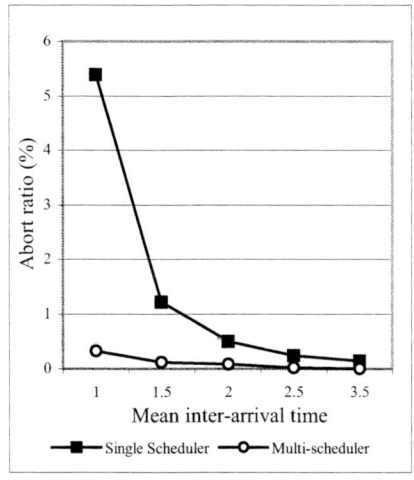

Fig. 2. Num_page = 125

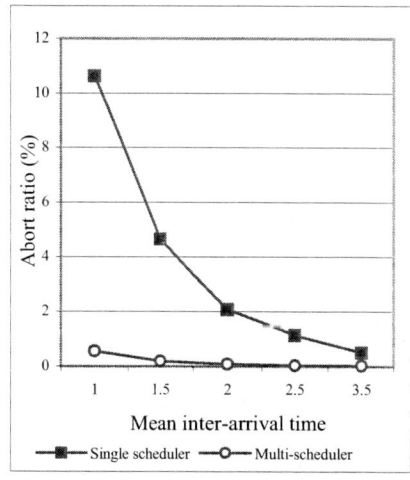

Fig. 3. Num_pages = 188

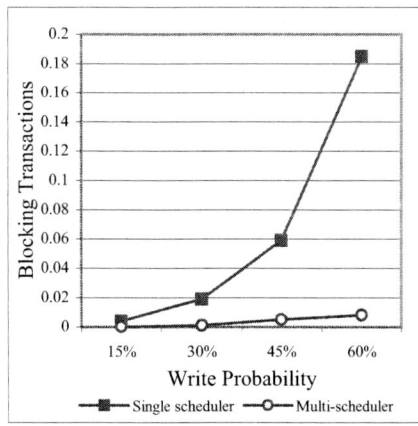

Fig. 4. Blocking Transactions

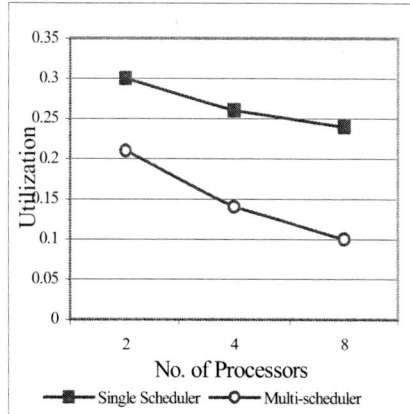

Fig. 5. Utilization

Table 1 shows the parameter settings for the experiments. Specific parameter setting values are again mentioned in the concerned experiment. OLTP (On-Line

Transaction Processing) workload is assumed for the system. We assume the 80-20 rule for data access. 80-20 rule states that 80% of the transactions access 20% of the data items and they are known as hotspots. Inter-arrival time for the transaction is assumed to be *exponential distribution* with mean *m*. Data is partitioned across all the PE's using *range partitioning* with equal number of data objects in each node.

4.2   Results and Discussions

Fig. 2 and 3 shows the performance results for 2-node case. Fig. 2 shows the variation of *abort ratio* with different loading condition for database size of 125 pages. We note that as the *mean* of inter-arrival time approaches to the maximum waiting time, abort ratio for both the strategies coincides. Under heavy loading conditions (i.e. with less inter-arrival time) data contention increases, thus the locking overhead also increases and consequently the abort ratio increases for single scheduler as well as for multi-scheduler strategy. We note that the rate of increase of abort ratio is much higher for single scheduler. The graph clearly shows the effect of locking overhead in single scheduler strategy is higher than multi-scheduler strategy.

Next we run the experiment under same environment by increasing the data volume by 1.5 times, that increases the number of pages per node from 125 to 188 (see fig. 3). The nature of the graph is same as that of fig. 2 but an interesting fact to note is that the rate of increase in abort ratio is less in multi-scheduler strategy. Under heavy loading conditions (inter-arrival time with $m = 1$) the increase in abort ratio for single scheduler is 97.22% but for multi-scheduler strategy the increase in abort ratio is 71.87%. Thus increasing the size of the database adversely affects the performance of single scheduler as discussed in previous sections.

Fig. 4 presents the relation between write-probability and percentage of blocking transactions. Number of nodes for this experiment was 4 and the mean of the inter-arrival time was 2.5 second. Depending on the type of transaction e.g. update, read-only etc., write-probability of the transaction may vary. Fig. 4 shows the effect of write-probability on the percentage of transactions that are blocked. We vary the write-probability from 15% to 60% and measure the percentage of blocking transactions for both single and multi-scheduler strategy. Observation shows a marginal difference between the two strategies at higher write-probability. At lower write-probability of 15% the number of blocking transactions for both strategies are approximately equal but with increasing write-probability the separation between the two strategies increases exponentially. Write operations acquire locks in exclusive mode and also tend to keep locks for longer duration. Hence, with increasing write-probability lock contention increases and centralized or single scheduler's performance deteriorates greatly.

The next experiment was motivated to test the affect of distributing the scheduling responsibilities on scale-up issues of high performance database systems. We vary the number of processors to 2, 4 and 8, then measure the average utilisation of the processors (see fig. 5). As expected the utilisation of the processors reduces in both the strategies but the rate of reduction in both the cases are different. Decrease in average percentage utilisation for the single scheduler case is 13% for 4 processors and 20% for 8 processors (percentage reduction is measured with respect to 2 processors). For multi-scheduler strategy the decrease in average percentage utilisation is 33% for 4 processors and 52% for 8 processors. Thus in multi-scheduler strategy, with increase in number of processors the average utilisation of processors reduces faster and the processor can be used to serve more transaction and thus achieve scale-up.

5   Conclusion

This paper aimed to evaluate and compare the performance of single and proposed multi-scheduler policies in high performance database applications. The motivation behind this work was to reduce the message and locking overheads of centralized scheduling scheme. Though the performance of both strategies is comparable under light workload, the performance of single scheduler deteriorates exponentially under heavy loading conditions. Thus, we can conclude that multi-scheduler algorithms show better chances of scalability and performance under heavy workload conditions. The reason for enhanced performance of multi-scheduler algorithm (TMCC in this case) is due to better distribution of work to individual nodes.

In future we plan to test the sensitivity of multi-scheduler strategy to different data distribution strategy e.g. hash partitioning and round robin partitioning. We also intend to investigate the direct relation of speedup and scaleup with the multi-scheduler strategy. The performance of the algorithm may also be examined for different loading conditions of the application such as decision support queries and mixed workloads (OLTP and decision support queries).

References

[1]   Bhide, "An Analysis of Three Transaction Processing Architectures", *Proceedings of 14th VLDB Conference*, pp. 339-350, 1988.

[2]   D.J. DeWitt, J. Gray, "Parallel Database Systems: The Future of High Performance Database Systems", *Communication of the ACM*, vol. 35, no. 6, pp. 85-98, 1992.

[3]   J. Gray, A. Reuter, Transaction Processing: Concepts and Techniques, Morgan Kaufmann, 1993.

[4]   M. Stonebraker, "The Case for Shared-Nothing", *IEEE Data Engineering*, vol. 9, no. 1, pp. 4-9, 1986.

[5]   P. Valduriez, "Parallel Database Systems: The Case For Shared Something", *Proceedings of the International Conference on Data Engineering*, pp. 460-465, 1993.

[6]   P. Valduriez, "Parallel Database Systems: Open Problems and New Issues", *Distributed and Parallel Databases*, vol. 1, pp. 137-165, 1993.

[7]   P. A. Bernstein, V. Hadzilacos, N. Goodman, *Concurrency Control and Recovery in Database Systems*, Addison-Wesley, 1987.

[8] T. Ohmori, M. Kitsuregawa, H. Tanaka, "Scheduling batch transactions on shared-nothing parallel database machines: effects of concurrency and parallelism" *Data Engineering, Proceedings, Seventh Intl. Conference on,* 8-12, pp: 210 –219, Apr 1991.

[9] K. Barker, "Transaction Management on Multidatabase Systems", PhD thesis, Department of Computer Science, The university of Alberta, Canada, 1990.

[10] T. Ozsu, P. Valduriez, "*Distributed and Parallel Database Systems*", *ACM Computing Surveys,* vol.28, no.1, pp 125-128, March 1996.

[11] M.T. Ozsu and P. Valduriez, editors. *Principles of Distributed Database Systems* (Second Edition). Prentice-Hall, 1999.

[12] CSIM, User's Guide CSIM18 Simulation Engine (C++ Ver.), Mesquite Software, Inc.

[13] S. Goel, H. Sharda, D. Taniar, "Multi-scheduler Concurrency Control Algorithm for Parallel Database Systems", *Advanced Parallel Processing Technology, Lecture Notes in Computer Science*, Springer-Verlag, 2003.

Gridscape: A Tool for the Creation of Interactive and Dynamic Grid Testbed Web Portals

Hussein Gibbins and Rajkumar Buyya

Grid Computing and Distributed Systems (GRIDS) Laboratory
Department of Computer Science and Software Engineering
The University of Melbourne, Australia
{hag,raj}@cs.mu.oz.au

Abstract. The notion of grid computing has gained an increasing popularity recently as a realistic solution to many of our large-scale data storage and processing needs. It enables the sharing, selection and aggregation of resources geographically distributed across collaborative organisations. Now more and more people are beginning to embrace grid computing and thus are seeing the need to set up their own grids and grid testbeds. With this comes the need to have some means to enable them to view and monitor the status of the resources in these testbeds. Generally developers invest a substantial amount of time and effort developing custom grid testbed monitoring software. To overcome this limitation, this paper proposes Gridscape – a tool that enables the rapid creation of interactive and dynamic testbed portals (without any programming effort). Gridscape primarily aims to provide a solution for those users who need to create a grid testbed portal but don't necessarily have the time or resources to build a system of their own from scratch.

1 Introduction

As we develop into an information dependant society, the amount of information we produce and the amount that we desire to consume continues to grow. As we continue to advance in this information society, we find the need for more sophisticated technology, faster computation and large-scale storage capacity to handle our information wants and needs on demand in real-time. Recently we have seen the rise of utilisation of resources distributed across the Internet (e.g., SETI@Home [1]) in order to solve problems that need large-scale computational resources. This paradigm is popularly known as grid computing [2], which has the potential of being able to deal with large-scale data and compute intensive problems and is opening the door to new innovative computing solutions to problems in both scientific (e.g., bioinformatics[22]) and commercial (e.g., portfolio pricing[23]) fields.

The components that make up computational grids include instruments, displays, computational resources, and information resources that are widely distributed in

S. R. Das, S. K. Das (Eds.): IWDC 2003, LNCS 2918, pp. 131-142, 2003.

location and are managed by various organisations. Grid technologies enable large-scale sharing of these resources and in these settings, being able to monitor any resources, services, and computations is challenging due to the heterogeneous nature, large numbers, dynamic behavior, and geographical distribution of the entities in which a user might be interested. Consequently, information services are a vital part of any grid software or infrastructure, providing fundamental mechanisms for monitoring, and hence for planning and adapting application behavior [3].

With the rise in popularity of grid computing, we see an increasing number of people moving towards grid enabling their work and their applications. Many people are now attempting to harness distributed resources and are setting up grid testbeds. Once testbeds have been set up, there is a need for some application or portal to enable the viewing and monitoring of the testbed's status. Gridscape, presented in this paper, aims to assist with this problem by assisting the creation of web based portals as well as making administering these portals an easy process.

The design aims of Gridscape are that it should:

- Allow for the rapid creation of grid testbed portals;
- Allow for simple portal management and administration;
- Provide an interactive and dynamic portal;
- Provide a clear and user-friendly overall view of grid testbed resources; and
- Have a flexible design and implementation such that core components can be leveraged, it provides a high level of portability, and a high level of accessibility (from the browsers perspective).

In the remainder of this paper we further identify the need for Gridscape and how it fits into the grid architecture. We then discuss the design and implementation as well as walking through an example of Gridscape's usage.

2   Related Work

As mentioned earlier, there is a definite need for software to monitor testbeds once they are created. However, because of the complex nature of grids, creating useful tools to gather and present resource information is a challenge. There are a number of implementations which currently exist, each with their own strengths and weaknesses. The majority of these efforts are either application specific portals or portal development toolkits. Here we discuss a few representative implementations and compare and contrast their design and development methodologies along with their strengths and weaknesses, clarifying the need for a tool such as Gridscape.

The most basic, and perhaps least useful type of implementation is application specific and uses HTML with static content to present resource status information. This type of implementation does not provide users with up-to-date, real-time feedback about the status of grid resources. This type of monitoring tool is easy to create, however limits the relevance and usefulness of the information provided and is also difficult to maintain or keep updated. Also, these types of portals tend to provide complete, unprocessed grid resource information data, which makes it hard to locate specific characteristic about any given resource, thus severely hindering its usefulness as a monitoring tool. An example of such an implementation is an early version of GRIDView, which is used to monitor the status of the US Atlas Grid Testbed [4].

Another, more sophisticated approach is to use dynamic content within HTML (such as with PHP). This allows for a real-time view of how the grid resources are performing, which is ideal for this type of tool. The NorduGrid [5] Grid Monitor is a good example of this, providing current load information as well as processed and user-friendly Globus MDS [3] (Metacomputing Directory Service, recently called Monitoring and Discovery Service) information. One feature lacking from this and the previous implementation is the availability of a spatial or geographical view of the resources. It is often useful to be able to have a visual picture of where your resources are located geographically. Again, the downside to this type of implementation is that it is tailored specifically to a particular testbed or particular needs, which means that this monitoring tool cannot be used to monitor any general testbed we may want to monitor.

An even more sophisticated tool can be produced with the use of technology such as Java and Java Applets. This approach has been taken in a number of instances, such as the new GRIDView monitoring tool for the US ATLAS Grid Testbed [6]. However, this implementation doesn't provide the user with immediate and concise information, it is again application specific.

Moving away from the application specific type of portal, we see a number of Grid portal development toolkits. They include GridPort [7], GPDK [9], and Legion Portal [10]. These toolkits assist in the construction of application specific portals; however they operate at a much lower level and aim to provide developers with libraries or low-level interfaces to Grid resources in order to assist in portal creation. For example, GridPort toolkit libraries/interfaces have been utilised in the development of NPACI HotPage[8] portal. Gridscape, on the other hand, requires no explicit programming effort in order to create a testbed portal.

Map Center [11] basically provides a web interface for querying resources (by issuing commands such as ping and grid-info-search) for status information. The command interface is bit low level, for instance, one need to supply attributes at LDAP (Lightweight Directory Access Protocol) syntax level for querying the MDS. Gridscape provides high-level and user-friendly portal interface—status of Grid resources is displayed on a geographic map on the testbed and they can be queried further for detailed information.

3   Architecture

The architecture of Gridscape and its interaction with various Grid components is shown in Figure 1. Gridscape itself consists of three components: web application, administration tool, and interface to grid information service.

Web Application. The web application consists of a customisable template portal which provides an interactive graphical view of resource locations and the ability to monitor its status and details, with the added ability of being able to submit queries to identify resources with specific characteristics.

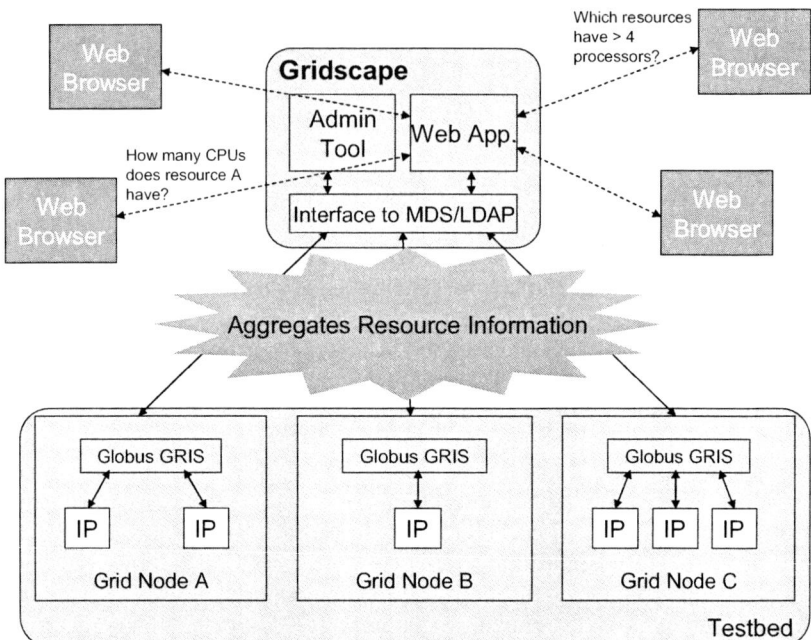

Fig. 1. Gridscape architecture

Administration Application. The administration tool provides the user with a simple and user-friendly way of customising and updating their personal testbed portal. Users are able to manage the resources to be used in the portal by adding, removing and editing their details.

Interface to Grid Information Service. The information provided by Gridscape is gathered from individual grid resources by accessing the Monitoring and Discovery Service (MDS) [3] provided by the Information Services component of the Globus Toolkit [12], which is run on them. MDS is designed to provide a standard mechanism for publishing and discovering resource status and configuration information. It provides a uniform, flexible interface to data collected by lower-level information providers.

Within Gridscape, "interface to MDS" component has been basically developed as a Java based class containing high-level methods that hide low-level details (e.g., LDAP protocols) of accessing MDS services. This level of separation of low-level MDS access mechanisms from other Gridscape components will ensure their portability. For example, if there is a change in MDS access protocols say from LDAP based to XML-based Web services, we can easily update our "MDS access interface" without the need of updating other components.

As the MDS services are utilised by Gridscape while gathering individual Grid node information, it seems logical we first discuss MDS components briefly in order to better understand Gridscape's interaction with them.

Globus MDS

The MDS reduces the complexity of accessing system information. This is achieved by having local systems use a wide variety of information-generating and gathering mechanisms, but users only need to know how to interact with MDS to access the information. MDS acts as a point of convergence between the large number of information sources and the large number of applications and high-level services which utilise them.

The MDS represents information in accordance with the Lightweight Directory Access Protocol (LDAP) [3]. LDAP is a set of protocols for accessing information directories.   LDAP, a Lightweight version of the old X.500 Directory Access Protocol, supports TCP/IP communication and is becoming the standard protocol when dealing with any directory information applications. Using an LDAP server, MDS provides middleware information in a common interface.

There are three components which make up the MDS hierarchy: Information Providers (IPs), the Grid Resource Information Service (GRIS), and the Grid Index Information Service (GIIS) [13]. At the lowest level there are Information Providers (IPs) which provide resource data such as current load status, CPU configuration, operating system type and version, basic file system information, memory information, and type of network interconnect. These IPs, interface from any data collection service, and report to GRIS. The GRIS runs on a resource and contains the set of information relevant to that resource, provided by the IPs. Individual resources can then be registered to a GIIS, which combines individual GRIS services to provide an overall view of the grid. The GIIS can be explored and searched to find out information about resources, as you would any index.

Gridscape's Interaction with MDS

Gridscape discovers the properties of individual resources of a given testbed by making MDS queries to individual GRIS installations. Results are sent back to Gridscape which caches these details for further use. Because Gridscape aims to be free of 3rd party tools such as a database, and because querying distributed resources continually is very costly, Gridscape caches the current status of the testbed and allows this store to be shared by any web browsers accessing the portal. The current status information held by Gridscape can be automatically updated periodically, or an immediate status update can be requested at any time.

The action of accessing the GRIS to collect details of individual resources allows Gridscape to behave as a GIIS, to a certain extent, in that it provides users with a collection of separate GRIS information from various resources, in order to provide more of a holistic view of a grid testbed.

4   Design and Implementation

The Gridscape web application is designed following the MVC (Model-View-Controller) based, Model-2 type architecture [14] shown in Figure 2. This architecture, which was developed for use with web applications implemented in technology such as Java Server Pages and Servlets, provides a means of decoupling the logic and data-structures of the application (server-side business logic code) from the presentation components (web pages)[15][16]. In order to make implementation easier, and enhance reliability, the Jakarta STRUTS framework [17] has been adopted. STRUTS provides a framework for building Model-2 type web applications.

With Gridscape, the Model component of the Model-2 architecture becomes the most interesting to investigate further. The reason for this is because this is where the main functionality is located within Gridscape. Also, because of the separation of the presentation, control and the business logic achieved with the application's architecture, we are able to leverage the Model component from the web application and re-use it in the Gridscape Administration Tool. We can see that in this way, other applications could be developed which also make use of the core functionality provided in the Model, by offering a new presentation and control or application layers which access these core components, as illustrated in Figure 3 .

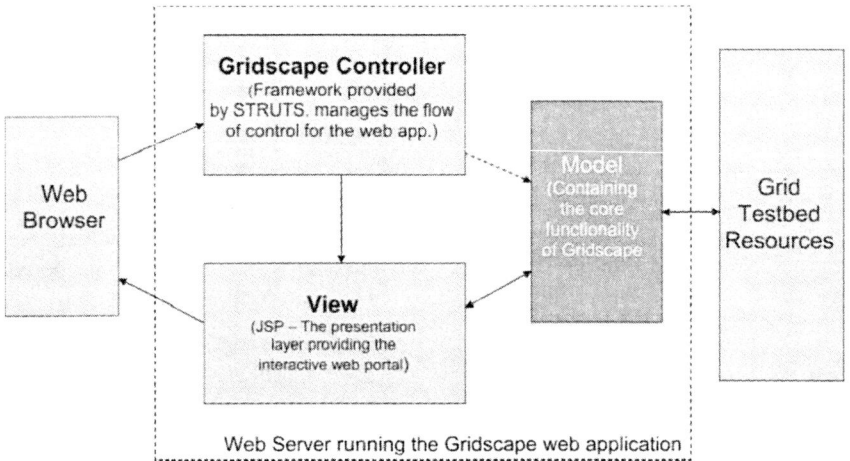

Fig. 2. MVC Model2 architecture of Gridscape implementation

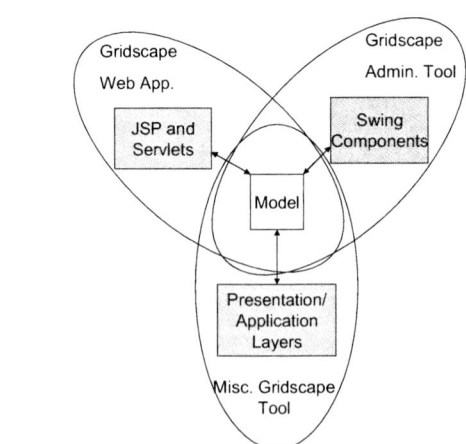

Fig. 3. Flexibility and reuse of the Model component

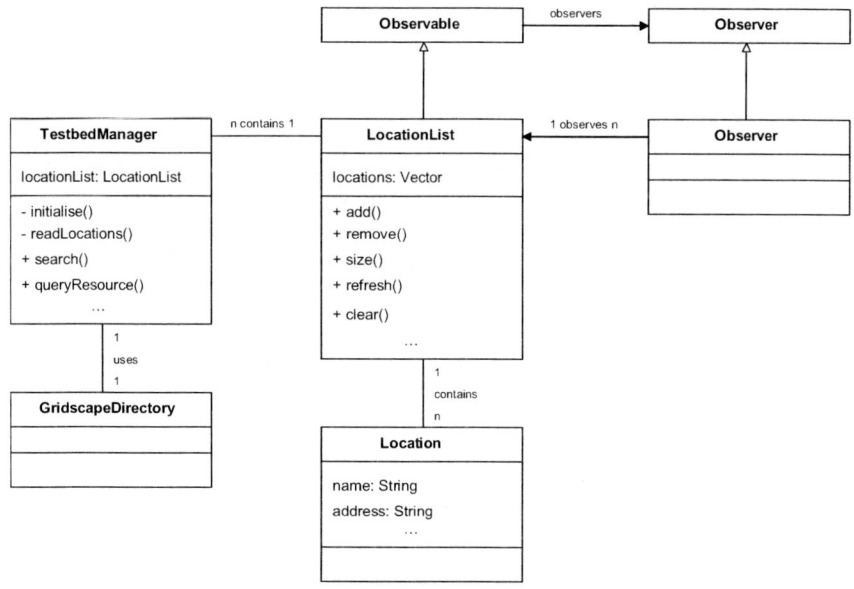

Fig. 4. Class diagram of the core classes of Gridscape's Model

4.1   The Model

So far we have identified the significance of the Model and seen its flexibility. The Model itself though consists of a number of important classes. In this section we will take a closer look at some of these classes, their properties, and how they interact with one another.

4.1.1 GridscapeDirectory

This class provides a convenient wrapper around the necessary elements of the naming and directory access packages of the core Java API. It provides us with an easy means of connecting to, and querying, the resources in the testbed. It is easy to see that this class will be invoked whenever we communicate with the testbed resources, whether it be through the web application or the admin tool.

4.1.2 Location

This low level class is used to represent and hold information about a particular resource. This information includes things such as the name, address, port number, as well as the MDS data which is gathered from the resource.

4.1.3 LocationList

As its name suggests, this class is used to hold a list of the various locations in a testbed. This class extends the Observable class provided by Java. Allowing this class to become observable means, that developing presentation layers or views which depend on this data, is made easy.

4.1.4 TestbedManager

This class manages the other components within the Model and is responsible for handling queries which are communicated from components outside of the Model. It handles the initialisation of the core of the application and handles duties such as searching by allowing other components to collaborate. It is interesting to note that while the TestbedManager contains an instance of the LocationList, it contains only a singleton instance. The benefit of this type of implementation is that through the web application, even though each client accessing the application is given a new instance of the TestbedManager, there is only one instance of the data. This means that information retrieved from testbed resources is cached, making the site more responsive, and ensuring that everyone is seeing the same up-to-date data.

5   Gridscape in Practice

Gridscape has already been used by a number of virtual organisations to create their Grid testbed portals for visualising and monitoring resources [21]. They include Australian Virtual Observatory and UK AstroGrid Collaboration, Belle Analysis Data Grid (BADG), and our own World-Wide Grid (WWG) testbed. In this section we will walk through the steps involved in creation of portal for your own Grid testbed using Gridscape and illustrate them with an example of creating a portal for WWG.

5.1   Deploying the Gridscape Web Application

To begin using Gridscape the user must first deploy the web application within their Jakarta Tomcat installation and also install the administering tool.

5.2   Creating Your Portal

Creating your own customised testbed portal with Gridscape simply involves customising the blank template portal which is provided with Gridscape. Gridscape supports intuitive GUI (see Figure 5) using which you can supply various elements of the testbed: a testbed logo, a map for displaying physical location of resources, and details of resources that are part of the testbed.

5.3   Customising Your Portal

Most of the details about your testbed are stored in configuration files which can also be edited manually. To make customisation easier, an administrating tool has been provided. Open the template from within the administrating tool to continue with customising your portal. Figure 5 shows a snapshot of Gridscape taken while creating a portal for the WWG testbed.

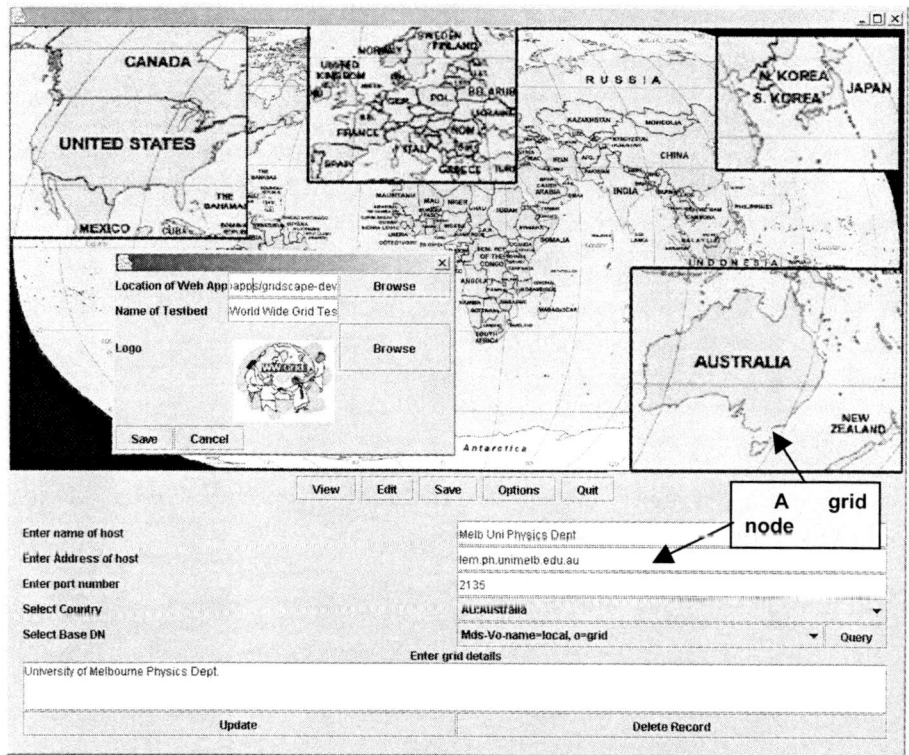

Fig. 5. A snapshot of Gridscape utilisation while creating the WWG portal

5.3.1 Changing Testbed Name, Logo and Other Details

These items are all customisable from the 'Options' menu. If a new logo is required for the web portal, for example, we can select the 'Change Logo' option. We are then presented with a dialog box which allows us to browse for a suitable image for our logo. Once the logo is selected we can save the selection and the changes will propagate immediately and directly to the web portal. This means that when we next visit the page, we will be able to see this change. This functionality is the same for all options. A small pop-up window shown in Figure 5 illustrates how one can supply testbed name, logo image file, and portal deployment location to Gridscape.

5.3.2 Managing Testbed Resources

The next step in setting up the portal is to tell Gridscape details about the resources which will be involved. The Gridscape Admin tool provides the user with two modes, an editing and a viewing mode. The editing mode allows the user to edit information regarding resources, while the viewing mode allows the user to simply browse and query existing resources. To add resources to the testbed portal we must enter 'Edit' mode. This can be achieved by choosing the appropriate mode from the 'Mode' menu.

Adding a New Resource

To add a new resource to the testbed, simply click the mouse in a vacant area on the map. Doing this will automatically create a new resource in your testbed. Position this resource in a desired location on the map by clicking and dragging this resource with the mouse. If the testbed is international, then you need supply a name of the country where the resource is physically located. Figure 5 shows an addition of a Grid node located in the School of Physics at Melbourne University.

Editing Resource Details

When a new resource is created, it is provided with the default property values. To change these properties we first need to select the resource by clicking on it with the mouse. Once selected, we can freely edit such details as its name, address and port number. Once completed, use the 'Update' button to store the changes.

Deleting an Unwanted Resource

If for some reason you need to remove a resource from the testbed, simply select the resource and use the 'Delete' button.

5.3.3   Querying Testbed Resources

Before saving your changes or viewing the web portal online, it is a good idea to go into 'View' mode and query the resources, by clicking on them with the mouse. This will give you confidence that the details you entered were correct and indicate the expected behaviour of your web portal.

5.4   Browsing the Testbed Portal

Once the customisation is complete, the testbed details are saved into a configuration file and deployed on the Web server. A snapshot of browsing and monitoring status of the WWG testbed resources through a portal created using Gridscape is shown in Figure 6. The portal can be accessed online by visiting the World Wide Grid Testbed website [19].

6   Conclusion and Future Work

Currently there are a number of unique applications designed to monitor, very specifically, details of only one grid testbed. This paper identifies the need for a tool of this nature - one able to automate the process of creating grid testbed portals. We propose Gridscape, a tool aimed to meet the needs of those who require a testbed portal but simply don't have the resources available to invest in creating their own software from scratch. Gridscape has the potential to provide users with any of the information made available through Globus MDS, and allows for quick and easy creation and administration of web based grid testbed portals.

We are planning to extend Gridscape to support live monitoring of application-level utilisation of Grid resources by integrating it with our Grid application management portal called G-monitor [20].

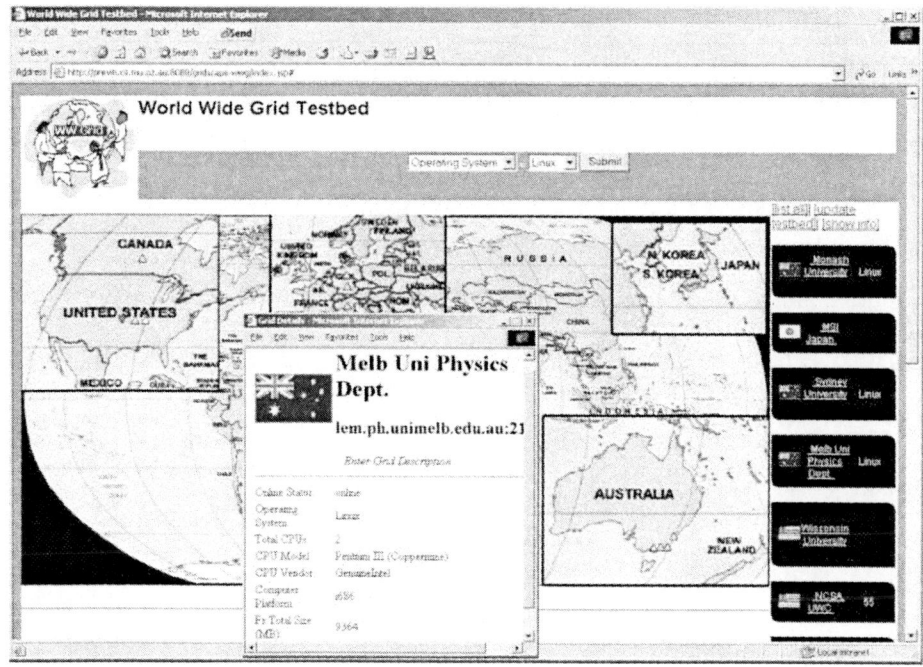

Fig. 6. A snapshot of browsing World Wide Grid testbed portal

Availability

The Gridscape software and user manual can be downloaded from the Gridbus project website: http://www.gridbus.org/gridscape/

Acknowledgements

We would like to acknowledge and thank Ankur Chakore, Yogesh Chadee and Rami Safiya, for their contributions in developing the initial portal creation tool called STAMPEDE [18] that served as an early seed for Gridscape. We would like to thank Anthony Sulistio for his comments on early drafts of the paper.

References

[1]   SETI@Home - http://setiathome.ssl.berkeley.edu/.
[2]   Foster and C. Kesselman (editors), *The Grid: Blueprint for a Future Computing Infrastructure*, Morgan Kaufmann Publishers, San Francisco, CA, USA, 1999.
[3]   MDS 2.2 User's Guide. http://www.globus.org/mds/mdsusersguide.pdf.
[4]   US ATLAS GRIDView (obsolete).
       http://heppc1.uta.edu/atlas/grid-status/index.html.

[5]   NorduGrid – Nordic Testbed for Wide Area Computing and Data Handling. http://www.nordugrid.org/.

[6]   US ATLAS GRIDView (current). http://www-hep.uta.edu/~mcguigan/applet/.

[7]   M. Thomas, S. Mock, J. Boisseau, M. Dahan, K. Mueller, D. Sutton, *The GridPort Toolkit Architecture for Building Grid Portals*, Proceedings of the 10th IEEE International Symposium on High Performance Distributed Computing, Aug 2001.

[8]   NPACI HotPage -- https://hotpage.npaci.edu/.

[9]   J. Novotny, *The Grid Portal Development Kit, Special Issue on Grid Computing Environments*, Journal of Concurrency and Computation: Practice and Experience, Volume 14, Issue 13-15, Wiley Press, USA, Nov.-Dec., 2002.

[10]   Natrajan, A. Nguyen-Tuong, M. Humphrey, M. Herrick, B. Clarke, A. Grimshaw, *The Legion Grid Portal*, Journal of Concurrency and Computation: Practice and Experience (CCPE), Volume 14, Issue 13-15, Wiley Press, USA, Nov.-Dec., 2002.

[11]   Map Center - An Open Grid Status Visualization Tool. http://ccwp7.in2p3.fr/mapcenter/.

[12]   The Globus Project. http://www.globus.org/.

[13]   Globus Toolkit 2.2, *MDS Technology Brief*, Draft 4 – January 30, 2003. http://www.globus.org/mds/mdstechnologybrief_draft4.pdf.

[14]   Govind Seshadri, *Understanding JavaServer Pages Model 2 architecture*. December, 1999. http://www.javaworld.com/javaworld/jw-12-1999/jw-12-ssj-jspmvc.html.

[15]   M. Foley, *STRUTS your stuff*, ZDNet Australia, 11 June 2002. http://www.zdnet.com.au/builder/program/java/story/0,2000034779,20265872,00.htm.

[16]   Q. Mahmoud, *Servlets and JSP Pages Best Practices*, March, 2003. http://developer.java.sun.com/developer/technicalArticles/javaserverpages/servlets_jsp/.

[17]   Jakarta STRUTS project homepage. http://jakarta.apache.org/struts/.

[18]   The STAMPEDE project – Helping utilise the power of many. http://members.optusnet.com.au/dgibbins/.

[19]   World Wide Grid Testbed Portal - http://previn.cs.mu.oz.au:8080/gridscape-wwg/.

[20]   M. Placek and R. Buyya, *G-Monitor: Gridbus web portal for monitoring and steering application execution on global grids*, Proceedings of the International Workshop on Challenges of Large Applications in Distributed Environments (CLADE 2002), In conjunction with HPDC 2003 symposium, June 21-24, 2003, Seattle, USA.

[21]   Gridscape-based Portals - http://previn.cs.mu.oz.au:8080/gridscape/.

[22]   R. Buyya, K. Branson, J. Giddy, and D. Abramson, *The Virtual Laboratory: Enabling Molecular Modeling for Drug Design on the World Wide Grid*, Journal of Concurrency and Computation: Practice and Experience, Volume 15, Issue 1, Wiley Press, Jan. 2003.

[23]   Crawford, D. Dias, A. Iyengar, M. Novaes, and L. Zhang, *Commercial Applications of Grid Computing*, IBM Research Technical Report, RC 22702, IBM, USA, Jan. 22, 2003.

Agent-Based Fair Load Distribution
in Linux Web Server Cluster

Kyeongmo Kang, MinHwan Ok*, and Myong-soon Park

Dept. of Computer Science and Engineering, Korea University
Seoul, 136-701, Korea
{kinghor,myongsp}@ilab.korea.ac.kr
panflute@korea.ac.kr

Abstract. Clustering the replicated Web servers is appropriate way to build a Web information system. An LVS(Linux Virtual Server) operating with software clustering technology provides the Web service based on Linux environment. The LVS is the centralized structure, which the load should be distributed to the real servers by a load balancer. A round-robin scheduling method or a least-connection scheduling method may cause a load imbalance amongst the real servers since such methods distribute the load to the real servers without considering the actual load of real servers. In this paper, we propose a fair load distribution scheme that can distribute the requests even fair than ABSS that was early proposed to solve the problem of scheduling methods in the LVS.

1   Introduction

These days Internet sites such as 'Lycos,' 'Excite,' or 'Yahoo' get hundred millions of requests from the clients worldwide.[1,5] This is four to five thousands of requests in one second and necessitates system scalability and availability. Famous digital libraries also suffer heavy amount of requests from worldwide clients. Clustering the server system is the popular solution nowadays. Many works on clusters support system scalability and availability.[2] Linux Virtual Server(LVS) is one such a system. The LVS is a centralized structure that consists of a load balancer in its front-end and real servers in its back-end. A load balancer controls the whole cluster system, and acts as the representative point of contact for clients and distributes requests to back-end nodes in the cluster system. The real server acts as the role of processing client requests that are actually allocated from the load balancer. Performance of the cluster system is determined by the scheduling method with which the load balancer divides client requests to the real servers.

The round-robin scheduling method and the least-connection scheduling method are usually used in LVS. The round-robin scheduling method distributes client re-

---

* Corresponding Author

S. R. Das, S. K. Das (Eds.): IWDC 2003, LNCS 2918, pp. 143–152, 2003.

quests to the servers sequentially. The least-connection scheduling method distributes client requests to a server that has least connections. These scheduling methods distribute client requests to the servers without considering the real load of real servers. Because of this, a load imbalance can occur between the real servers. If the load is concentrated in a particular real server due to the load imbalance between the real servers, it can bring a high latency time or may not service the client requests. Accordingly a successful service may be provided by sophisticated load balancing between real servers.

The Bit Project No.61 is a scheduling method which considers the real load of real servers.[3] However, the load balancer has lots of communication overhead due to the fact that the load balancer sends packets requesting load information to real servers and receiving the response packets from real servers.[1,3] As the number of real servers are increasing, the communication overhead of the load balancer is also more increasing. Therefore the single-point-of-bottleneck problem of the load balancer is also becoming more evident. The single-point-of-bottleneck problem also causes low scalability. The ABSS, the previous work of this paper, solved these problems.[1] The ABSS uses an agent-based scheduling method employing CPU utilization as the real load of real servers. In Web information systems, most tasks are searching for information including archives and viewing the information. It usually takes longer the time a person seek to find out wanted information from a result window, than the time a server prepares the result window. Moreover each window opened reflects imminent task, i.e. future searching any other archive, or the past one for the server. Each window occupies a server memory. Viewing the archive by a person takes much longer time than processing the archive content by a server. More importantly, a large number of viewing the archive of complex content may exhaust the server memory.

From this respect, we learn that using memory utilization can be rather precise than using CPU utilization to distribute load of the requests fairly in Web information cluster. The rest of the paper is as follows. In Section 2 we depict the Web information cluster and talks about the LVS with its scheduling methods. Section 3 describes the previous work, ABSS, in detail. In Section 4 we present the proposed load distribution scheduling. Section 5 shows the simulation results and discusses. We conclude in Section 6.

2   System Organization

The model of Web information system, in this paper, is a cluster structure, constituted Linux cluster of replicated Web servers. The cluster structures of replicated Web servers provide scalability with ability to respond quickly to requests from clients. Each replicated Web server presents an identical, entire Web site. Replicated Web servers need publish consistency among them[6], however, this paper makes a premise that the Web servers are consistent in publishing the Web pages.

The Web information cluster is based on Linux Virtual Server or LVS[7]. It has a centralized structure thus primary connection point to the system is the load balancer. Many Web server clusters are implemented in centralized structure due to its performance merit.

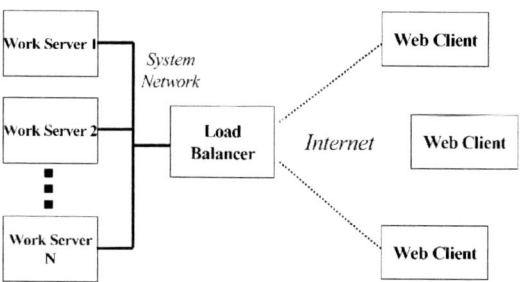

Fig. 1. Web Information System

2.1   Linux Virtual Server

The Linux Virtual Server (LVS) is a highly scalable and highly available server built
on a cluster of real servers. The LVS consists of the load balancer and real servers as
is typical in centralized structures. The architecture of the cluster is transparent to
end-users, and the users see only a single virtual server. The real servers are interconnected by high-speed LAN. At the front-end is a load balancer, which schedules requests to the different servers and makes parallel services of the cluster appear as a
virtual service on a single IP address. At the back-end are real servers, which actually
process client requests.

Figure 1 illustrates system organization of the Linux Web information cluster. We
call the real servers as work servers in this paper.

The load balancer schedules the request to a work server that is selected by the
following scheduling method. Each work server is a replicated Web server connected
to one database server. The work servers send a response to the client directly or
through the load balancer after it processes the request.

2.2   Scheduling Methods of LVS

The scheduling method greatly affects the performance of the whole cluster system. It
also affects the scalability and the reliability of the whole cluster system, and decreases the response time for the client request by evenly distributing the requests
over all work servers. The scheduling method can be divided into two methods. One
is the method that considers the real load information of work servers. The other is the
method that doesn't consider.

Round-robin scheduling is the method that doesn't consider the real load information. All the requests are treated as equals regardless of their loads. In the Web information cluster when the load balancer distributes client requests to the work servers, it
uses information from the TCP layer or below. That is, the load balancer doesn't know
the contents of the request from the client since it processes the requests with an IP
address or a port number only. The least-connection scheduling method and the agent-
based scheduling method are appropriate to this restriction.

The least-connection scheduling algorithm directs the network connections to the
work server with the least number of established connections. The increment and dec-
rement of live connections can be counted by detecting SYN(initializing a session)

and FIN(terminating a session) flags in the TCP packets sent from the client side respectively. That is, the established connection is considered as the real load of a work server. All the requests are treated as equals regardless of their load.

The agent-based scheduling method is presented in 'Bit Project No.61'.[3] An agent is installed in the work servers and in the load balancer. Throughout the action of the agent, the load balancer gains the load information of the work servers each time and saves it in their own Load Information Table. When the request is received, the load balancer selects a work server that has the least load for offering the service, and so is able to achieve a sophisticated load balancing. However the load balancer has a lot of communication overhead since it sends to the work servers the packets that have requested the load information and also receives response packets from the work servers. As the work servers increase, the communication overhead of the load balancer is further increasing. It compounds the single-point-of-bottleneck problem of the load balancer. Our earlier works proposed the agent based sophisticated and scalable scheduling method that supplements this problem.[1]

3   Previous Works

3.1   Web Information Cluster of Web Servers

Many works has proposed load sharing schemes for clusters of replicated Web servers. For Alexandria digital library systems an adaptive load sharing scheme that uses CPU and I/O utilization was proposed on heterogeneous work servers.[8] Assigning load could be redirected to other work server in the cluster. However since predicting the cost of a database searching query exactly was difficult, estimation about database access was omitted from I/O, resulting in diminished load sharing. Migration of request among work servers was another disadvantage since migrated request should suffer longer response time. A model of an adaptive load balancing scheme considers overall server resource usage was presented,[9] on heterogeneous work servers. The model is time consuming, since it requires gathering overall resource usage from each work server and calculating recursive functions. This may introduce additional delay into each request delivery.

For rapid distribution of requests the cluster of homogeneous work servers is preferred. Gathering load information of each work server should be light-weight and calculation in selection of the work server should not be complex since long computation may introduce additional delay. The following scheduling method has such characteristics.

3.2   Agent-Based Sophisticated and Scalable Scheduling Method (ABSS)

Various model of the cluster system has been devised.[10] The dispatcher-based architecture showed the best performance in the article, and ABSS is proposed for dispatcher-based architecture. The architecture of ABSS scheduling method is illustrated in Figure 2. The agent runs in the load balancer and work servers. The Load Information Packet itinerates between work servers and gains the load information of work

servers. The load balancer distributes client requests to work servers using the real
load information of work servers.

The basic composition of the ABSS scheduling method is illustrated in Figure 3.
The load balancer consists of a scheduler part and an agent part. The scheduler performs the role that decides which work server will process a client request referring to
the Load Information Table. The Server Management module of the load balancer
performs the role that decides the order among the work servers when the work server
is added or removed. The Initiate module creates a new Load Information Packet(LIP)
and calls the Communication module. The Communication module sends the LIP
packet to the first work server. The Communication module of the work server gets a
LIP and calls the Load Information Process module. The Load Information Process
module has the load information of the work server and adds it to the LIP packet.
Then the Load Information Process module calls the Communication module again.
The Communication module sends the LIP to the next work server. The LIP packet
gathers the load information of each work server while itinerating between the work
servers. The Communication module in the load balancer gets the LIP packet from the
last work server. The Process module copies the LIP to its own Load Information Table. And the load balancer distributes the client requests to the work servers referring
the load information of work servers.

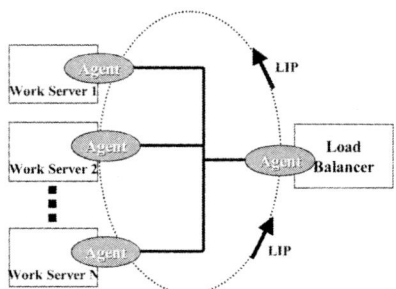

Fig. 2. Architecture

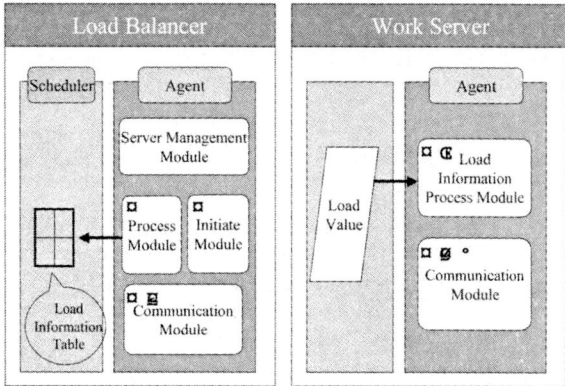

Fig. 3. Agent Module

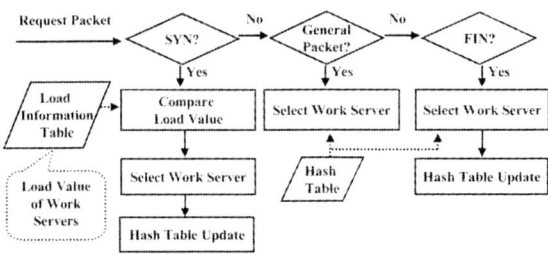

Fig. 4. Algorithm

Figure 4 illustrates the algorithm that processes each client request packet in the load balancer. When a request packet sent from the client is entered in the load balancer, the load balancer classifies it into three. The first packet means the beginning of one request and it is called the SYN packet. The last packet means the end of one request and it is called the FIN packet. The rest is the general packet. When the load balancer receives the SYN packet, it selects the work server that has the least load referred to the Load Information Table and sends the request packet to the selected work server. Then the load balancer stores the scheduling result at the Connection Hash Table. If the request packet is the general packet, the load balancer finds the work server that has offered the service previously through the Connection Hash Table. If the request packet is the FIN packet, the load balancer removes the entry from the Connection Hash Table.

The ABSS uses the CPU utilization as a standard that decides the load of work servers. However there is difficulty in using the CPU utilization as the real load information of work servers for two reasons. The first reason is that the CPU utilization just tells the past record. If it takes t to arrive at each work server, then it takes $n*t$ the LIP to be delivered to the load balancer. The second reason is that the CPU utilization changes frequently. The CPU utilization is not stable during the execution of a process. Moreover the required CPU clock time is different with each instruction. The CPU utilization is the indefinite value during the execution of the process. Thus CPU utilization is not suitable for the load information of the work servers.

4   Proposed Scheduling Method

The memory utilization doesn't show the throbbing of CPU utilization. Therefore we consider the correct load information of work servers than the CPU utilization that is not stable.

The BSD class' network code is assigned three types of memory, *socket, inpcb* and *tcpcb*.[12] The socket information is correlated with the communication link. The inpcb has the information correlated with an IP protocol control block, transport layer. The tcpcb has the information correlated with TCP control block, TCP. The assigned memories are kept until the connection is released. Also the memory utilization expresses the utilization of present memory unlike the CPU utilization.

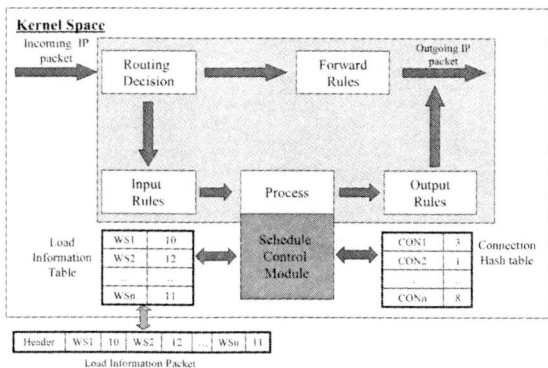

Fig. 5. Internal Implementation

Figure 5 illustrates the internal implementation of the load balancer. The load balancer creates or updates the Load Information Table with the Load Information Packet that has the load information of work servers. And the load balancer maintains the Connection Hash Table using the connection information. The load balancer performs the following scenario.

In case of SYN packet the load balancer received, it delivers the packet to the work server after selecting the work server to process the request referring the Load Information Table. In case of general packet the load balancer delivers the packet to the corresponding work server after referring the Load Information Table. Finally, in case of FIN packet the load balancer delivers the packet to the work server after referring Load Information Table and deletes the relevant entry from the Connection Hash Table.

5   Performance Evaluation

5.1   Simulation Conditions

We suppose one load balancer and four work servers in the simulation. The transmission time of Load Information Packet between two nodes is assumed 0.3ms. We conceived an idea from the results of ping program in the Linux environment. RTT in the results were less than what we assumed but the size of Load Information Packet is bigger than the size of ping packet. CPU-intensive requests and Memory-intensive requests are assumed in about the same numbers.[13]

The simulation was conducted during ten seconds. 60 requests are generated per second. The CPU requirement and the memory requirement of each request are chosen at random with some variation. Particularly we defined the average processing time for one request to be 67*ms*. The Web server receives a maximum of 2,000 requests per second [11], and they are served by four work servers. However 60 requests were served per second in the simulation. While the simulation processes one request, 33.33 requests are processed in reality. Therefore, the average processing time is 1/500 second in reality. Thus we defined 33.33/500(≈0.067) seconds as average processing time in the simulation.

The reason why we give the CPU requirement and memory requirement in each request is to compare the fairness of load distribution by *Fair-Ready*, the proposed scheduling, with that by ABSS.

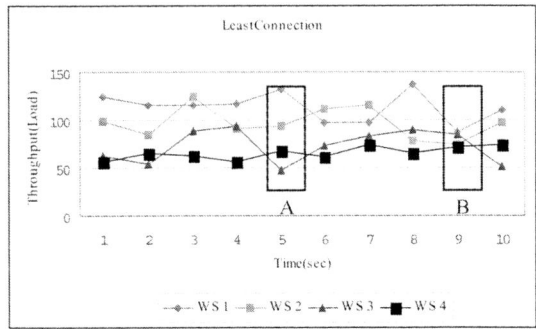

Fig. 6. Least Connection Scheduling

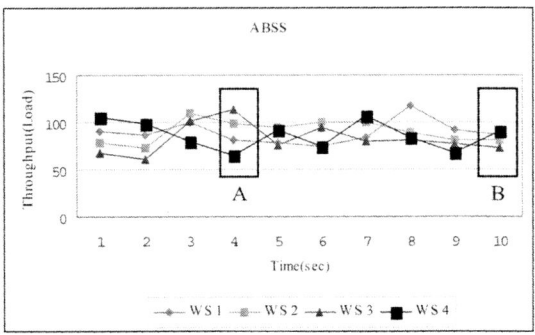

Fig. 7. ABSS Scheduling

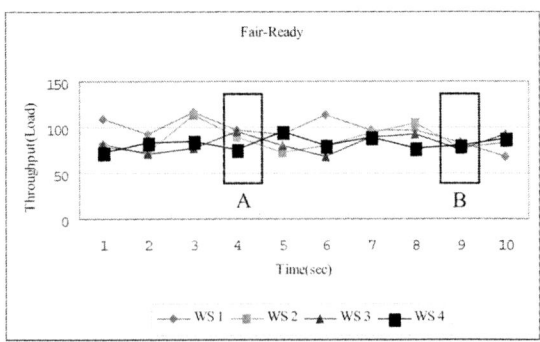

Fig. 8. Fair-Ready Scheduling

Table 1. Average Standard Deviation between Work Servers

Scheduling method	Average standard deviation
Least Connection(10)	21.87
ABSS (10)	11.73
Fair-Ready (10)	9.88
Least Connection(20)	38.26
ABSS (20)	16.73
Fair-Ready (20)	16.23

5.2   The Results

Three scheduling methods were evaluated. The first, least-connection scheduling, is the method that considers the number of connections as load information of the work server. The second, ABSS, is the method that considers the CPU utilization, and the third, Fair-Ready, is the method that considers the memory utilization.

The above figures show the results respectively of the least-connection, of the ABSS and of the Fair-Ready. As narrower the difference of throughput among the work servers, the load balancer has distributed the load closer to fairness. The part that is marked by 'A' displays the widest difference of throughput among work servers for the total simulation time. On the contrary the part that is marked by 'B' displays the narrowest difference of throughput among the work servers. The result of ABSS and Fair-Ready has narrower differences of the throughput among work servers than the result of least-connection. Further the result of Fair-Ready has narrower differences of the throughput among work servers than the result of ABSS. Table 1 shows the average standard deviation among the work servers for the least-connection, the ABSS and the Fair-Ready scheduling method.

In table 2, we can see that the standard deviation of Fair-Ready is less than those of least-connection and ABSS, respectively. It is still true the range of CPU requirement and memory requirement fluctuation is wider, (20) in the table. This implies that the Fair-Ready scheduling method is the even fair load distribution scheme than the other two methods.

6   Conclusion

Linux LVS is adopted for clustering replicated Web servers in order to build a Web information system. We proposed the even fair scheduling method than previous ones by using the memory utilization and the proposed method gains the real load information of work servers through the Load Information Packet and selects a work server in scheduling with the memory utilization. In the simulation, our proposed method showed even fair load distribution among the work servers than that in ABSS. This means infrequent changes of memory utilization become more accurate metric than frequent changes of CPU utilization. Therefore, the Fair-Ready scheduling method should help the Web information cluster provide steady service under heavy request traffic.

References

[1]   Y.-H. Shin, S.-H. Lee, H.-K. Baik, G.-H. Kim, M.-S. Park: Agent-based So-
       phisticated and Scalable Scheduling Method. In: Proceedings of Third Interna-
       tional Network Conference. (2002)
[2]   Linux Virtual Server open project sites. http://www.linuxvirtualserver.org
[3]   High Available Cluster System. Bit Project No.61. (2001)
[4]   D. Adresen, T. Yang, V. Holmedahl, O.H.Ibarra: SWEB: Towards a Scalable
       World Wide Web Server on Multicomputer. In: IEEE Proceedings of IPPS '96.
       (1996) 850-856
[5]   M. F. Arlitt, C. L. Williamson: Internet Web Servers: Workload Characteriza-
       tion and Performance Implications. Trans. Networking, Vol.5 No. 5. IEEE
       (1997) 631-645
[6]   R. C. Burns, D. D.E. Long: Efficient Data Distribution in Web Server Farm.
       Internet Computing, Vol 5 No. 4. IEEE (2001) 56-65
[7]   B. H. Klovning, E. Kure, O.: A Comparison of Load Balancing Techniques for
       Scalable Web Servers. Network, Vol 14 No. 4. IEEE (2000) 58-64
[8]   H. Zhu, T. Yang, Q. Zheng, D. Watson, O. H. Ibarra, T. Smith: Adaptive Load
       Sharing for Clustered Digital Library Servers. In: IEEE Proceedings of Interna-
       tional Symposium on High Performance Distributed Computing. (1998) 235-
       242
[9]   J. Aweya, M. Ouellette, D. Y. Montuno, B. Doray, K. Felske: An Adaptive
       Load Balancing Scheme for Web Servers. Int. J. Netork Mgmt. 12 (2002) 3-39
[10]   V. Cardellini, M. Colajanni, P. S. Yu: Dynamic Load Balancing on Web-server
       Systems. Internet Computing, Vol. 3 No. 3. IEEE (1999) 28-39
[11]   http://support.zeus.com/doc/tech/linux_http_benchmarking.pdf
[12]   http://home.postech.ac.kr/~jysh/research/IDS/survey/synki ll/synkill.html
[13]   http://www.microsoft.com/korea/msdn/library/dnw/mt/html/tunewmserver.asp

Location Management by Movement Prediction
Using Mobility Patterns
and Regional Route Maps

R. K. Ghosh[1], Shravan K. Rayanchu[2], and Hrushikesha Mohanty[3]

[1] Department of CSE, IIT-Kanpur, Kanpur 208016, India
rkg@cse.iitk.ac.in
[2] Department of CSE, IIT Guwahati, Guwahati 781039, India
rayanchu@iitg.ernet.in
[3] Department of CIS, University of Hyderabad, Hyderabad 500046, India
hmcs@uohyd.ernet.in

Abstract. In this paper we argue that in most of the cases the movement pattern of a mobile host repeats itself on a day-to-day basis, but for the occasional transient deviations. Taking the spatio-temporal properties of a mobile host into account, we propose a new location management scheme. The scheme achieves the near optimal routing as it bypasses the default reliance on the routes through the home agent for most of the calls made to a mobile host. It uses the mobility pattern of the mobile host to predict the cell location of that host. Transient deviations ranging from 5-30% are tackled by tracking down a host efficiently with the help of a regional route map which is the physical route map of a small neighbourhood of the last known location of that host. The performance of the proposed scheme is evaluated with respect to varying values of call-to-mobility ratio (CMR), and found to be quite good even for transient deviations ranging upto 30%.

1   Introduction

In this paper we propose a new location management scheme by taking the spatio-temporal localities of mobile hosts into account. The strategy works on the idea that the movement patterns of most mobile hosts get repeated on day-to-day basis, but for occasional transient deviations. Thus, by modeling the mobility pattern of a mobile host at home agent and replicating the same at some chosen network sites it would be possible to predict the likely location of that mobile host. The transient deviations, if any, can be tackled by localizing the search to the neighbourhood of the last known location of a mobile host. We use the physical route map of the neighbourhood region of the last known location of a mobile host to capture its spatial locality assuming that the mobility of the host almost surely be guided by the region route map. An update is not required as long as a host moves within the area under a Mobile Switching Center (MSC) and conforms to its mobility pattern. When a mobile host deviates, it is tracked down at its new location using the regional route map. An update for

S. R. Das, S. K. Das (Eds.): IWDC 2003, LNCS 2918, pp. 153–162, 2003.

new (deviated) position is made at the MSC where it is found. The proposed
scheme have following advantages: (i) it takes into account the call stability as
the patron hosts [3] are identified and the mobility pattern is stored at the patron
stations to achieve optimal routing; (ii) it takes into account the call locality,
i.e., the calls arising from the stations that are under the same MSC as the cell
in which the mobile host currently resides find the pattern stored at that MSC
and are routed directly bypassing the lengthy route through the home agent;
(iii) it improves over the one in [4], where the patron service is invoked for every
global move, by invoking the patron service only when the host undergoes global
deviations. (iv) the paging costs are optimal and update cost is zero, when the
host conforms to the mobility pattern; (v) it is better than [2] which switches to
conventional schemes even for a transient deviation.

The rest of the paper is organized into 7 sections. The focus of section 2 is
on the attempts to capture spatio-temporal localities of mobile hosts for efficient
location management. Section 3 deals with the models of mobility patterns. An
overview of the proposed location management scheme is given in section 4.
Section 5 provides formal algorithmic specification of the scheme along with
justification and analysis. The results of simulation appear in section 6; section 7
concludes the paper.

2   Location Tracking Using Spatio-Temporal Locality

The performance of the location management schemes depends heavily on the
subscriber mobility patterns. To analyze the performance of certain selected
schemes, two mobility models, namely, activity-based mobility model and random mobility model were used in [6]. Very few schemes take into account the
spatio-temporal properties of the host movement pattern [6]. All such schemes
use static host profiles to approximately trace the location of the host at any
given time. When a call is made, initially the callee's profile is used to find the
expected locality. But if the callee is not found at that locality, they use the triangular routing scheme to track the callee. This approach works only for hosts
with strictly unchanging schedules. Moreover, many schemes [1, 2] assume particular topologies of the cells — e.g., circular, square, or hexagonal — in order
to simplify analysis. In [5], a multilayer neural network was used to model the
host mobility pattern. It was observed that the average accuracy of prediction
for uniform patterns was 93% and 40-70% for regular patterns. The scheme proposed in [2] also uses a neural network model to capture the mobility pattern of
the host. It attempts to reduce the paging and the update costs by using this
pattern to predict the location of the mobile host. So, though move stability exhibited by the host is accounted for, the schemes fail to take into consideration
the factors like the frequency and the source of calls, and the call stability in
designing and evaluating location management strategies. Thus the inefficiency
due to triangle routing follows. In summary, the techniques used so far tend to
decrease the cost of either the move or the call in the expense of other. Moreover,

the scheme switches to conventional schemes (zone based or movement based) too often, even when the deviations are temporary.

3   The Mobility Model

The mobility patterns of majority of hosts can be modeled using the three models, namely, (i) the Traveling Salesman (TS) model, (ii) the Boring Professor (BP) model, (iii) the Pop-Up (PU) model.

The stations where there is a high probability of finding the host at certain predetermined times as Boring Professor (BP) stations. The host moves at approximately same times everyday between two BP stations. All the stations on the path between two BP stations are referred to as trajectory (TY) stations. Typically, a host makes a transient deviation to the locations that are close to its one of its BP stations; and such deviations are occasional. For example, consider the itinerary of a mobile user shown in figure 1. The home of the user is at station A, he works at a location which is denoted by station B; and the club he visits daily is at a station C. Everyday, the user starts from home at 9:00 AM, reaches his office around 9:30 AM, stays in the office till 5:00 PM and then he visits a club at 5:30 PM, He returns home by 8:00 PM starting from the club at 7:30 PM. The mobile host's user spends a significant amount of time at home, office and the club. Thus, these stations are considered as the BP stations. As the user travels from home to office, office to club, club to home, he mostly moves along the same route. All the stations in this route are the TY stations. Sometimes the user visits a shopping center at station X and barber at station Y. These stations are the PU stations for the mobile host. The PU stations account for the occasional transient deviations of the host from its mobility model. In figure 1, a transient deviation to a hospital on the way from home to the office is depicted.

For a majority of the mobile hosts for most of the time, the movement profile repeats itself from day-to-day basis. The mobility pattern is more or less stable but for the occasional transient changes and rarely permanent changes in the pattern. Thus, if we can tackle these transient deviations in an efficient way, then we could use the mobility pattern of the hosts to achieve minimal costs for the paging as well as for the updates.

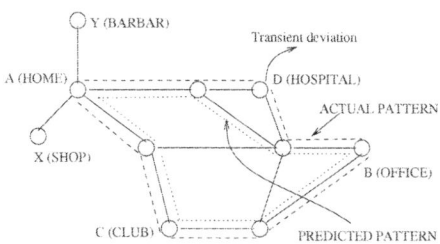

Fig. 1. Graph mobility model

4   The Proposed Scheme

In our scheme, a cellular network is divided into a few big location areas (BLAs). Each BLA, consisting of about 100 cells, is under a higher level MSC. The MSC has the access to the location information databases in the network, which are used to store the location and service information for each registered mobile host of PCS network. We enhance the functionality of the MSC by introducing special mobility functions; and MSC has a unique regional route map (RRM) associated with it.

An RRM is a bounded degree graph connected graph $G = (V, E)$ that represents the physical road connectivity between the cells under an MSC. Each cell is represented by a node. An edge between nodes A and B indicate a direct route between the cells represented by these nodes. Note that A and B must be adjacent cell if an edge exists between them. The weight $W(A, B) = t_{min}(A, B)$ denotes the minimum time required to move from cell A to cell B. The MSC uses this regional route map in order to track down the hosts which have undergone transient deviations.

The mobility pattern of a host can be obtained by monitoring the host's movements for a period of time or the information can be gathered from the host's user in advance. This pattern is replicated at the home station and other MSCs (if any) which come under the span of the mobility pattern. The concept of patron host, introduced in [3] is used here. For each mobile host, the set of source hosts from where the majority of traffic for the host has originated are called patron hosts. We call the MSCs under which one or more patron hosts reside as patron stations. It is not necessary to store the entire mobility pattern at the patron stations, it is only sufficient to know under which MSC the host resides at a particular point of time. Since the number of patron stations is very less in number compared to the number of patron hosts; it would not result in much storage overhead.

When a call request arrives at the patron station, the information from the callee's mobility pattern stored at the patron station is used and the request reaches the MSC under which the callee is present. From the complete mobility pattern, which is stored at this MSC, it finds the cells in which the callee is expected to be present. In the majority of the cases, the callee is successfully tracked down. But if the callee is not found, then the MSC use locally stored route map to track down the callee. When a call request arrives at the home station or other MSC at which the pattern is stored, the same approach is used to track the callee. This optimization is also possible for non-patron hosts residing at the patron stations or at the home station. But if a call originates from a host, outside these stations, then the request is first directed to the home station and then the same process follows. Thus, triangle routing is avoided for most of the calls.

An update is not required when the host moves locally (within an MSC) if it conforms to the mobility pattern. When a call is made to a host, if the MSC finds that the host does not conform to the pattern (local transient deviation), it uses the route map to track down the host. It also updates the new deviated position only at the MSC under which the host is present. For the subsequent calls, when the request arrives at the MSC, it finds out that the callee has deviated and thus a pages the new location. If the callee is still not found then, the hunt may begin from the deviated location or the location at which the callee was supposed to be found at that time. If the host moves into a cell that comes under a different MSC, it is checked whether the host conforms to the mobility pattern. If so, then no update takes place. Now, if there is a call for this host, then the request would indeed arrive at the new MSC as the pattern indicates it and the callee can be tracked down. But if it is found that the host moves globally and does not conform to its pattern i.e., it moves to a BLA in which the pattern is not stored, then the patron service is invoked. The home station and the patron stations are informed about this deviation. Hence, for the subsequent calls which arrive at the patron station (or the home station), it is understood that there is a global deviation. If the new BLA is a neighbour of the home station then the route map of the home station can be copied there so as to carry out the hunt efficiently as in the earlier case. But, if the new MSC is not a neighbour of the home station, then the route map is not copied. This is because, the deviation is no longer transient and the host has deviated a lot from his mobility pattern. Effectively, each MSC also has the route maps of its neighbouring BLAs to assist them in case of global deviations of hosts from their mobility patterns. If the host deviates to BLAs which are not the neighbours of home station, the hunt can still be carried out; but the host mobility pattern will not be used in assisting the MSC. Whenever there is a call, the callee can be tracked down with the help of RRM of the new BLA. The update would made periodically so as to bound the region of paging. However, it is expected that the host would resume its pattern at the earliest, so that the previous scheme can be used again.

The important idea behind the scheme is to diminish the effect of deviations in the host mobility in order to hide the local deviations completely from the rest of the network and to confine the effects of global deviations to those hosts which are more likely to call again, thereby reducing the updates and at the same time achieving optimal paging costs for the majority of cases.

5   The Hunt Algorithm

Let $S(t) = \{C_1, C_2, \ldots, C_k\}$ be the set of cells, in which the callee is most likely to be found at that time t. This set is obtained from the mobility pattern of callee stored in the MSC in which he currently resides. On an average, this set would consist of 1-3 cells. Suppose that the callee gets a call at time t_i. There are two cases depending on whether the callee has deviated or not by a variable $flag\_dev$ stored at the MSC. If a deviation has been recorded for the callee at a time $t_{i-1} \le t_i$, then the cell in which the callee was found is also stored.

Case 1. If the callee has not deviated at some time t_{i-1} before (i.e., $flag\_dev$ = 0), then the set of cells $D = S(t_i)$ in which the callee is most likely to be found are paged. This set is obtained from the mobility pattern of the callee stored in the MSC in which the callee currently resides.

Case 2. If the callee has already deviated (i.e., $flag\_dev = 1$), then the cell *prev_cell* in which the callee was previously found is paged. In this case, $D = prev\_cell$.

In both the cases if the callee is found, the location of the user is relayed to the caller. If the callee is not found then the hunt begins.

We define a reachable cell R of B, as a cell which can be reached from cell B within a time T. For this cell R to be reachable from B within time T it is necessary that $t_{min}(B, R) \leq T$. We also define remaining time t_r, as the maximum time for which the user can still move after reaching R. Hence, $t_r = T - t_{min}(B, R)$. Note that a cell is not reachable if $t_r < 0$. Extending the definition further, a reachable cell R from a set of cells X, is a cell that is reachable from any of the cells of X within time T and the remaining time $t_r = \max(T - t_{min_{R \in X}}(C, R)) = T - \max_{C \in X}(t_{min}(C, R))$, where C is any cell of X from which R is reachable within T.

Now the callee can only be in the cells reachable from D within the time $T = t_i - t_{i-1}$. Let Q be the set of ordered pairs of ⟨`reachable cells, remaining time`⟩. Initially $Q = \{(x, T) \mid x \in D\}$. Since this is a transient deviation, there is a strong possibility that the callee would resume its movement pattern at the earliest. Taking this into consideration a cost for all cells is calculated and then the *best* cell is obtained by finding the cell whose cost is minimum among all the cells of Q. The best cell is then paged. If the callee is found then, the new position of the callee is updated and the time t_i is also noted. A variable $flag\_dev$, which represents whether the callee has deviated, is updated or reset. In the case 1, if it is noted that the callee has deviated from his mobility pattern then $flag\_dev$ set to 1; and in case 2, if the callee is found to resume his movement pattern, then the information is updated by resetting $flag\_dev$ to 0. On the other hand if the callee is not found then, the *best* cell is removed from Q and its neighbouring cells which are 'reachable' from Q but are not yet paged are added to the set Q. Again the *best* cell among the cells of Q is found and the algorithm then continues till the callee is found. Note that the callee will definitely be found before all the elements of Q are exhausted. The specification of the hunt algorithm, as discussed above, appears below.

Algorithm 1: Hunt Algorithm

```
     // Executed by MSC under which callee resides or by its home station
01   if (flag_dev = 0) then
02       T = bound;
03       Page all cells in D = S(t_i);
         // Let C_i be the reply of paging from BSs of the paged cells.
         // C_i = null implies callee not available in any of the paged cells.
04       if (C_i ∈ S(t_i)) then return C_i endif
05   else
06       T = t_i − t_prev_call;
07       if (T is very large) then
             // The callee may have returned to pattern
```

```
08              T = bound;
09              Page all cells in D = S(ti);
10        else
11              Page D = prev_cell;
12              if (Ci ∈ D) then return Ci endif
13        endif
14  endif
15  Q = {(x, T) | x ∈ D};
16  repeat
            // For estimating the cost of the cells to be paged a separate
            // function (see Algorithm 2) was used. Efficiency of search will
            // improve with an improved cost estimation function.
17        best_cell = min(x,Tx)∈Q{Cost(x)};
18        Page best_cell;
19        if (Ci = best_cell) then
20              if (Ci ∈ S(ti)) then flag_dev = 0 else flag_dev = 1 endif
21              prev_cell = Ci; // Update
22              t_prev_call = ti;
23              return Ci;
24        else
25              R = { (x, Tx) | x is a reachable neighbour cell of best_cell
                              not paged yet and Tx is the remaining time};
26              Q = Q ⋃ R;
27              Remove best_cell from Q; // Remove already paged cell.
28        endif
29  until (Q = φ)
```

The estimation of cost for paging cell during search at line 17 of hunt algorithm has not been specified so far. We used the following algorithm for estimation of cost for paging the cell.

Algorithm 2: Estimation of Cost

```
01  P = {j | j ∈ S(t), where ti − δ < t < ti + δ};
02  η1 = |S(ti)|/(|P| + |S(ti)|);
03  η2 = |P|/(|P| + |S(ti)|);
04  // n is total number of cells under the MSC.
05  for (j = 1 to n) do
06      Wp(j) = min∀x∈P{tmin(j, x)};
07      Ws(j) = min∀x∈S(ti){tmin(j, x)}
08      Cost(j) = η1Wp(j) + η2Ws(j);
09  endfor
```

5.1   Justification and Analysis

The set of cells P in which the callee is expected to be found between the time $t_i - \delta$ and $t_i + \delta$ is determined from the mobility pattern. The parameter

δ specified either by the callee or a default value is used. As the deviations are transient in nature, the callee must have deviated some time ago; and would resume his movement pattern some time later. The parameter δ is an estimate of this interval. Since the P is the set of cells which the callee may visit in the interval $[t_i - \delta,\, t_i + \delta]$, it is likely that the callee would be: (i) either in one of these cells but his arrival at the predicted cell is delayed due to his busy schedule, traffic, etc.; (ii) or the callee has actually deviated from one of these cells to a nearby PU station and would be returning back to one of these cells again. Hence, more the closer a cell is to P, more is the possibility that the callee be found in that cell. Thus, the cost of each cell in the MSC is found on the basis of the sets P and $S(t_i)$. The cells are then paged in the increasing order of their cost. For a cell k in the MSC, $W_p(k) = \min_{\forall x \in P}\{t_{min}(k, x)\}$ and $W_s(k) = \min_{\forall x \in S}\{t_{min}(k, x)\}$. And the cost of cell k, $Cost(k) = \eta_1 W_p(k) + \eta_2 W_s(k)$. $W_p(k)$ represents the proximity of the cell k from the set of cells P. A small value of W_p indicates that the cell is close to the set of cells P. For all cells in P, $W_p = 0$. For a small value of W_p, the cost is small and the cell is paged early. Similarly, W_s represents the proximity of a cell from the set $S(t_i)$, where the callee should have been found at t_i. The number of cells in P depends on the mobility pattern of the callee and the value of δ chosen. If P is large, then we should avoid paging all cells in P before paging the other cells in proximity of $S(t_i)$. That is, W_s must be given more weightage for large P. But for a cell k in P, the cost is $\eta_2 W_s(k)$, as $W_p = 0$. Hence, η_2 must be proportional to $|P|$. Since the other cells nearer to $S(t_i)$ must be paged before paging cells of P, η_1 must be 0 for these cells. That is, the determining factor must be W_s. It may so happen that the callee is supposed to be at his work place at time t_i, thus he would be staying at the same place during $[t_i - \delta,\, t_i + \delta]$. Here $P = S(t_i)$, that is the callee resides in a BP station during this time. Hence, $\eta_1 = |S(t_i)|/(|S(t_i)| + |P|)$, $\eta_2 = |P|/(|S(t_i)| + |P|)$.

When $|P| \gg |S(t_i)|$, $\eta_1 \approx 0$ and $\eta_2 \approx 1$ and $Cost \approx W_s$. Thus, in this case the cells in the proximity of $S(t_i)$ are paged first. No additional preference is given to the cells in P. But if P is relatively small, then it would not cost us much to page the cells in P first and then the cells in the vicinity of $S(t_i)$. So additional preference is given to cells in P. Whereas if, $P = S(t_i)$ then $\eta_1 = \eta_2 = 0.5$. Thus, cost of each cell would be equal to $W_s = W_p$. Thus the cells in the vicinity of the BP station ($S(t_i)$) are paged first. After the calculation of the cost is done, the hunt algorithm can proceed by finding the best cell starting with a cell from D. Any improvement on design of cost function will improve the search and paging.

In case 1, $D = S(t_i)$. The paging starts from the nearest reachable neighbouring cells of D and then gradually proceeds farther away from the pattern to page all the possible reachable cells within the bound T till the callee is found. Depending upon the value of $|P|$, the cells in P or the cells nearer to $S(t_i)$ are paged first. If $|P|$ is large, all the cells nearer to $S(t_i)$ are paged first and then we gradually move away from $S(t_i)$. Whereas if $|P|$ is small, then the cells of P are paged first and then we gradually move away from P to page all the reachable cells. In case 2, the paging starts from $D = prev\_cell$ and then its reachable neighbour cells nearest to $S(t_i)$ or P, (depending upon the value of $|P|$) are paged first. In any case, the search starts from D proceeds towards the pattern till the bound is reached. Then, the other cells farther away from the pattern are searched till the callee found.

6   Simulation Results

The performance of the proposed scheme was evaluated for a MSC with 50 cells. The road connectivity i.e., the edges between the nodes and their weights were randomly generated. The total simulation time was 200 units. The calls for a host were also randomly generated within this interval. The scheme was simulated for a number of mobility patterns with varying degree of deviations. The deviations were randomly generated and the average search and the update cost were calculated for varying call-to-mobility ratios (CMRs). For the purpose of simulation, $S(t)$ is assumed to contain only one cell; and δ was taken to be 15 units. The total number of moves made by the host is taken to be 10. The plots in left half of figure 2 show the average number of cells paged for different CMRs and varying percentages of deviations. Whereas, the right half of the figure shows the average number of updates per call for different CMRs and varying percentages of deviations.

It is found that on the average only 1-2 cells are required to be paged. The paging cost is considerably less even for 30% deviation in the mobility pattern for the hosts with very high CMR. This is because, the host gets calls very frequently and thus would not be very far away from the position where it received the last call. Thus, the bound T would be much less for many cases. Consequently the paging cost is very small. But with a very low CMR, the cost would be high at large percentage of deviations. For very high CMRs, the update cost is found to

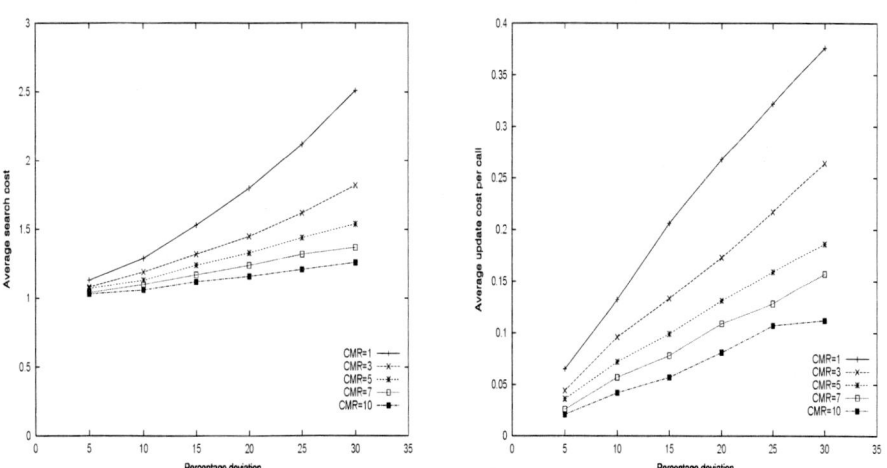

Fig. 2. The search and update costs

be very less. When the CMR is 10, the average update cost is found to be 0.021 per call. That is, the update takes place every 50 calls. Even for a deviation of 30%, the update takes place once for every 10 calls. But for 30% deviation with CMR=1, one update is required approximately after every 3 calls. However, for a deviation of 10% even for a very low CMR value, on the average one update takes place for every 10 calls.

7   Conclusion

In this paper we have proposed a scheme for locating a mobile host taking its spatio-temporal properties into account. The scheme uses the host mobility pattern and the regional route map of the of the last known location of the mobile host to determine the candidate cells where the target host may be found. The performance of the proposed scheme is found to be acceptable even when there are transient deviations upto 30% over the known host mobility patterns. The deviations are tackled using a best-first branch and bound search to locate the candidate cells for likely locations of the destination. The bound function is modeled to restrict the locality search space by using appropriate weighted functions of mobility pattern and regional route map.

References

[1] A. Bar-Noy, I. Kessler, and M. Sidi. Mobile users: To update or not to update ? *Wireless Networks*, 1(2):175–195, July 1995.

[2] Goutam Chakraborty. Efficient location mangement by movement prediction of the mobile host. In *IWDC, LNCS*, volume 2571, pages 142–153, 2002.

[3] G. Cho and L. F. Marshall. An efficient location and routing scheme for mobile computing environments. *IEEE Journal on Selected Areas in Communications*, 13(5):868–879, June 1995.

[4] A. Hac and Y. Huang. Location update and routing scheme for a mobile computing environment. *International Journal of Network Management*, 10:191–214, July-August 2000.

[5] B. P. Vijay Kumar and P. Venkataram. Prediction based location management using multilayer neural networks. *Journal of IISc*, 82:7–21, 2002.

[6] A. A. Siddiqui and T. Kunz. The peril of evaluating location management proposals through simulations. In *Proceedings of 3rd International workshop on Discrete algorithms and methods for mobile computing and communications*, pages 78–85, August 1999.

Towards "Always-Connected" Wireless Network: A Framework of Infrastructureless Networks

Zhe Guang Zhou, Prawit Chumchu,
Sanchai Rattananon, and Aruna Seneviratne

School of Electrical Engineering and Telecommunications
The University of New South Wales, Sydney, Australia
{zheguang,chumchu,san,a.seneviratne}@mobqos.ee.unsw.edu.au

Abstract. The increasing research in ad-hoc networking enables the formation of transient peer communities for asynchronous information exchange. This has made it possible to offer network connectivity and a wide range of network services with no fixed infrastructure present. However, most ad-hoc network research has focused on the former, i.e. the provision of network connectivity. In this paper, we address the latter, the provision of services through asynchronous message exchanges. We propose a framework that provides virtual connectivity for mobile users while they are in regions without any wireless infrastructure support. Through the mathematical analysis and simulations, we show that with the virtual connectivity provided through the use of couriers, it is able to offer a number of valuable services with an acceptable level of quality.

1 Introduction

Today's mobile devices such as Personal Digital Assistants (PDAs), smart phones and sub-notebook computers have significant processing power, multimedia and storage capabilities. In parallel, there have been significant advances in the field of wireless local area communications with the standardization of WLANs. These developments have given rise to increased research efforts into ad-hoc networking and the provision of Internet services through wireless "hotspots". These researches to date have focused on the provision of connectivity to a group of peer nodes for communication amongst themselves and extending the range of the Internet. However, wireless network infrastructure has limitations in terms of coverage area. It is impossible to cost-effectively provide a network for all subscribers with continuous connectivity. Therefore, most wireless networks today have small coverage which is surrounded by regions that either have no network connectivity. However, from the standpoint of the mobile user, the ideal scenario is to have network connectivity anywhere, anytime.

Technological advances in mobile devices and wireless networking make it possible to form transient peer communities to exchange information using ad-hoc network. With the formation of transient peer communities, it is possible to provide asynchronous message exchange, which can be used as the basis for offering a wide range of best-effort services with no infrastructure available. A couple

S. R. Das, S. K. Das (Eds.): IWDC 2003, LNCS 2918, pp. 163–173, 2003.

of such applications have been proposed by [4, 9]. Although these proposals show the viability of using transient peer communities for some specific applications, they do not demonstrate their general viability. In this paper, we address this by proposing a framework that provides a *virtual connection* for mobile users in areas with no wireless infrastructure. Through our framework, mobile users are able to enjoy everyday network applications, such as collecting email, and sending/receiving instant messages or receiving small amounts of multimedia data with an acceptable level of quality. Through mathematical analysis, we show the general viability of transient peer community formation and validate our analysis with simulations.

The rest of this paper is organized as follows. Section 2 presents the related work. Section 3 and Section 4 present the proposed framework and the analysis. Section 5 presents the simulation results. Section 6 presents the conclusion.

2   Related Work

Infostation [7] is proposed as an information island for the mobile users (mobile nodes, MNs). The Infostation operates as a small base-station with high bandwidth but small coverage. MNs will experience two different bandwidths: 1) high bandwidth when there is an Infostation; 2) low bandwidth when there is only normal base-station. It provides high bandwidth connection to the MN in order to enable partial downloading and information buffering. Subsequently, after the MN comes close to another Infostation, they are able to resume and/or complete the transaction. Iacono et al [2] proposed a similar approach specifying that all Infostations belong to an organization. This scheme provides services with intermittent high quality connection. A file is divided into segments to be transmitted by several Infostations along the movement path of the MN. It is assumed that the paths of the MNs are known by the system and the MNs are traveling at constant velocities. These assumptions simplify their simulation model. In reality, the segmentation greatly affects transmission since any lost segment will result in complete failure of the transaction and waste the resources transferring the remaining segments.

Kodeswaran et al [9] proposed a framework that uses a third party for resuming multimedia data in a continuous format while the MN is in the area without network coverage. It is assumed the destination of the data requesting MN is known by the user's schedule such as appointment books. This is impractical since most users are not willing to offer their personal appointment information to the system to determine their movement details. Moreover, there is no data duplication considered in their analysis since it is assumed that the movement trajectory is accurately determined by the system. This is not a valid assumption since uncertainties exist with any of the current location management schemes. Therefore, to resume multimedia data, the memory size requirement of the third parties will increase dramatically if data duplication is considered. Furthermore, for large multimedia files, not only the data segments must be considered, but also the success of all resumed segments and their arrival order.

3   Proposed Framework

In order to address the problems of general viability, we propose a framework that provides *virtual connection* in the physically disconnected area for the mobile users. In this area, mobile users are able to receive emails, instant messages or small amounts of multimedia data with an acceptable level of quality. It exploits the formation of the transient peer community with ad-hoc network. The viability is shown through the mathematical analysis and simulations.

3.1   Framework Descriptions

In our framework, the entire operational environment is divided into two areas (regions). The first is the *connected area*: the area that has wireless infrastructure. In this region, mobile users are able to access the network services under the coverage of one or more access routers (ARs). The second is the *disconnected area*: the area that does not have any wireless coverage for mobile users. Two possible scenarios will form this type of region: 1) wireless connectivity might exist but not provide public accessibility for anonymous mobile users due to security and resource management issues; 2) there is no wireless infrastructure at all. Examples for the former are private companies or residential areas and for the latter are private or public parks.

We assume all network infrastructures in the connected areas are owned by one enterprise. Furthermore, we also assume that there exists a large client population who requires network services in the disconnected areas. An example for this kind of system could be a metro network. Each metro station represents a connected area as it contains a wireless infrastructure. All wireless network systems in the metro stations are connected to the metro company via a fixed backbone network. The wireless systems in those metro stations are not able to cover the entire geographical area of the metro system due to the limited coverage range. Those areas surrounding the metro stations are the disconnected areas as mentioned before, with large client population and normally they do not have network services while traveling. Another example is a highway system. Again, the highway system belongs to one enterprise. Each entry/exit point and the toll booth are the connected areas and elsewhere are the disconnected areas. In these environments, it is possible to provide network services for the mobile users in the disconnected areas by using some of other mobile users of the system as couriers — *courier clients*. These couriers are the mobile users who are able to create a transient peer community with other mobile user(s) in the disconnected area. Therefore, these couriers can carry information from/to the mobile user(s) in the disconnected area. The transient peer community is formed by setting up an ad-hoc network between the normal mobile user and the courier within radio communication range. The delivered information can be emails, instant messages and small chunks of multimedia data etc. Although the physical network connection does not exist in the disconnected area, the mobile users still have network services. Thus, we refer to this type of connection as *virtual connectivity* shown schematically in Figure 1.

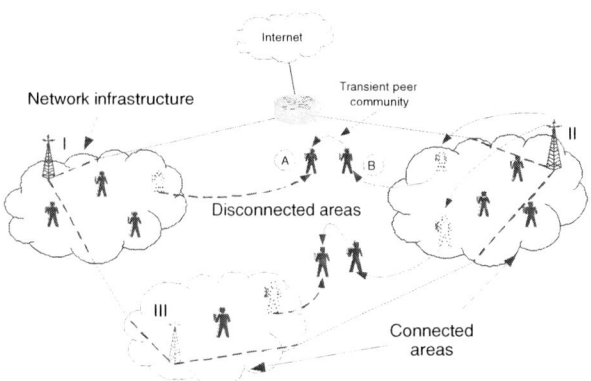

Fig. 1. Framework Structure

The details of the framework description for Figure 1 are as follows. Network I and network II are under the same administration domain. Network I knows that Mobile Node (MN) A leaves it, moving towards network II with the aid of mobile movement tracking techniques. While MN A is on its way, some application data (referred to as the message or data) such as an email, needs to be delivered to it. Since it had an association history with network I, network I is able to forward this data to network II. Meanwhile network I also buffers the message in case MN A returns to its coverage area. Network II doesn't have any association information with MN A, and therefore considers it as an incoming MN. It first buffers this message, and then selects one of the MNs, MN B, under its coverage to act as the *courier*. The selection decision could be random or performed using a movement prediction algorithm. After the selection, network II embeds this data into the selected courier without any notification to the user. It is only the communication between network II and the mobile device of the selected courier. The courier travels independently along its path. When this courier leaves network II and meets MN A on the way, the transient peer community can be established and the data can be delivered. There are three possible cases in which this delivery will not occur. Firstly, MN A returns to its original network. In this case, the data buffered in network I will be delivered. Secondly, MN A enters the coverage of network II while the data is being forwarded to network II. In this case, after MN A establishes an association with network II, the data can be transmitted. Thirdly, unsuccessful delivery by the MN B if it doesn't meet MN A on its way, or if MN B does not travel in the opposite direction of MN A since the movements of MN A and MN B are mutually independent. In this case, it is similar to the previous case as the data is buffered in network II.

3.2   Framework Design

Accurately identifying the best courier client is the most critical process for this framework to be successful. There are also a number of system issues that need to be resolved, such as the resource requirements, security etc. In this paper, we will concentrate on the courier identification process and discuss it in Section 4. The other issues are briefly discussed here for completeness.

Incentive, Security and Transient Peer Formation. Firstly, the incentives for becoming a courier could be simply handled by subscription policies. For example, lower subscription rate for people who are willing to become couriers or making it a mandatory subscription requirement. Secondly, security issues and resource requirements could be handled through the use of a SIM card similar to those used in GSM system. All subscribers will be provided a wireless card similar to the SIM card which has sufficient buffer. The SIM card will provide necessary authentication mechanism. The storage size for all MNs is equal and constantly dedicated. This is easier for the system to perform resource management. Finally, the transient peer community is created by ad-hoc network.

Message Format, Size and Duplication Handling. The delivered data is numbered by sequential mechanisms and with a TTL (Time-to-Live) field in the header of the message. The former is used by the receiver to discard duplications and the latter is used for the courier device to reclaim the buffer space for unsuccessful deliveries. Data segmentation could be one of the solutions for large messages. However, the system needs to ensure the successful delivery of all segments. This will result in an increasing number of couriers required and hence increases the number of duplications. Thus, in this paper, the message delivered is an entire enveloped package for the MN without segmentations. A threshold value is set for determining the delivery method (see Section 4.2). If the message size is less than it, then the full message will be delivered in one go, otherwise, a notification will be delivered. This notification informs the user that there is a message for him. In this case, the notification message will contain the header of the original message such as the sender, the subject etc. The received mobile user will decide the importance of this message and decide to find out a nearby network facility for receiving the complete message. Furthermore, network II discards the buffered message when MN enters its coverage. Network I discards the buffered message by the notification from network II. Due to limited space, the protocol developments are not discussed in this paper.

Hence, the viability of this scheme depends on accurately identifying couriers from a given user population and the formation of transient peer communities.

4   Courier Client Identification

This section presents the mathematical analysis of the viability of this system. For simplicity, we only consider two connected areas and a disconnected area in between, as shown in Figure 2. Firstly, we assume there are K possible paths from network I to network II. MN A is moving away from network I towards network II along the shortest path. Secondly, we also assume that the residence

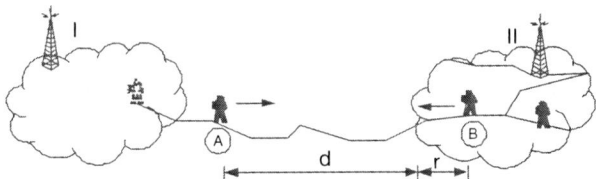

Fig. 2. Framework Analysis

time of the MNs in network II is exponentially distributed thus the probability of leaving network II is given by Poisson distribution as shown in [6]. Finally, as discussed before, the courier selection is not necessary if MN A enters the coverage of network II before the message is forwarded to network II. Therefore, we further assume that the distance, d, is sufficient for the creation of a transient peer community and to exchange the data that is being carried by the courier client, MN B. It must be larger than the distance, r, which is the distance for the courier client, MN B, must travel to leave network II.

4.1   Random Selection

Assume that there are a total of M mobile users in network II. Network II randomly selects n clients as potential couriers for delivering the message to MN A with the condition $1 \leq n \leq M$. We only need to consider the probability that at least one courier successfully deliver the message to MN A as messages contain sequence numbers that enable MN A to detect and discard duplications.

In this scenario, we need to compute the probability of at lease one courier which is moving in the correct direction, given that n MNs are selected as the couriers and j of them are leaving network II, for $0 \leq j \leq n$. By the Law of total probability, this can be done as follow. The probability that out of n MNs that are being selected, j of them will leave network II, is given by the Binomial distribution. Since there is only one shortest path, if all paths have an equal leaving probability, then the probability of success to leave on the correct path is $1/K$. Hence, the probability of at least one successful delivery is:

$$P(success\_random) = \sum_{j=0}^{n} \left[1 - \left(1 - \frac{1}{K} \right)^j \right] \binom{n}{j} P^j (1 - P)^{n-j} \qquad (1)$$

Simplifying the above equation by Binomial theorem yields: $P(success\_random)$ $= 1 - \left(1 - \frac{P}{K} \right)^n$, where P is the probability of each courier leaving network II, and K is the total number of paths. Without loss of generality, if all paths have unequal leaving probability, then the probability of successful delivery is:

$$P(success\_random) = \sum_{j=0}^{n} \left[1 - (1 - P_{dir,c})^j \right] \binom{n}{j} P^j (1 - P)^{a-j}$$

where $P_{dir,c}$ is the probability that the courier will be traveling in the correct direction. This equation is similar to Equation (1). These equations show that the probability of at least one successful delivery depends on four factors. First is the number of MNs, n, being selected. The second is the number of MNs, j, leaving network II. The third is the probability of each courier leaving network II. The final is the probability of the leaving MN moving in the correct direction.

4.2 Successful Delivery Period and Message Size Determinations

In addition to Section 4.1, we are also interested in the probability of the message being delivered within some expected period, T. We need to find that the probability of the minimum time period for delivery being less than the expected time. This expected time, T, must be in the range of the minimum and maximum possible delivery time. This probability is given by:

$$P(min(t_1, t_2, \ldots, t_m) \leq T) = 1 - P(t_1 > T, \ldots, t_m > T), for \ 0 \leq m \leq j \leq n$$

Since it is assumed the residence time is exponentially distributed, hence:

$$P(t_{min} \leq T) = 1 - [1 - F_t(T)]^m = 1 - e^{-m\lambda T}, for \ \lambda > 0, t \geq 0, 0 \leq m \leq j \ (2)$$

where $t_1, t_2, \ldots, t_m$ are the times for m couriers to deliver the message. $F_t(T)$ is the cumulative distribution function (cdf) of time t. t_{min} is the time for the first successful delivery which is the minimal delivery time. The mean of exponential distribution is $\mu = 1/\lambda$. Therefore, to complete Equation (2), the time T needs to be deduced empirically from the statistical data for the particular environment.

Furthermore, the size of the delivered message is another concern in our framework. The maximum message size depends on two essential parameters. Firstly, the longest distance that the communication can be held between two MNs, which is $min(R_A, R_B)$, where R_A and R_B are the coverage radius of MN A and the courier respectively in the case that they are using different mobile devices with different coverage ranges. Secondly, it is their velocities, v_A and v_B, respectively. Therefore, the period for the transient peer communications is:

$$t_{trans} = \frac{2min(R_A, R_B)}{v_A + v_B} - t_{setup}$$

where t_{setup} is the ad-hoc network setup time for two MNs. Then the maximum message size, S, is the product of the data rate, D, and the period t_{trans} (i.e. $S = D \times t_{trans}$). For example, in the highway, the MN A and the courier are all traveling with a constant velocity 110km/h (the maximum speed limit in Australia). Given that the coverage radius of a mobile device is about 160m with IEEE802.11b technology and the processing time to establish the ad-hoc network is about 100-600ms [3]. Then the total time for the transient communication is 4.64s. Theoretically, the maximum number of bytes that can be exchanged between two network interfaces is 6.38MB. This value includes all signaling packets during the communication. We performed a simple TCP performance test using

PCATTCP [1] and the ORiNOCO^{TM}wireless network cards which have a data rate of 11Mbps. The average data transmission rate is 4.58Mbps for both transmitter and receiver. Hence, the total massage size is 2.65MB. Although this value is also affected by other factors such as the packet length and the error rate, it gives an indication for computing the threshold value for determining the size of the message. The coverage radius used in here is obtained from the ORiNOCO^{TM}specifications. Nowadays, the coverage of the wireless cards is often greater than 160m. Therefore, the value of the threshold could be larger. Moreover, a slower traveling speed will also result in a larger threshold.

4.3   Selection with Movement Prediction

When mobile movement trajectory prediction is used, the probability of success depends only on the accuracy of the prediction algorithm applied. The movement direction of the couriers can be highly predictable in the metro and highway examples mentioned before. Movement prediction can be achieved in a number of different ways within the coverage areas that has one or more ARs [5, 8, 10]. The probability of successful delivery with trajectory prediction is:

$$P(success\_prediction) = 1 - (1 - P_{acc})^n \quad for \ 0 \leq n \leq Y \leq M \qquad (3)$$

where P_{acc} is the probability of accuracy of the prediction algorithm. Notice that the number of couriers selected, n, here is different. The condition for n is less than or equal to the total number of MNs that are estimated or predicted movement in the correct direction, Y, from a total mobile population, M, in network II. The analysis can be further investigated, if guarantees of delivery are provided. The requirement for guaranteed delivery can be satisfied as long as the number of selected couriers satisfies the following. Suppose P_{acc} is the percentage of accuracy of a given mobile trajectory prediction algorithm or mechanism. Then the delivery is guaranteed if:

$$M(1 - P_{acc}) < n \leq Y \leq M \qquad (4)$$

where, $M(1 - P_{acc})$ indicates the total number of couriers that are predicated moving in the correct direction but the predictions are wrong due to the accuracy of the prediction mechanisms. As long as condition (4) is satisfied, the probability of successful delivery is one (i.e. delivery guaranteed). Otherwise, Equation (3) holds for the probability computations. Obviously, the higher the value of P_{acc}, the higher the overall probability. Moreover, when the mobile trajectory algorithm is involved, the movement paths of the MNs are known. Therefore, the delivery period can be computed by using the traveling distance and velocity relationships of the MNs. Similarly the maximum message size depends on the coverage distance of the wireless devices being used and the speeds of the MNs.

The analysis suggests the probability of success depends on the accuracy of the movement prediction. If the prediction has a high degree of accuracy, then the value of $1 - P_{acc}$ is small. Raising this value to a power of n will make this value smaller, so the probability of success is higher. Comparing to the previous descriptions, the success rate for random selection depends on the probability of leaving and the probability of leaving in correct direction. As the product of these two probabilities is small, the failure probability is high. This will cause a lower probability of successful delivery. Therefore, the probability of success with mobile tracking is larger than or equal to the random selection.

To further improve our computations, we can also assume that the movement of MN A is unknown, and network I only knows MN A has left its coverage and it is in a disconnected area. Then suppose $P_{A,k}$ is the probability of MN A moving out from any arbitrary direction (the kth way) over a total of K directions. Since the direction MN A moved out is mutually independent from the movements of all other clients in network II, the overall probability of successful delivery for random selection is simply the product of the $P_{A,k}$ and $P(success\_random)$ for independent events. Similarly, the overall probability of successful delivery with movement prediction can be determined as well. The only difference is that, instead of using $P_{A,k}$, the probability of prediction accuracy is used.

5   Simulation and Results

The validity of the framework was also evaluated through simulations. The simulation model was developed in Java. The MN population is 300 in network II. Each MN has a random coordinates in the initial state. During the simulation, each MN moves in a random direction with a random velocity. For each particular n MNs selected, the simulation was carried out for 200 times. The simulation results are shown in Figure 3. Figure 3(a) and 3(b) are the plots for the random selection and the selection based on the mobile trajectory prediction with different accuracies respectively.

In Figure 3(a), the center curve was obtained using the analytical model presented in the previous section. We repeated the complete simulation several

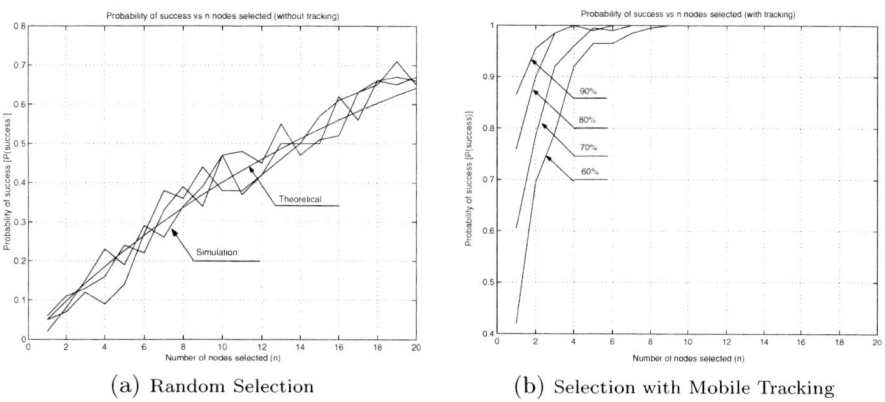

(a) Random Selection   (b) Selection with Mobile Tracking

Fig. 3. Simulation Results

times. All results show similar behavior. Figure 3(a) only shows three of them. From the figure, it can be seen that the simulation validates the theoretical results. The probability of success increases as the number of courier clients selected increases. The variations of the simulation results are only due to the number of simulations that were carried out (200 times) for each particular n. Increasing this number of simulations will result in much closer correspondence of simulation and analytical results. As expected, the results show that the probability of successful delivery, with movement prediction, is higher than random selection. This is clearly illustrated in Figure 3(b). For the mobile trajectory prediction, while the condition of Equation (4) is satisfied, delivery is guaranteed since the probability of success is equal to one. For $n \leq M(1 - P_{acc})$, the probability curves follow the exponential functions as expected. Comparing the results with mobile prediction and without movement prediction, for 60% movement prediction accuracy, nine courier clients are required for guaranteed delivery. This can achieve only 35% of successful probability for random selection. An accuracy of 60% is very low for movement prediction compared to the accuracy of the proposed prediction algorithms. For higher accuracy, the number of courier clients required to guarantee delivery is lower. A lower number of couriers selected implies a lower number of duplications. This directly affects the efficiency of the system.

6   Conclusion

This paper proposes a framework that enables the provision of asynchronous communication for users in disconnected areas, without additional communication infrastructure. The viability of the proposed scheme, which uses some clients as couriers to other clients were demonstrated by developing a mathematical model and simulations. This confirms the intuitive reasoning that successful message delivery could be achieved through "intelligent" selection of couriers. It can be done by employing a mobile trajectory prediction algorithm. As there are numerous mobile trajectory prediction mechanisms, especially in special environments, the proposed scheme can be effectively realized. In future research, we will improve our framework for delivering larger amounts of data and investigate the probability of successful delivery when data segmentation is required.

Acknowledgements

This work was carried out through an Australian Research Council Linkage Postgraduate Research Award with Singtel Optus Communications, Australia. The authors wish to thank Dr P J Cooke, Stephen Herborn, Chak Tong Kam and Rayson Chan for their invaluable helps.

References

[1] "TCP Performance Testing Tool for Windows, PCATTCP Test TCP Utility V2.01.01.06", www.pcausa.com/Utilities/pcattcp.htm, 2003. 170

[2] A. L. Iacono and C. Rose, "Bounds on file delivery delay in an infostations system", In Proceedings of the IEEE VTC, 2000 164

[3] A. Mishra, M. Shin and W. Arbaugh, "An Empirical Analysis of the IEEE 802.11 MAC layer Handoff Process", 2002. 169

[4] C. Becker, M. Bauer and J. Hahner, "Usenet-on-the-fly - Supporting Locality of Information in Spontaneous Networking Environments", Workshop on Ad hoc Communications and Collaboration in Ubiquitous Computing Environments, 2002. 164

[5] M. M. Zonoozi and P. Dassanayake, "User Mobility Modeling and Characterization of Mobility Patterns", IEEE JSAC, 1997 170

[6] N. Banerjee and S. K. Das, "Analysis of Mobile Multicasting in IP-based Wireless Cellular Networks", IEEE ICC, 2002. 168

[7] R. H Frenkiel, B. R Badrinath, J. Borras and R. D Yates, "The Infostations Challenge: Balancing Cost and Ubiquity in Delivering Wireless Data", IEEE Personal Communications, 2000. 164

[8] R. Hsieh, ZG. Zhou and A. Seneviratne, "S-MIP: A Seamless Handoff Architecture for Mobile IP", IEEE INFOCOM, 2003. 170

[9] S. B Kodeswaran, O. Ratsimor, A. Joshi, T. Finin and Y. Yesha, "Using Peer-to-peer Data Routing for Infrastructure-based Wireless Networks", PerCom, 2003. 164

[10] ZG. Zhou, A. Seneviratne, R. Chan and P. Chumchu, "A Software Based Indoor Relative Location Management System", In Proceedings of IASTED WOC, 2002. 170

Multi-path Transmission Algorithm
for End-to-End Seamless Handover
across Heterogeneous Wireless Access Networks*

Shigeru Kashihara, Katsuyoshi Iida, Hiroyuki Koga,
Youki Kadobayashi, and Suguru Yamaguchi

Graduate School of Information Science, Nara Institute of Science and Technology
Ikoma-shi, Nara, 630-0192, Japan
{shige-ka,katsu,koga,youki-k,suguru}@is.aist-nara.ac.jp

Abstract. In mobile networks, new technologies are needed to enable
mobile hosts to move across various kinds of wireless access networks.
In the past, many researchers have studied handover in IP networks.
In almost all cases, special network devices are needed to maintain the
host's mobility. However, in these technologies, mobile hosts cannot move
across different wireless access networks without closing the connection
and degrading the goodput. To overcome these, we propose here a multi-
path transmission algorithm for end-to-end seamless handover. The main
purpose of this algorithm is to improve the goodput during handover by
sending the same packets along multiple paths, reducing unnecessary
consumption of network resources. We evaluate our algorithm through
simulations and show that mobile hosts gain a better goodput.

1   Introduction

In future mobile networks, a Wireless Local Area Network (WLAN) hotspot
service will be available at many places including transportation waiting lounges
and coffee shops, while cellular service will also exist. Although cellular service
provides a wide coverage area, it is relatively expensive, narrowband and has
unstable connectivity. On the other hand, the WLAN hotspot service is relatively inexpensive, wideband and has stable connectivity, but its coverage is
very limited. In this environment, we believe some new technologies are needed
to enable mobile hosts to move across various kinds of wireless access networks
that are based on different wireless access technologies and operated by different
service providers. In this paper, we focus on end-to-end seamless handover to
avoid connection severance and degradation of communication quality, which are
unacceptable for real-time communications including video streaming and Voice
over IP (VoIP).

---

* The work of K. Iida was supported in part by grants from the Support Center
for Advanced Telecommunications Technology Research (SCAT), Japan. The work
of Y. Kadobayashi was supported in part by the Telecommunication Advancement
Organization of Japan.

S. R. Das, S. K. Das (Eds.): IWDC 2003, LNCS 2918, pp. 174–183, 2003.

To accomplish this, we propose an end-to-end seamless handover mechanism in which mobile hosts can roam across different wireless access networks without closing the connection and degrading the communication quality. Our approach does not require any special network devices such as the Home Agent (HA) of Mobile IP [1, 2]. Instead, the mobile host needs to coordinate with its corresponding host to realize mobility on an end-to-end basis.

To end-to-end seamless handover, we employ the Stream Control Transmission Protocol (SCTP) [3] as an end-to-end protocol. This is because we think SCTP is an appropriate research platform for end-to-end protocols, since it has numerous functions that have arisen from many years of research in transport protocols. As one such function, SCTP has a multihoming function which can handle multiple network interfaces. We make use of this multihoming function to avoid connection closure.

Our main contribution in this paper is a multi-path transmission algorithm. For end-to-end seamless handover with minimum consumption of network resources, our algorithm has two modes: a single-path transmission mode and a multi-path transmission mode. Single-path transmission is the normal mode. If the transmission path becomes unstable, a mobile host switches to multi-path transmission to avoid degradation of communication quality. Moreover, after one of the transmission paths returns to the stable state, the mode returns to single-path transmission to reduce unnecessary consumption of network resources.

2   Related Work

Mobile communication technologies in the Internet are being investigated by a number of researchers. Most of this research is classified into network assisting mobility support (NAMS). In NAMS, each mobile host is communicating with a network device to handle the location of mobile hosts. A list of NAMS is the following: Mobile IP [1, 2], Hawaii [4], Cellular IP [5], and Hierarchical Mobile IP [6]. All of them need some special network devices to support mobility. Since the deployment of NAMS technologies has been just begun, it will be a long time before global deployment is complete. Moreover, special network devices are needed to maintain states in terms of locations of mobile hosts, causing high cost. For the above reasons, the notion of end-to-end assisting mobility support (EAMS) has been proposed.

In EAMS, each mobile host coordinates with its corresponding host to maintain mobility of both hosts. Thus, EAMS does not need special network devices to maintain mobility; instead, it needs a special end-to-end protocol for this purpose. Examples of EAMS are TCP migration [7] and Multimedia Multiplexing Transport Protocol (MMTP) [8]. TCP migration, which has been proposed by Snoeren and Balakrishnan, is a modification of normal TCP, allowing mobile hosts to change IP addresses. When a mobile host changes its own IP address, it will send a Migrate SYN packet to notify the change of IP address to its corresponding host. On the other hand, MMTP, which has been proposed by Magalhaes and Kravets, is a novel transport protocol for real-time communications on mobile hosts. Due to the nature of wireless links, wireless access networks provide relatively small bandwidth. To create a virtual channel with more bandwidth, MMTP is designed to aggregate the available bandwidth from multiple channels. However, no existing EAMS supports seamless handover.

3   Multi-path Transmission Algorithm for End-to-End Seamless Handover

First, we define end-to-end seamless handover. Since the conventional terminology of "handover" is assumed as NAMS, the terminological combination of "end-to-end" and "seamless handover" may be confusing. Our definition is this: seamless migration of a mobile host from one network to another in EAMS systems. Since existing techniques of seamless handover are designed only for NAMS, we need to develop a new technique.

3.1   Motivation for Choosing SCTP

This section describes why we chose SCTP as our base protocol. The SCTP is a novel transport protocol designed for both reliable and unreliable data transmissions. One major advantage of SCTP is that it has a multihoming function which can switch between different network interfaces without the connection begin closed. We think this is a basic function to realize end-to-end seamless handover.

Moreover, there is an enhancement of SCTP, called ADD-IP [9]. ADD-IP enables mobile hosts to dynamically add/delete IP addresses without loss of the connection. This function should be used in conjunction with our algorithm. Since our main focus in this paper is the multi-path transmission algorithm, we do not go into the details of ADD-IP.

Our terminology related to SCTP is as follows. *Path* is a combination of source and destination addresses. Due to the multihoming function of SCTP, one transport connection may contain multiple paths. Among these paths, *primary path* is the path actively chosen for data transmission. The other paths are called *backup paths*.

3.2   Modifications to SCTP

Since the current standard of SCTP in RFC2960 provides only reliable data transmission, as does TCP, some modifications are needed for supporting real-time communications. Here is a list of modifications to SCTP.

1. Disable the congestion control mechanism and the retransmission mechanism
2. Periodically send HEARTBEAT packets to all path
3. Change the method of increasing and resetting the error counter
4. Change the sending interval for HEARTBEAT packets
5. Implement the multi-path transmission algorithm

The first, second and third modifications are for supporting real-time communications. Since the SCTP in RFC2960 is optimized for non-real-time communications, it has a congestion control mechanism and a retransmission mechanism. However, these two mechanisms are obstacles for real-time communications, so that disabling them is the first modification.

We next explain the second modification. The SCTP is a HEARTBEAT mechanism to test reachability. Hosts periodically send HEARTBEAT packets to backup paths to test reachability. After the corresponding host receives the packet, it sends a HEARTBEAT Acknowledgment (HEARTBEAT-ACK) packet to the sender host. However, since HEARTBEAT packets are sent only to the backup paths in RFC2960, the primary path's reachability is not tested. Therefore, the second modification is that hosts send HEARTBEAT packets to all paths.

Moreover, hosts maintain a parameter called an error counter that is a metric of reachability for each path. In RFC2960, the error counter for a path is increased by 1 when a data/HEARTBEAT packet sent through the path is timed out. Conversely, the error counter for the path is reset to 0 when a host receives a data/HEARTBEAT acknowledgment packet. However, since this condition cannot appropriately give changes of reachability for each path in a wireless access network, we change the way of increasing and resetting the error counter, as the third modification:

- When a host sends a HEARTBEAT packet along the path, the error counter for the path is increased by 1.
- When a host receives a HEARTBEAT-ACK packet along the path, the error counter for the path is reset to 0.

For the fourth modification, we then describe our HEARTBEAT interval (H) calculation, which is given by

$$H = HB.Interval \times (1 + \delta) \tag{1}$$

where $HB.Interval$ is a constant, and δ is a random value, uniformly distributed between -0.5 and 0.5, to reflect a fluctuation of loads of computers and networks.

The fifth modification is our key proposal. We describe it in the following section.

3.3 Multi-path Transmission Algorithm

In this section, we describe our multi-path transmission algorithm. In wireless access networks, many packet losses degrade the quality of real-time communication, so that more redundant packet transmission is one possible implementation of seamless handover. However, redundant packet transmissions consume network resources; thus, we need to reduce unnecessary packet transmissions as much as possible. In order to achieve these two contradictory goals, we provide two modes: a single-path transmission mode and a multi-path transmission

```
[Primary Path]                      [All paths]
  send(HB);                           receive(HBack);
  ErrorCount++;                       if(mode == SinglePath){
  if(ErrorCount > MT){                    ErrorCount = 0;
      mode = MultiPath;             }else if(mode == MultiPath){
  }else if(ErrorCount > PMR){            ErrorCount = 0;
          state = inactive;              if(seqnum == HBack_seqnum - 1){
  }                                          StabilityCount++;
                                             if(StabilityCount > ST){
                                                 mode == SinglePath;
                                             }
[Backup paths]                           }else{
  send(HB);                                  StabilityCount = 0;
  ErrorCount++;                           }
  if(ErrorCount > PMR){                }
      state = inactive;             seqnum = HBack_seqnum;
  }

  (a) Sending HEARTBEAT              (b) Receiving HEARTBEAT-ACK
```

Fig. 1. Pseudo code of multi-path transmission algorithm

mode. The modes change according to an error counter for each path, with only a small consumption of network resources.

When the quality of the primary path is degraded significantly, the host switches to multi-path and activates one of the backup paths. In this mode, the host sends the same packet to two paths simultaneously. After one of the paths becomes stable, the host chooses the stable path as the new primary path and switches back to single-path.

Fig. 1 illustrate pseudo codes of the multi-path transmission algorithm. Fig. 1 (a) is executed for each path at every interval of H that is calculated by Eq. (1). The timing of this execution is different for each path. Fig. 1 (b) is executed when a mobile host receives a HEARTBEAT-ACK packet.

We use an error counter as an indicator to switch to multi-path. There is a threshold for the error counter called *Path.Max.Retrans* (*PMR*). *PMR* is a threshold to indicate that the primary path is inactive due to some network trouble. In addition to this threshold, we newly provide *Multi-path.Threshold* (*MT*), which is a threshold for the error counter to switch to multi-path. This process is shown in Fig. 1 (a). Note that *MT* must be much smaller than *PMR*.

To switch back from multi-path to single-path, we provide a stability counter and *Stability.Threshold* (*ST*) as, respectively, a new counter and a new threshold. The stability counter is activated only in the multi-path transmission mode. The initial value of the stability counter is 0. The stability counter is increased by 1 when two consecutive HEARTBEAT-ACK packets are received. To confirm this, we add a sequence number to HEARTBEAT and HEARTBEAT-ACK packets. When the stability counter exceeds *ST*, which means the path has become stable, the host then switches back to single-path, shutting down the

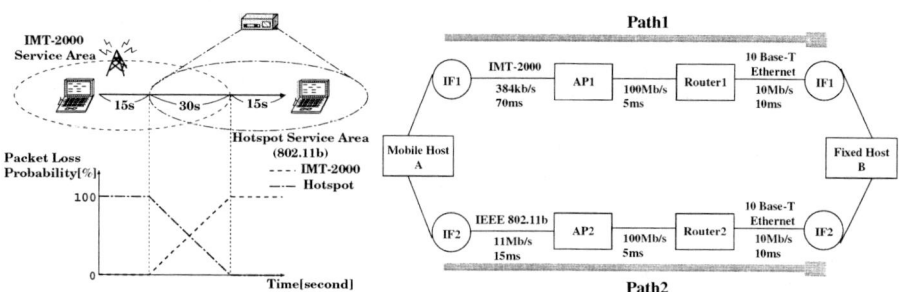

Fig. 2. Mobile scenario   **Fig. 3.** Network topology

operation of the stability counter. The stability counter is maintained in both paths currently in use, choosing the better path when the mobile host switches to single-path. The reset condition of the stability counter is when two consecutive HEARTBEAT-ACK packets cannot be received. This process is illustrated in Fig. 1 (b).

4 Simulation Experiments

In this section, we do some simulation experiments to get some basic evaluations. The simulator used to implement our algorithm is the Network Simulator version 2 (NS-2) [10] with SCTP module [11].

4.1 Simulation Setting

Since our objective is seamless handover across various kinds of wireless access networks, we consider the following scenario as shown in Fig. 2. Mobile host A has two different wireless network interfaces including an IMT-2000 card as a cellular service and an IEEE 802.11b card as a WLAN hotspot service. At first, mobile host A is only within the IMT-2000 service area. After that, the mobile host moves towards the WLAN Hotspot service area.

We then show our network topology in Fig. 3. Mobile host A communicates with a corresponding fixed host B. Fixed host B has two wired network interfaces. IF1 of mobile host A and IF1 of fixed host B form Path1. Similarly, IF2 of mobile host A and IF2 of fixed host B form Path2. Both paths are contained by one SCTP connection. The network capacity of IF1 of mobile host A is 384Kb/s according to the assumption of IMT-2000. On the other hand, that of IF2 is 11Mb/s. The network capacity of the rest of the links is illustrated in Fig. 3.

The moving speed of mobile host A is walking velocity. When it goes into the overlapping area of two wireless access networks, the packet loss rate of IF1 increases linearly whereas that of IF2 decreases linearly as a function of time, as illustrated in Fig. 2.

Table 1. Simulation parameters

$HB.Interval$	0.1, 0.3, 0.5s
$Path.Max.Retrans$	79, 25, 15
$Multi\text{-}path.Threshold$	1, 2, 3
$Stability.Threshold$	1–10

Our assumed application is real-time video streaming at a rate of 300kb/s. The streaming is uni-directional from mobile host A to fixed host B.

Time lines of all the simulations follow Fig. 2. Simulation start time is 0s. Between 15s and 45s, mobile host A is within the overlapping area. Simulation ends at 60s.

4.2   Results

Parameter Selection As listed in Table 1, SCTP with Multi-path Transmission Algorithm (MTA) has the four parameters, while it has three important performance metrics: goodput, total multi-path transmission period, and communication overhead caused by data/HEARTBEAT packets. Among four parameters, $HB.Interval$, MT and ST are especially important. Here, we first would like to determine $HB.Interval$ to simplify the rest of the discussion.

The communication overhead caused by HEARTBEAT packets can be calculated from the following equation,

$$\text{Overhead} = \frac{\text{HEARTBEAT packet size} \times 8}{HB.Interval}, \qquad (2)$$

where the HEARTBEAT packet size is a constant 60 Bytes. Therefore, this metric is affected by only the $HB.Interval$. If we choose 0.1s as the $HB.Interval$, the overhead becomes 4.8kb/s, which is unlikely to be an acceptable value.

Fig. 4 shows goodput as a function of ST. The MT value is set to 2. Goodput is the total number of bytes received by the receiver when mobile host A is in the overlapping area. We can see from Fig. 4 that a small value for the $HB.Interval$ gives us high goodput in general. When we select 0.3s as the $HB.Interval$ value, this results in an overhead of 1.6kb/s. This would be acceptable in the IMT-2000 environment at a rate of 384kb/s. Note that the simulation results listed in this section are average values from 100 experiments for each set of parameter values.

We then select a value for MT. We show the impact of ST and MT on goodput in Fig. 5. $HB.Interval$ is set to 0.3s. Goodput calculation is followed by the same way as in Fig. 4. Fig. 5 shows that small MT results in high goodput. Fig. 6 illustrates the impact of ST and MT on the total multi-path transmission periods, which indicates the total length of the redundant transmission period. The figure shows that large MT reduces this period. According to Figs. 5 and 6, we select 2 as MT.

Next we select a value for ST according to Figs. 5 and 6. Large ST results in slight improvement of goodput as well as an increase of the total multi-path

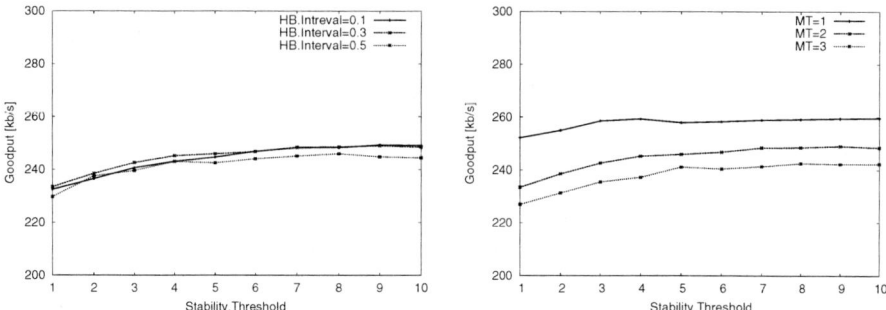

Fig. 4. Goodput versus ST and HB. **Fig. 5.** Goodput versus ST and MT
Interval

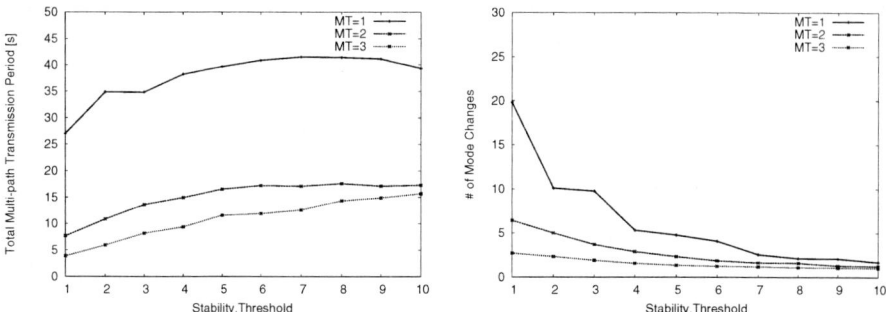

Fig. 6. The total multipath transmis- **Fig. 7.** The number of mode changes
sion period versus ST and MT versus ST and MT

transmission period. Among these relationships, we select 2 as ST to limit the
total multi-path transmission period to about 10s.

For the complementary purpose, we show the simulation results for the number of mode changes from single-path to multi-path in Fig. 7.

Finally, we determine PMR from

$$PMR = \frac{\text{decision period}}{HB.Interval} - 1 \qquad (3)$$

where the decision period is the time until complete disconnection of a path. For
example, PMR becomes 25 if we choose 8s as the decision period.

Comparisons We compare goodput for SCTP with and without MTA. All the
parameter values for SCTP with MTA follow the results of parameter selection
in the previous section. A summary of parameter selection is listed in Table 2.
Figs. 8 and 9 illustrate throughput and goodput performance of SCTP with and
without MTA, respectively. Fig. 8(a) shows there is a time period in which both

Table 2. Parameter values for SCTP with MTA and without MTA

	with MTA	without MTA
HB.Interval	0.3	0.3
Path.Max.Retrans	25	2
Multi-path.Threshold	2	–
Stability.Threshold	2	–

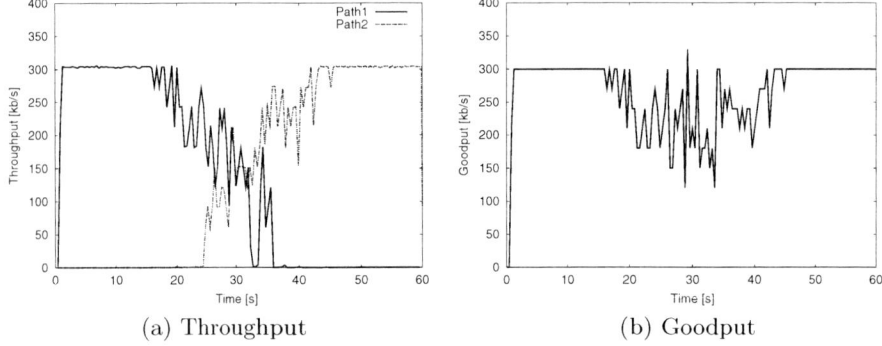

(a) Throughput   (b) Goodput

Fig. 8. Throughput and Goodput performance of SCTP with MTA

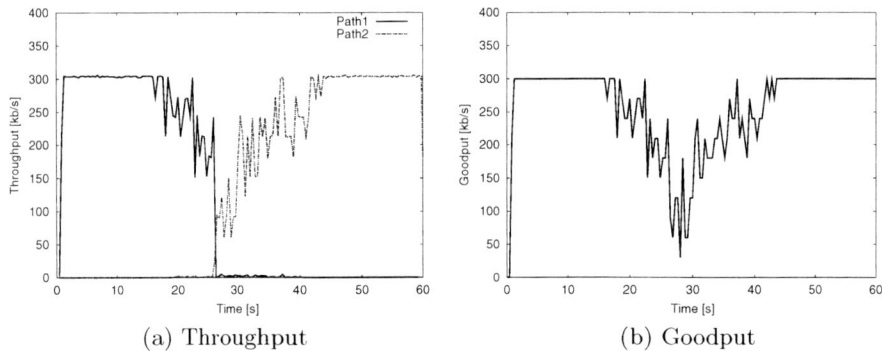

(a) Throughput   (b) Goodput

Fig. 9. Throughput and Goodput performance of SCTP without MTA

paths are sending the same packets, while Fig. 9(a) does not have such period. As a result, goodput of SCTP with MTA is better than that without MTA. The average goodput of SCTP with MTA is 238.5kb/s, and that of SCTP without MTA is 215.7kb/s. The difference between these two values is about 20kb/s. One thing we would like to emphasize is that Fig. 9(b) includes a drastically decreased goodput period, less than 100kb/s, which is not an acceptable performance. On the other hand, SCTP with MTA shows that goodput is kept high (i.e., more than 150kb/s) at almost all times.

5   Conclusion

In this paper, we have proposed a multi-path transmission algorithm (MTA) for end-to-end seamless handover across various kinds of wireless access networks. The aim of the MTA is to improve goodput during handover by implementing an end-to-end assisting handover. To realize our algorithm, we have modified SCTP which can handle multiple network interfaces. In the MTA, mobile hosts have two modes: single-path and multi-path transmission modes. If a mobile host detects that the network condition of the primary path becomes unstable, it switches to multi-path transmission mode to avoid quality degradation, and sends the same packets to multiple paths simultaneously. After that, if one of the transmission paths becomes stable, the mobile host switches back to single-path transmission mode to reduce unnecessary consumption of network resources. Through simulations, we have shown that our multi-path transmission algorithm gives us high goodput while limiting unnecessary consumption of network resources.

Acknowledgements

We would like to thank Prof. Ian R. L. Smith for his help in the proof reading of this paper and the many corrections of English. We also would like to thank Mr. Kosuke Hata for his usuful comments on this paper.

References

[1] C. Perkins, Ed., "IP mobility Support," *IETF RFC 2002*, October 1996. 175
[2] D. Johnson, et al., "Mobility Support in IPv6," *IETF Internet-Draft*, draft-ietf-mobileip-ipv6-21.txt, February 2003. 175
[3] R. Stewart, et al., "Stream Control Transmission Protocol," *IETF RFC2960*, October 2000. 175
[4] R. Ramjee, et al., "HAWAII: A Domain-Based Approach for Supporting Mobility in Wide-area Wireless Networks," *Proc. IEEE ICNP'99*, October 1999. 175
[5] A. Valkó, "Cellular IP: A New Approach to Internet Host Mobility," *ACM Comp. Commun. Rev.*, Vol. 29, No. 1, January 1999. 175
[6] E. Gustafsson, et al., "Mobile IPv4 Regional Registration," *IETF Internet-Draft*, draft-ietf-mobileip-reg-tunnel-07.txt, October 2002. 175
[7] A.C. Snoeren and H. Balakrishinan, "An end-to-end approach to host mobility," *Proc. 6th ACM/IEEE International Conference on Mobile Computing and Networking (Mobicom)*, August 2000. 175
[8] L. Magalhaes and R. Kravets, "MMTP: Multimedia Multiplexing Transport Protocol," *Proc. ACM first Workshop of Data Communications in Latin America and the Caribbean (SIGCOMM-LA 2001)*, April 2001. 175
[9] R. Stewart, et al., "Stream Control Transmission Protocol (SCTP) Dynamic Address Reconfiguration," *IETF Internet-Draft*, draft-ietf-tsvwg-addip-sctp-07.txt, February 2003. 176
[10] The Network Simulator – ns-2 –; http://www.isi.edu/nsnam/ns/. 179
[11] Protocol Engineering Lab; http://www.pel.cis.undel.edu/. 179

Global Mobility Management: A Three Level Architecture for Next Generation Wireless Networks

Iti Saha Misra[1], Mohuya Chakraborty[2], Debasish Saha[3], Amitava Mukherjee[4],
Billawdeep Deb[1], and Bitihotra Chatterjee[1]

[1] Department of Electronics and Telecommunication Engineering
Jadavpur University, Kolkata 700032, India
itimisra@cal.vsnl.net.in
{stolidcraze,bitihotra_ju22}@yahoo.com
[2] Netaji Subhash Engineering College, Garia, Kolkata 700084, India
mohuyacb@yahoo.com
[3] MIS & Computer Computer Science Group
Indian Institute of Management (IIM)
Calcutta, Joka,, D.H.Road, Kolkata 700 104, India
ds@iimcal.ac.in
[4]IBM Services Calcutta, Salt Lake, Kolkata 700 091, India
amitava.mukherjee@in.ibm.com

Abstract. In this paper, we propose a three level global mobility model that supports IP mobility. The limitations of the existing protocols vis-à-vis their advantages in dealing with different types of mobility management schemes provide an insight to incorporate their positive features into a proposed Global Mobility Model as an extension to existing protocols. This three level model introduces a Global Mobility Agent to handle interdomain mobility. The concept is in conformation with the three level UMTS architecture and is an effort to map it to the IP mobility scenario. Simulation results, using ns-2, show that the model reduces the frequency of location updates and handover delays (especially for intradomain changes which are inevitable for mobility in large networks) with an efficient signaling mechanism. An enhanced soft handover scheme has also been proposed, which not only reduces the possibility of false handovers but also takes care of packet duplicacy and misordering.

1 Introduction

An essential feature of macromobility protocols is to ensure efficient mobility management of mobile nodes (MNs) from one domain to another as well as within a single domain. The former is called interdomain (or global) mobility and the latter is known as intradomain mobility. The aim is to provide a network architecture that provides seamless and fast connectivity as the mobile node moves from one domain to another.

S. R. Das, S. K. Das (Eds.): IWDC 2003, LNCS 2918, pp. 184-193, 2003.

Visualizing the present day scenario, we find that there are quite a large number of IP mobility protocols that support intradomain and interdomain mobility management. Notable among them are Mobile IPv6 (MIPv6) [1], Hierarchical MIPv6 (HMIPv6) [2] and TeleMIP [3]. A great deal of research work has been done regarding micromobility management [4, 5, 6]. But, to the best of our knowledge, methods of improving macromobility management have been little dealt with. The limitations of the existing protocols are as follows.

i)   In MIPv6, location updates are always generated whenever the MN changes a subnet in the FN. In situations with an extremely large population of MNs, the signaling overhead increases enormously.

ii)   MIPv6 has large handover latency, if the MN and HA or CN are widely separated. Data losses also take place until the handover completes and a new route to the MN is established.

iii) Since MIPv6 standard requires the mobile to change the care-of-address at every subnet transition, it is difficult to reserve network resources all along the path between the CN and the mobile.

iv)   In HMIPv6 location update latency increases as the hierarchy level increases.

v)   Though TeleMIP architecture offers several advantages over other schemes, it does not support QoS management since the global reservation terminates at the MA and not at the MN itself.

Understanding the problems of existing protocols we are motivated to design a protocol that supports global mobility. We propose here a three level Global Mobility Model (GMM) that combines the positive aspects of the existing protocols with some enhancements in the handover procedure. This technique reduces the signaling overhead and location updates at global level granularity up to HA. GMM supports global mobility by considering a three level architecture consisting of Foreign Agents (FAs) at the lowermost level, MAs at the intermediate level and Global Mobility Agents (GMAs) at the topmost level. The model is an extension of TeleMIP, which uses a two level atchitecture. In GMM, an MN obtains three care-of-addresses (CoAs). Whenever an MN enters a subnet, it obtains a local CoA (LCoA) of the FA, a regional CoA (RCoA) of the MA and a global CoA (GCoA) of the GMA. A GMA is located at the boundary of two domains and handles a large number of MAs. Within a subnet the LCoA of the MN does not change. It changes only when the MN moves to a different subnet handled by the same MA. The change of MN's point of attachment with MA changes its RCoA. The GCoA of the MN changes only when it moves to a different domain or to a new subnet handled by an MA that is not under the control of the existing GMA. It is this GCoA with which the MN is registered with the HA. So there is a decrease in the frequency of location updates and handover latency in this model in comparison with other protocols.

The route optimization in GMM enables any CN to send packets directly to MN via GMA. HA is made responsible for sending binding updates to CN.

A new enhanced handover management scheme has been proposed which mainly uses the concept of proactive handover using bicasting [7]. Bicasting is advantageous when an MN switches between two domains several times (this type of MN movement is known as ping-ponging). In HMIPv6 handover is done from the mobility anchor point. In GMM, bicasting is done from the MA unlike MIPv6 where bicasting is done from the HA. Bicasting from HA produces scalability problems and higher signaling overhead. The problem of scalability is eliminated in this model by using a load-balancing algorithm.

The paper is sectionized as follows. After the introduction in Section 1, Section 2 proposes and discusses the architecture of the proposed Global Mobility Model (GMM). The mobility management for the GMM is discussed in section 3. A performance analysis of the existing protocols and that of the GMM is made in section 4. Finally section 5 concludes the paper with some highlights on GMM and future works.

2   Proposed Global Mobility Model (GMM)

GMM that takes care of macromobility (intradomain) and global mobility (interdomain) is shown in Fig. 1. A new node called GMA is introduced for this scheme at the network layer granularity higher than that of a subnet, thus reducing the generation of global location updates. By limiting intradomain location updates to the GMA, the latency associated with intradomain mobility is reduced. The three level mobility management scheme allows the use of private addressing and non-IP mobility management within the provider's own domain.

This model follows a hierarchical structure. At the top level lies a boundary-interworking unit known as Global Mobility Agent (GMA). The GMA is placed at the boundary between two administrative domains. It is connected to the HA of an MN through the Internet. A domain is divided into a large number of subnets. Each subnet has an FA. There are several MAs in a domain handling a number of FAs. This can be done using any load-balancing algorithm [9]. As soon as an MN enters into a subnet, a request is sent for a new CoA to the FA. The FA in that subnet assigns an LCoA to the MN after authentication validation. The FA then sends proper route control messages to the MA. MA allocates an RCoA to the MN through FA. A GCoA is then allocated to the MN via the selected route from GMA to MN via MA and FA handled by the GMA. It is this GCoA with which the GMA registers with the HA of the MN after the MN has moved away from its home area.

3   Mobility Management in Global Mobility Model

Mobility management of GMM consists of routing and handover technique. Improved handover technique employed here reduces the possibility of false handover and provides fast and seamless roaming. Route optimized routing overcomes the latency of triangular routing.

3.1   Routing

The GMM uses the concept of route optimization to eliminate the drawbacks of triangular routing [1]. The protocol has been designed to make HA responsible for providing binding updates to any concerned correspondent node at foreign networks. The correspondent nodes need to be sure of the authenticity of the binding updates. In short there are four major steps involved in route optimization in GMM.

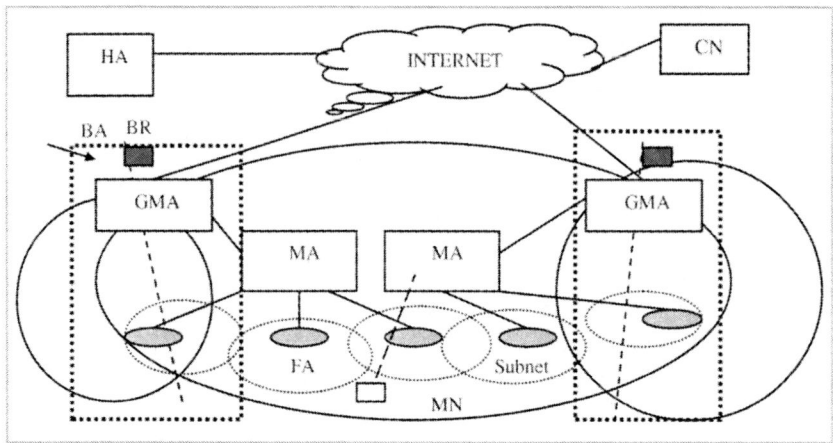

Fig. 1. Proposed Global Mobility Model

i)   A binding warning control message sent to the HA indicates the unawareness of a CN, about MN's CoA.

ii) A binding request sent by CN.

iii) A binding update sent by HA containing MN's current GCoA.

iv) A binding acknowledgement sent by CNs to HA for smooth handover s.

Packets may be sent by the CN to GMA of the foreign domain without interacting with HA. GMA sends the packets to the intended destination MN through MA and FA. Here route optimization technique up to GMA level has been adopted instead of FA as in MIPv6. The main reason for this is to avoid security problems. In cases where fast and seamless handover predominates over secured data transmission, a more efficient route optimization technique up to the FA level may be adopted. The additional steps involved after the initial route optimization up to GMA are as follows.

i)   A binding warning control message sent to the GMA indicates the unawareness

ii) of a CN about MN's LCoA.

iii) A binding request sent by CN.

iv) A binding update sent by GMA containing MN's current LCoA.

v)   A binding acknowledgement sent by CNs to GMA for smooth handover s.

3.2   Handover Procedure

For handover an MN analyses the router advertisement sent by the FA periodically to determine its new point of attachment. The FA sends router advertisement on receipt of request from the MN in the form of router solicitation. The router advertisement contains information to create a new LCoA for the MN. The uniqueness of the link-local address on the new link is verified first by the MN. It then performs duplication address detection (DAD) on its link-local address in parallel with its communications.

Then, it uses either stateful [8] or stateless [9] address auto-configuration to create its new LCoA.

In order to have an improved handover technique, we define a boundary area BA, as shown in Fig.1. It is the residing area of the mobile node when it sends the registration request to the target domain. There is a boundary register (BR) located within the BA associated with the GMA to maintain a location information database of all MAs under the control of the GMA. The handover procedure employed here is based on bicasting [7] and provides a fast handover technique [10].

For intradomain handover, an MN receives a trigger from target FA indicating that it is about to perform a handover. It then requests fast handover to the current MA, which then sends a valid address (RCoA, LCoA) for the new subnet under a different MA to the MN and address validation request to the destination MA. The destination MA then controls if the address is unique in that subnet and sends a validation result to the current MA. If the address is valid then the current MA forwards the authorization to the MN in both subnets. The MN can use this new address and can send a binding update to the HA and CNs via GMA.

The process of bicasting is applied for interdomain handover. Bicasting allows the MN to simultaneously register with several MAs. This is called proactive handover. Handover actually takes place before the MN has actually changed its domain. All packets intended for the mobile node are then duplicated in several potential localizations. But bicasting performed by the HA suffers from scalability problem and produces higher signaling overhead. By producing local bicasting from MA, these problems have been eliminated to a large extent.

The main drawback of proactive or soft handover is false handover. The GMM uses an enhanced bicasting scheme to eliminate this problem. In this model a mobile node may request bicasting in its local registration when its distance from the boundary is less than the handover distance dab, determined by equation (1). This scheme guarantees that the mobile node completes the handover procedure in an area that is within a threshold distance away from the boundary of the two domains, A and B.

In order to determine the BA that is dependent on the mobile node's handover distance, a dynamic handover policy is developed to decide the handover distance on a per-user basis [11]. Under this scheme, the handover distance, dab for the mobile node moving from A to B is based on velocity v of the mobile node.

If a mobile node is moving very fast, it may arrive at the new domain in a short time. Thus, the handover distance should be long so that the mobile node can send the registration request before it enters the new domain B. Taking this into consideration, we may establish the handover distance of the mobile node as:

$$d_{ab} = d\,(v_{ab}\,/\,v) \tag{1}$$

where, d is the constant of proportionality, which is equal to the minimum handover distance from the boundary and depends upon the system parameters like area of the domain etc. and v_{ab} is the average velocity of all the mobile nodes roaming in that domain.

Table 1. A comparative chart of location updates

Protocol	Mobile IPv6	HMIPv6	TeleMIP	GMM
Global location updates *(up to HA)*	P*N	P*(N/R)*L	P*(N/R)	P*[N/(R*K)]

4   Performance Analysis

This section deals with the performance analysis of GMM with respect to location update, handover delay and signaling overhead using ns-2 [12] Linux code.

4.1   Location Update

Let us consider that the network of Fig. 1 has N subnets, M MAs, G GMAs and P mobile nodes. Each subnet has an FA. Each MA can handle R subnets where R is very large. Each GMA can handle K MAs.

Performance gain will be more compelling if R and K are large and a mobile node spends a significant period of time within the domain controlled by a single MA and a single GMA. It is assumed that N/R – M i.e., an MA handles all subnets within a single city. It is also assumed N>>M and M>>G. Table 1 shows the number of location updates generated by P mobiles as they visit N subnets one by one using different mobility protocols.

Here L is the number of levels of access routers per domain in HMIPv6. Fig. 2 shows the plot of location updates as number of mobile nodes (P) per domain increases for different types of protocols under study. For GMM, location update remains at a very low value even with a very high value of P unlike others, which show an increase with P. So GMM has been observed to be advantageous over other models with respect to global location updates up to HA.

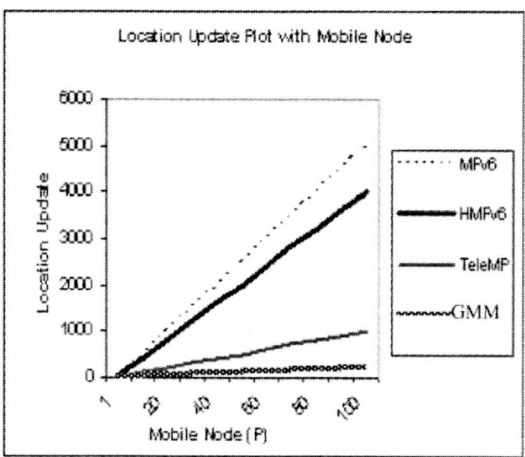

Fig. 2. Number of location updates with mobile nodes Parameters: N=50, R=5, M=10, G=2, K=5, L=4

4.2   Handover Delay

To study handover delay we simulated the various protocols and also the proposed architecture in ns-2 with required standard enhancements. Table 2 shows a comparative chart of handover delay of various protocols. It is observed that in MIPv6, whenever an MN changes a subnet in the FN, a location update is generated up to HA resulting in a very high handover latency of 204.462 ms. In our analysis, it is obvious that a change of MA produces a change of subnet and a change of GMA produces a change of MA. So, though the concept of MA and GMA are not present in MIPv6, yet for analysis purpose, we have assumed that changes of MA and GMA cause the MN to contact HA for location update leading to a delay of 204.462 ms.

In HMIPv6, an assumption of L=4, leads to a handover delay of 40.924 ms for an FA change which is larger than that of TeleMIP and GMM. When an MN changes its point of attachment with MA, HMIPv6 and TeleMIP show a delay of 282.082 ms and 224.925 ms respectively both of which are quite high as compared to 24.694 ms delay of GMM. This is a drastic change (handover delay improves by 89% over TeleMIP architecture for an MA change) and is very advantageous in large networks where an MN can change MAs very often. This would mean lower packet losses and faster handovers on MA change.

In GMM, location updates up to HA occur only when an MN changes its point of attachment with the GMA located at the boundary of two domains. This feature of GMM is found to be advantageous especially for real time service applications when the number of MNs per domain increases leading to a higher inter-subnet mobility. For HMIPv6 and TeleMIP the idea of GMA is not present. But from similar argument mentioned above, it follows that for GMA change, both HMIPv6 and TeleMIP have to contact HA leading to a delay of 282.082 ms and 224.925 ms respectively. For GMA change it is observed that the handover latency for GMM is higher than that of TeleMIP. But as the frequency of updates for GMM is less and in cases where frequency of GMA changes is very less, the overall effective handover latency is less in GMM compared to other protocols as seen from Fig. 2.

Table 2. A comparative chart of handover delay (Results from a standard simulation)

Protocol	Handover Delay (in ms)		
	For FA change	*For MA change*	*For GMA change*
MIPv6	204.462	204.462	204.462
HMIPv6	40.924	282.082	282.082
TeleMIP	14.231	224.925	224.925
GMM	14.231	24.694	255.619

Table 3. A comparative chart of signaling overhead

Protocol	Signaling Overhead		
	Local per hop	Global per hop	Total in the network
MIPv6	0	2*L3/T1	2*L3*NHA/T1
HMIPv6	2*L*L2/T1	2*L3/T2	2*L*L2*NMA/T1 + 2*L3*NHA/T2
TeleMIP	2*L2/T1	2*L3/T2	2*L2*NMA/T1 + 2*L3*NHA/T2
GMM	2*L1/T1 + 2*L2/T2	2*L3/T3	2*L1*NMA/T1+ 2*L2*NGMA/T2 + 2*L3*NHA/T3

4.3   Signaling Overhead

A comparison of the signaling overhead associated with GMM with that of MIPv6 and TeleMIP has been made in accordance with [13] as shown in Table 3.

Here, L1, L2 and L3 are the sizes of subnet, domain and global registration packets having values of 54, 50 and 46 (in bytes) respectively. As L3 does not contain subnet specific CoA and domain specific CoA and L2 does not contain subnet specific CoA so L3<L2<L1. NMA, NGMA and NHA are average number of hops from MN to MA, MN to GMA and MN to HA having values of 2, 3, and 5 respectively (arbitrary values). T1, T2 and T3 represent average duration for which MN remains in a subnet, in a domain and in a network respectively. Also M and N are the number of MAs and FAs respectively. T1 and T2 depend on network topology and mobility pattern of MN. But for the sake of simplicity in our analysis we have used T2 being N times T1 and T3 being M times T2.

The total signaling overhead for the entire network has been plotted in Fig 3. As expected the total signaling overhead is tremendously low. It also goes down when an MN stays longer in a subnet. It can be seen that the signaling overhead of the network remains almost the same as that of TeleMIP but is advantageous as the proposed scheme reduces handover delay significantly in the process.

5   Conclusion

GMM is a highly efficient scalable model particularly in case of very large networks. But as the network size increases, the problem of scalability creeps in. This model is an efficient solution to the scalability problem. The three level scalable GMM architecture for global or interdomain mobility management introduced in this paper, results in far-reaching improvement in handover latency for intradomain changes. The global signaling overhead up to HA is also reduced enormously. The handover scheme of the proposed model also reduces false handovers.

Since obtaining IP addresses in a commercial environment can be very expensive, GMM approach is very flexible to network service providers allowing then the use of private address pool. Providers may use a two level hierarchical addressing scheme under GMA for their networks.

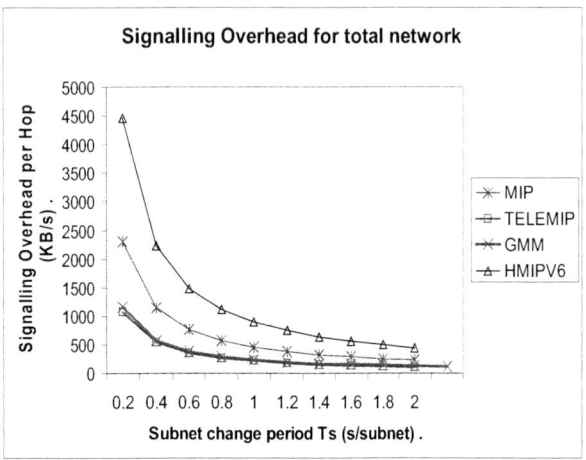

Fig. 3. Comparison of total signaling overheads

An efficient use of the model could be in organizations spread over multiple campuses. The GMAs could be placed at each of the campuses. This would reduce the handoff latency of the MN when it changes MAs in the same administrative domain (in this case a campus). The management scheme would be highly effective when there are huge numbers of MNs and MAs in the same administrative domain.

A number of open issues remain. Most notably however is a framework to support QoS guarantees in GMM infrastructure. Work is in progress to meet this requirement.

References

[1]   D. Johnson and C. Perkins, "Mobility Support in IPv6," IETF draft, draft-ietf-mobileip-ipv6-15.txt. July 2001.
[2]   H. Sollman et al.., "Hierarchical MIPv6 Mobility Management," IETF draft, draft-ietf-mobileip-hmipv6-05.txt, July 2001.
[3]   Subir Das, Archan Misra, Prathima Agrawal, and Sajal K. Das, "TeleMIP: Telecommunications-Enhanced Mobile IP Architecture for fast Intradomain Mobility," IEEE Personal Communications, Aug 2000, pp 50-58.
[4]   Andrew T. Campbell, Javier Gomez, Sanghyo Kim, Chieh-Yih Wan, Zoltan R. Turanyi and Andras G. Valko, "Comparison of IP Micromobility Protocols," IEEE Wireless Communication, Feb. 2002, pp 72-82.
[5]   M. Chakraborty, I. S. Misra, D. Saha, A. Mukherjee, "A Comparative study of existing protocols supporting IP mobility", Proceedings of World Wireless Congress 2003, Sanfrancisco, USA, May 2003.
[6]   S. Deering and R. Hinden, "Internet Protocol Version 6 (IPv6) Specification," IETF RFC 2460, Dec. 1998.

[7]   N. Montavont and T. Noel, "Handover Management for Mobile Nodes in IPv6 Networks," IEEE Communications magazine, August 2002, pp 38-43.

[8]   J. Bound et al., "Dynamic Host Configuration Protocol for IPv6 (DHCPv6)," IETF draft, draft-ietf-dhc-dhcpv6-24.txt, Oct. 2001.

[9]   S. Thomson and T. Narten, "IPv6 Stateless Address Autoconfiguration," IETF RFC 2462, Dec 1996.

[10]   G. Dommety et al., "Fast Handovers for Mobile IPv6," IETF draft, draft-ietf-mobilrip-fast-mipv6-03.txt, July 2001.

[11]   F. Akyildiz and W. Wang, "A Dynamic Location Management Scheme for Next Generation Multi-tier PCS Systems", IEEE Trans. On Wireless Communications, Jan 2002, pp 178-189.

[12]   ns-2 home page, http://www.isi.edu/nsnam/ns.

[13]   S.Das, A.McAuley, A.Dutta, A.Mishra, K.Chakraborty, S.K.Das, "IDMP:an intradomain mobility management protocol for next generation wireless networks, IEEE Wireless Communications, June 2002, pp.38-46.

Distributed Certified E-mail System for Mobile Users

Jong-Phil Yang[1], Chul Sur[1], and Kyung Hyune Rhee[2]

[1] Department of Computer Science, Pukyong Nat'l Univ.
599-1, Daeyeon3-Dong, Nam-Gu, Pusan 608-737, Republic of Korea
{bogus,kahlil}@mail1.pknu.ac.kr
[2] Division of Electronic, Computer and Telecommunication Engineering
Pukyong Nat'l Univ.
599-1, Daeyeon3-Dong, Nam-Gu, Pusan 608-737, Republic of Korea
khrhee@pknu.ac.kr

Abstract. We propose a new certified e-mail system which is based on the traditional cryptographic schemes and server-supported signature scheme for fairness and confidentiality of message. Since we intend to minimize the computational overhead of mobile device on public key algorithm, our proposed system becomes to be suitable for mail user who uses mobile devices such as cellular phone and PDA. Moreover, the proposed system is fault-tolerant, secure against mobile adversary and conspiracy attack, since it is based on threshold cryptography on server-side.

Keywords: Certified E-mail, Mail security, Secret sharing

1 Introduction

In recent years, e-mail has become an essential communication tool for business as well as academic area. Due to easy and convenient communication over e-mail, many people and businesses are moving into on-line transactions and the Internet access becomes more commonplace in everywhere. However, the Internet does not provide all the services required by business communication model such as secure, reliable and fair electronic exchange. The Certified e-mail is a different solution from the existing secure e-mail systems such as PGP and S/MIME [17]. It must provide *fairness*: at the end of exchange, it must be guaranteed that either each party has received what it expects to receive or neither party has received anything useful [11].

In this paper, we present a new certified e-mail system which is called **RE-CEM**(REliable Certified E-Mail system). One goal of our proposal is to reduce the users' computational overhead in mobile system environment through server-supported signatures[12][18]. Another goal is to provide reliability and security against mobile adversary and conspiracy attack through threshold cryptography.

The rest of the paper is organized as follows. The next section describes preliminaries to induce the main idea of the paper. Section 3 outlines the certified e-mail system used by our system. We analyze and evaluate the proposed system in Section 4. Finally, we have conclusions in Section 5.

S. R. Das, S. K. Das (Eds.): IWDC 2003, LNCS 2918, pp. 194–204, 2003.

2   Preliminaries

Server-Supported Signature: N.Asokan [12] had proposed server-supported signatures scheme for reducing the users' computational overhead on public key algorithm. In this scheme, a server performs digital signature on users' behalf. It is possible to provide security services such as non-repudiation both of origin and receipt on a signature. Moreover, if the server is regarded as a TTP(Trusted Third Party), it can guarantee fair-exchange between a sender and a receiver. X.Ding [18] had proposed the variation of [12].

Threshold Cryptosystems: When users depend on a single server for their cryptographic operation, the configuration of the server becomes to be simple. However, the single server will be a main target of malicious adversaries. When the single server is compromised, the whole cryptographic operation of it must be stopped. To overcome this situation, we can make use of threshold cryptosystems for developing a more robust and reliable server system.

In the case of (n, t)-threshold signature scheme [5], $n \geq 2t + 1$, there is a server system which consists of n servers. There is one secret/public key pair for the server system. At the beginning, a TTP computes n secret shares s_i, $1 \leq i \leq n$, from the secret key, and securely distributes s_i to each server. It allows any subset of $t+1$ servers out of n to generate a signature with the secret key, but it does not allow the creating of a valid signature if only t or fewer servers participate in the protocol. For the purpose of corrupting the whole server system, an adversary has to corrupt at least $t + 1$ servers and obtains their secret shares [1, 13].

Certified E-mail: Almost all certified e-mail requires a TTP as a mediator for fair-exchange of e-mail. Recently, many authors, such as G.Ateniese [6], K.Imamoto [8], J.Zhou [7] and B.Schneier [4], have proposed certified e-mail systems. Certified e-mail systems can be classified into on-line protocols and optimistic protocols according to their involvement of TTP. There are some desirable properties for certified e-mail: *fairness, authentication, integrity, non-repudiation*. Fairness means that both a sender and a receiver can obtain the result which each user desires, or neither of them does. Specially, fairness is the most important property in certified e-mail.

3   RECEM(REliable Certified E-Mail) System

3.1   System Model

In this section we introduce our system and a communication model.

Secure Indexing Server: SIS(Secure Indexing Server) is a trusted authority, and it issues a *credential* which is used for supporting user authentication, and securely saves some information for users' signature.

Distributed Mail Delivery: DMD(Distributed Mail Delivery) is implemented by a set of n MD(Mail Delivery)s, $n \geq 2t + 1$, each runs on a separate processor in a network. There is one service public/secret key pair in DMD. It is used for signing a message on users' behalf. The service secret key is not held by any MDs for obvious reasons. Instead, n different shares of the service secret key are distributed and stored on each MD, and threshold cryptography is deployed to construct signatures on a message.

User: A person who wants to send certified e-mail.

Communication Model: We assume that our communication model on server-side is composed of a set of n MDs and SIS. They are connected by a complete network of authenticated and dedicated multicast channel; by authenticated and dedicated we means that if a server multicasts a message, it is received by every other server and recognized as coming from the server. These assumptions allow us to focus on a high-level description of the protocol.

When an user who wants to send certified e-mail, he/she sends a request to single MD in DMD, the MD becomes a *delegate* for the user. The delegate must collaborate with SIS and $n - 1$ MDs in DMD for performing cryptographic operation. We use several traditional cryptographic schemes such as digital signature and cryptographic one-way hash function, etc. For simplifying the description of this paper, we assume that the used traditional cryptographic schemes are secure against well-known attacks.

3.2   Notations

We introduce some notations that are used for describing our protocol.

- S, R : the identities of sender and receiver, respectively.
- C : the information that explains a message M.
- MD_i : the identity of i-th Mail Delivery, where $1 \leq i \leq n$.
- NRT : the non-repudiation token. This is signed by DMD.
- SK : a session key for symmetric cryptosystem. It is used during a single session. The encrypted message M with session key SK is represented as $[M]_{SK}$.
- $h_X()$: one-way collision resistant hash function for user X. Users should personalize the hash function. For example, this can be always done by including their unique name as an argument: such as $h(X, M)$, where M is a message.
- $H(M)$: the message digest of a message M using one-way collision resistant hash function H.
- K_X : a randomly chosen secret key from the range of $h_X()$.
- K_X^i : a user X's $(n - i)$-th signing key. Based on K_X, the user X computes the hash chain $K_X^0, K_X^1, \cdots, K_X^n$, where

$$K_X^0 = K_X, K_X^i = h_X^i(K_X) = h_X(K_X^{i-1})$$

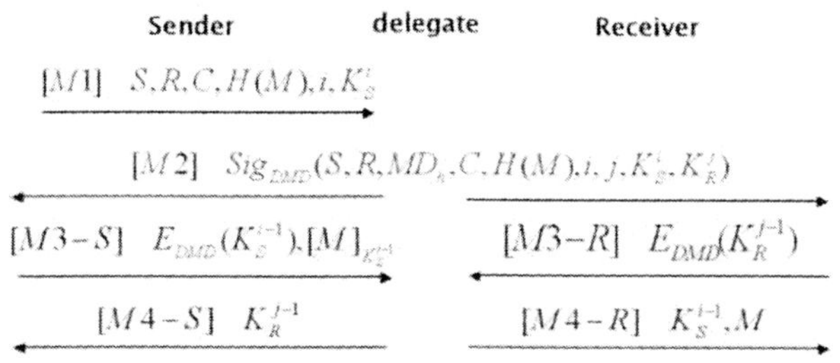

Sender   delegate   Receiver

$[M1]$ $S, R, C, H(M), i, K_S^i$

$[M2]$ $Sig_{DMD}(S, R, MD_a, C, H(M), i, j, K_S^i, K_R^j)$

$[M3-S]$ $E_{DMD}(K_S^{i-1}), [M]_{K_S^{i-1}}$ $[M3-R]$ $E_{DMD}(K_R^{j-1})$

$[M4-S]$ K_R^{j-1} $[M4-R]$ K_S^{i-1}, M

Fig. 1. Basic Mail Delivery Protocol

K_X^n constitutes X's *root signing key*, the current value of i is *signature counter*, K_X^i is X's *current signing key*.

- $Sig_X(M)$: a digital signature of a message M with a user X's secret key.
- $E_X(M)$: an encryption of a message M with a user X's public key.
- Cre_X : a user X's *credential*, which is issued by SIS.

$$Cre_X = Sig_{SIS}(X, n, K_X^n, SIS)$$

3.3   Basic Mail Delivery Protocol

In this section, we introduce basic mail delivery protocol in RECEM. We assume a situation such that each MD_i $(1 \leq i \leq n)$, already has its secret share s_i for the service secret key of DMD. Fig. 1 shows a basic mail delivery protocol in RECEM.

[**Step 0**] To participate in RECEM, each user X randomly generates K_X, and computes $K_X^n = h_X^n(K_X)$. X submits the root signing key K_X^n to SIS for a credential. SIS issues a credential for X and publishes the X's credential to a directory service.

[**Step 1**] A sender S who wants to send certified e-mail hashes a mail message M, and sends $S, R, C, H(M), i, K_S^i$ as $[M1]$ in Fig. 1 to a $MD_h, (1 \leq h \leq n)$ in DMD.

[**Step 2**] The MD_h which receives $[M1]$ from S becomes a *delegate*. It verifies the current signing key K_S^i based on the root signing key in the S's credential Cre_S, i.e., checks that $h_S^{n-i}(K_S^i) = K_S^n$. For generating a *candidate NRT*, it obtains the signature counter(j) and signing key(K_R^j) of the receiver(R) from SIS. The delegate configures a message consisting of $S, R, MD_h, C, H(M), i, j, K_S^i, K_R^j$, denoted by α for convenience, and multicasts it to $n - 1$ $MD_{k \neq h}$ $(1 \leq k \leq n)$. Each MD_k including the delegate computes partial signature $PS_{S_k}(\alpha)$ for α with its secret share s_k. All MD_k except the delegate send their partial signatures to the delegate as responses. For generating a valid signature of DMD, the delegate needs at least $t + 1$ correct partial signatures. Therefore, the delegate chooses $t + 1$ partial signatures, and computes $SIG_{DMD}(\alpha)$. If the computed value is invalid, the delegate tries to compute $SIG_{DMD}(\alpha)$ again with another sets of partial signatures. Finally, the delegate generates a candidate

NRT, $Sig_{DMD}(S, R, MD_h, C, H(M), i, j, K_S^i, K_R^j)$, and then sends it to both S and R as $[M2]$ in Fig. 1.

[**Step 3**] After receiving $[M2]$ in Fig. 1, S and R perform as followings:

- S verifies the received candidate NRT. If the verification is successful, S computes the *next signing key* K_S^{i-1}, and encrypts it with the service public key of DMD. S sends $E_{DMD}(K_S^{i-1})$ with $[M]_{K_S^{i-1}}$ to the delegate as $[M3-S]$ in Fig. 1.
- At the beginning, R reads C in $[M2]$ of Fig. 1. After reading C, if R wants to receive certified e-mail from S, R verifies the received candidate NRT. If the verification is successful, R computes the next signing key K_R^{j-1}, and encrypts it with the service public key of DMD, and sends $E_{DMD}(K_R^{j-1})$ to the delegate as $[M3-R]$ in Fig. 1.

[**Step 4**] The delegate multicasts $E_{DMD}(K_S^{i-1})$ and $E_{DMD}(K_R^{j-1})$ to the others $MD_{k \neq h}(1 \leq k \leq n)$ for decryption of the encrypted next signing keys with the service public key of DMD. Each MD_k including the delegate computes partial decryption $PD_{s_k}(K_S^{i-1})$, $PD_{s_k}(K_R^{j-1})$ with its secret share s_k. Except the delegate, all MD_k send their partial decryptions to the delegate as responses. Additionally, all MDs in DMD send $PD_{s_k}(K_S^{i-1})$, $PD_{s_k}(K_R^{j-1})$ and the identity of the delegate(MD_h) to SIS. SIS stores the received information for resolving potential disputes. For decryption, the delegate needs at least $t + 1$ correct partial decryptions. Therefore, the delegate chooses $t + 1$ partial decryptions, and decrypts K_S^{i-1}, K_R^{j-1}. By using the decrypted K_S^{i-1}, the delegate decrypts $[M]_{K_S^{i-1}}$. Finally, the delegate sends K_S^{i-1}, K_R^{j-1} to SIS. SIS checks the validity of the next signing keys for S and R as followings:

$$h_S^{n-i+1}(K_S^{i-1}) = K_S^n, h_S(K_S^{i-1}) = K_S^i$$

$$h_R^{n-j+1}(K_R^{j-1}) = K_R^n, h_R(K_R^{j-1}) = K_R^j$$

- If the verification is successful, SIS replaces the signature counter i by $i - 1$ for S, and the signature counter j by $j - 1$ for R. SIS stores K_S^{i-1} as the current signing key for S, and K_R^{j-1} for R. Then, SIS sends a message of "protocol proceed notification" to the delegate.
- If the verification fails, SIS sends a message of "protocol fail notification" to the delegate.

[**Step 5**] If delegate receives "protocol proceed notification", it sends K_R^{j-1} to S as $[M4 - S]$ and K_S^{i-1}, M to R as $[M4 - R]$ in Fig. 1. If the delegate receives "protocol fail notification", it stops basic mail delivery protocol.

[**Step 6**] Finally, S and R perform verification steps as followings:

- **Sender** : S checks whether the received K_R^{j-1} is the preimage of K_R^j in the candidate NRT. If the check is successful, S obtains NRT which R cannot repudiate the receipt of mail message.

$$SIG_{DMD}(S, R, MD_h, C, H(M), i, j, K_S^i, K_R^j), K_R^{j-1}$$

Finally, S records K_S^i as already used value by replacing the signature counter i by $i-1$.

- **Receiver** : R checks whether the received K_S^{i-1} is the preimage of K_S^i in the candidate NRT and whether the received message M is the preimage of $H(M)$ in the candidate NRT. If two checks are successful, R obtains NRT which S cannot repudiate the sending of mail message.

$$SIG_{DMD}(S, R, MD_h, C, H(M), i, j, K_S^i, K_R^j), K_S^{i-1}$$

Finally, R records K_R^j as already used value by replacing the signature counter j by $j-1$.

If there are any problems during [Step 6], a dispute will be occurred and a resolution procedure will be necessary to resolve the dispute.

RECEM is appropriate for threshold RSA [14, 16]. However, the schemes based on discrete logaritms also can be applicable [9, 10]. When it is implemented on mobile users, we suggest that the service public key of DMD is 3, i.e. $e = 3$ to minimize the computational overhead for mobile users. There are some methods to overcome the security weakness caused by using a small encryption exponent in [3]. However, by using a small exponent, we can minimize the computational overhead of users who verify a signature or encrypt a message through the service public key of DMD.

3.4   Dispute Resolution

In this section, we classify disputes or attacks into four-scenarios, and explains how to solve each problem.

Case-1: When a sender repudiates his/her e-mail that was sent.

- A receiver submits NRT and mail message M to an arbiter. Then, the arbiter who works together with SIS can verify as followings:
  1. The signature in NRT by DMD is valid.
  2. The current signing key of the sender in SIS is the same as the next signing key in NRT.
  3. The $H(M)$ value in NRT is the hash value of the mail message M.
- If at least one of these checks fails, then the arbiter judges the sender is correct. However, if these checks are all successful, the sender is allowed to the opportunity to repudiate the e-mail by providing a different NRT corresponding to the same current signing key.

Case-2: When a sender does not receive $[M4-S]$, which is a proof that a receiver has successfully received a corresponding e-mail.

- According to the basic mail delivery protocol, the delegate $MD_h (1 \leq h \leq n)$ performs threshold decryption with the others $n-1$ $MD_{k \neq h} (1 \leq k \leq n)$ after receiving $[M3 - S]$ and $[M3 - R]$. In the case of performing threshold decryption, all MDs in DMD send $PD_{s_k}(K_S^{i-1})$, $PD_{s_k}(K_R^{j-1})$ and the identity of the delegate(MD_h) to SIS. Therefore, it is impossible that the delegate does not send the decrypted K_S^{i-1}, K_R^{j-1} to SIS. So, SIS possesses the correct K_S^{i-1}.
- To resolve the dispute, the sender submits the candidate NRT to an arbiter. The arbiter who works together with SIS can verify as followings:
  1. The signature in NRT by DMD is valid.
  2. The current signing key of the sender in the candidate NRT is a hash of the current signing key of the sender in SIS.
- If these checks are successful, the arbiter judges that the delegate was compromised and did not send the next signing key of the receiver maliciously. Therefore, the arbiter makes SIS send K_R^{j-1} to the sender.

Case-3: When a receiver does not receive $[M4 - R]$, which is a proof that a sender has successfully received a corresponding e-mail.

- Basically, the solution for resolving the dispute is the same as case-2. If the receiver is correct, the arbiter make the sender or the delegate send mail message to the receiver.

Case-4: When fair exchange fails by conspiracy between a user(sender or receiver) and the delegate.

- Due to the strength of threshold signature scheme, it is only possible for an attacker to forge a signature and derive a failure of fair exchange, when the attacker compromises at least $t+1$ mail deliveries.
- Example, when a conspiracy attack between a sender and the delegate occurs.
  - The sender does not send an encrypted mail message $[M]_{K_S^{i-1}}$ in $[M3 - S]$. That is, the sender only sends $E_{DMD}(K_S^{i-1})$ as $[M3 - S]$. The delegate performs threshold decryption for decrypting $E_{DMD}(K_S^{i-1})$. The delegate sends $[M4 - S]$ to the sender, and sends only K_S^{i-1} as $[M4 - R]$ to the receiver or none. Consequently, the sender successfully obtains NRT for the receiver in spite of without sending the mail message.
  - According to [Step 6] in the basic mail deliver protocol, the receiver requests a dispute resolution to an arbiter. The method for resolving this dispute is the same as case-2 and case-3.

3.5   Simplified Approach for Confidentiality of Mail Message

The basic mail delivery protocol introduced in section 3.3 does not provide the confidentiality for mail message. Therefore, we introduce a simple method for confidentiality based on DH key agreement protocol. Fig. 2 shows the enhanced protocol for confidentiality. From the following literature, we only focus on the changed parts at basic mail delivery protocol in section 3.3.

[**Step 0**] SIS selects a large prime p and generator g of $Z_p^*(2 \leq g \leq p - 2)$, and publishes them to users. Each user X chooses a secret $x \in {}_R Z_{p-1}$, and computes $y = g^x \bmod p$ for generating DH key-pair. Each user X randomly generates K_X, and computes $K_X^n = h_X^n(K_X)$. X submits the root signing key K_X^n to SIS. SIS issues a credential for X:

$$Cre_X = Sig_{SIS}(X, n, K_X^n, g^x, SIS)$$

we introduce some additional notations which are used in this section:

- x_i : the DH secret key of a user i.
- y_i : the DH public key of a user i. That is, $y_i = g^{x_i} \bmod p$.
- T_i : the local timestamp value of a user i.

[**Step 1**] S generates a timestamp value(T_S) based on the local system clock, and sends $[M1]$ in Fig. 2 to a $MD_h, (1 \leq h \leq n)$ in DMD.

[**Step 3**] After receiving $[M2]$, S verifies the received candidate NRT. If the verification is successful, S computes a session key(SK) for secure communication with R by using DH public key($y_R = g^{X_R} \bmod p$) in the R's credential, DH secret key(X_S) of S and T_S.

$$SK = H(y_R^{X_S \cdot T_S} \bmod p) = H(g^{X_R \cdot X_S \cdot T_S} \bmod p)$$

S computes the *next signing key* K_S^{i-1}, and encrypts it with the service public key of DMD. S sends $E_{DMD}(K_S^{i-1})$, $[M]_{SK}$ to the delegate as $[M3 - S]$.

[**Step 4**] & [**Step 5**] The delegate cannot see the mail message M, because it is encrypted with SK which can be calculated by only S and R. If the delegate receives "protocol proceed notification" message from SIS, the delegate only sends $[M4 - S]$ to S, and $[M4 - R]$ to R.

[**Step 6**] R checks "Is the received K_S^{i-1} is the preimage of K_S^i in the candidate NRT?". If the check is successful, R computes a session key(SK) for secure communication with S by using DH public key($y_S = g^{X_S} \bmod p$) in S's credential, DH secret key(X_R) of R and T_S.

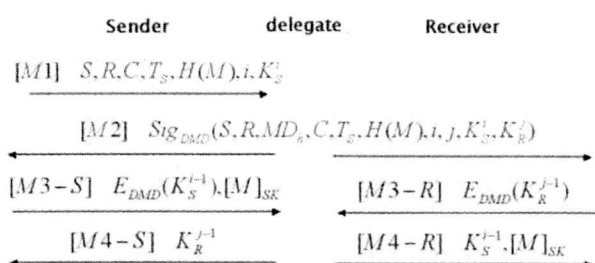

Fig. 2. Enhanced Protocol for Confidentiality

$$SK \; = \; H(y_S^{X_R \cdot T_S} \; mod \; p) \; = \; H(g^{X_S \cdot X_R \cdot T_S} \; mod \; p)$$

By using SK, R decrypts $[M]_{SK}$ and checks "Is the received message M is the preimage of $H(M)$ in candidate NRT?". If the check is successful,R becomes to obtain NRT which S cannot repudiate the sending of mail message and receives the mail message M.

4   Security Evaluation

The security of RECEM wholly depends on the security of the service secret key of DMD. Therefore, it is possible to use *proactive secret sharing scheme* to make RECEM more secure against mobile adversary [2, 15]. By using proactive secret sharing, we can periodically and securely update the secret share of each $MD_h, (1 \leq h \leq n)$, and recover the compromised MD_h. In [18], authors introduced a basic solution for denial-of-service attack, and it can be also applicable to our scheme. RECEM guarantees some desirable properties that were introduced in section 2, and provides additional security services:

– **Fairness** : By using server-supported signatures scheme, fairness is provided between a sender and a receiver, if DMD which supports users' signatures is correct.
– **Authentication** : Users can authenticate each other through candidate NRT and credential.
– **Confidentiality** : In section 3.5, we introduced a simplified approach for confidentiality.
– **Non-repudiation** : Through NRT of a sender and a receiver, it is impossible to repudiate his/her own activities successfully.
– **Attack against a malicious MD** : It is impossible for a single mail delivery to forge or delete a message successfully.
– **Attack against conspiracy between a user and a MD** : To forge or delete a message successfully, a user must conspire with at least $t + 1$ mail deliveries.
– **Fast revocation** : The fast revocation means the revocation for a signature ability of a user. When a user's signing key is compromised, SIS can revoke the user's credential and delete the user related information in SIS on instant. Consequently, DMD does not perform digital signature on behalf of users.
– **More secure signature** : Since DMD digitally signs a message on behalf of users, it is possible to use a more strong RSA key-pair without a burden of users for computational overhead.

5   Conclusion

A new certified e-mail system with low computational overhead for mobile users are proposed. The scheme is also reliable and secure against mobile adversary and conspiracy attack. Our proposal is suitable for users who want to send their secure e-mails by using cellular phone or PDA with limited computing power or battery. The communicational efficiency and implementation of the proposed scheme will be deployed for the future works.

Acknowledgements

This work is a product of University Research Program supported by Ministry of Information & Communication in Republic of Korea.

References

[1] A. De Santis, Y. Desmedt, Y. Frankel and M. Yung. "How to share a function securely". In Proceedings of the 26th ACM Symposium on the Theory of Computing, pages 522-533, Santa Fe, 1994. 195

[2] A. Herzberg, S. Jarechi, H. Krawczyk, and M. Yung "Proactive secret sharing or: How to cope with perpetual leakage". Advances In Cryptology-Crypto'95, the 15th Annual International Cryptology Conference, Proceedings, volume 963 of LNCS, page 457-469   202

[3] Alfred J. Menezes, Paul C. van Oorshot, Scoot A. Vanstone "Handbook of Applied Cryptography", 1997, CRC Press   199

[4] B. Schneier and J. Riordan. "A certified e-mail protocol". 13th Annual Computer Security Applications Conference, pages 100-106, Dec. 1998.   195

[5] D. Malkhi and M. Reiter. "Byzantine quorum systems" Distributed Computing, 11(4):203-213, 1998   195

[6] G. Ateniese, B. d. Medeiros and M. T. Goodrich. "TRICERT: A Distributed Certified E-Mail Scheme". In ISOC 2001 Network and Distributed System Security Symposium(NDSS'01), San Diego, CA, USA, Feb. 2001.   195

[7] J. Zhou and D. Gollmann. "Certified electronic mail". In Computer Security-ESORICS'96 Proceedings, pages 55-61. Springer Verlag. 1996.   195

[8] Kenji Imamoto, Kouichi Sakurai. "A Certified E-mail System with Receiver's Selective Usage of Delivery Authority". INDOCRYPT 2002, LNCS 2551, pp. 326-338, 2002.   195

[9] L. Harn, "Group oriented (t, n) digital signature scheme". IEE Proceedings-Computer and Digital Techniques, 141(5):307-313, September 1994   199

[10] M.Cerecedo, T.Matsumoto, H. Imai, "Efficient and secure multiparty generation of digital signatures based on discret logarithms". IEICE Transactions on Fundamentals of Electronics, Information and Communication Engineers, E76-A(4):532-545, April 1993   199

[11] M.Franklin and M.Reiter. "Fair exchange with a semi-trusted third party". In Proc. ACM Conference on Computer and Communications Security. 1997.   194

[12] N. Asokan, G.Tsudic, M.Waidner, "Server-Supported Signatures". European Symposium on Research in Computer Security , September 1996.   195

[13] P.Gemmel. "An introduction to threshold cryptography". in CryptoBytes, a technical newsletter of RSA Lab. Vol. 2, No. 7. 1997.   195

[14] R. Gennaro, S. Jarecki, H. Krawczyk, and T. Rabin. "Robust and efficient sharing of RSA functions". In Advances in Cryptology-Crypto'96, LNCS 1109, pp. 157-172, 1996   199

[15] S. Jarecki. "Proactive Secret Sharing and Public Key Cryptosystems". Master thesis. MIT. 1996. [14] Victor Shoup,   202

[16] Victor Shoup, "Practical threshold signatures", in Proc. Eurocrypt 2000   199

[17] William Stallings "CRYPTOGRAPHY AND NETWORK SECURITY : Principles and Practice" Second Edition, Prentice-Hall   194

[18] X. Ding, D. Mazzocchi and G. Tsudik "Experimenting with Server-Aided Signatures", 2002 Network and Distributed Systems Security Symposium (NDSS'02), February 2002.   195, 202

Models of Self-Stabilization and Sensor Networks

Ted Herman[*]

University of Iowa
ted-herman@uiowa.edu

Abstract. The advent of large-scale sensor networks highlights problems of fault tolerance and scale in distributed system, motivating designs that autonomously recover from transient faults and spontaneous reconfiguration. Self-stabilization is an attractive approach for such problems, however the standard model of research for self-stabilizing algorithms does not suit ad hoc networks of wirelessly communicating sensor nodes. The paper surveys some standard models of self-stabilization and relates these models to a sensor network. Challenges and opportunities are for integrating self-stabilization with sensor networks are illustrated with examples.

1 Introduction

Advances in microscale and nanoscale technology offer the promise of inexpensive components that measure physical quantities and trigger reactions to what they measure. Such components could be manufactured in large number and deployed in vast networks. Applications include environmental monitoring, detection of intrusion, and guiding vehicles through unmapped terrain. Distributed control of systems built from such components requires that sensing devices be coupled with communication and computing capabilities. For these sensor networks, unlike the vision of ubiquitous computing,[1] deployment could be *ad hoc* and temporary because typical applications for sensor networks have limited objectives.

Sensor networks need to autonomously react to the failures of sensors, disruptions in communication, and asynchronous reconfiguration events. Self stabilization is a general method, based on the abstraction of an unknown initial system state, by which systems autonomously adapt to arbitrary change of environment, configuration, and internal state. Most of the literature of self-stabilization research is expressed in a high-level model of processes that communicate using shared variables, often using large granularity of computation in each atomic step.

This paper attempts to reconcile the standard model of self-stabilization with a sensor network model. A detailed model of a sensor network is given in §2, then

---

[*] Research supported by NSF CCR-9733541 and DARPA contract F33615-01-C-1901.
[1] The term *ubiquitous computing* refers to a technology in which computational power is accessible in nearly all the appliances and infrastructure of daily human life.

S. R. Das, S. K. Das (Eds.): IWDC 2003, LNCS 2918, pp. 205–214, 2003.

§3 reviews the standard model and notation for self-stabilization. Integration of §2 and §3 is the topic of §4. Two sections, §5 and §6 show how it can be better to depart from the standard model of self-stabilization when designing for sensor networks.

2   Networks of Sensors

The sensor network for this paper is built from a set of nodes which use low-power radio to communicate. In contrast to heterogeneous systems, which can have powerful transmitters in base stations, all the nodes are intended to have the same computational and communication capabilities, and there exists no external backbone network or message repeater facility. Distribution of the nodes could be *ad hoc*, and even for some kind of planned distribution, it could be that a fraction of the nodes are unavailable due to malfunction or dead batteries. We suppose that each node p can communicate with a subset of nodes, called the *neighbors* of p, determined by the range of the radio signal. In the wireless model, transmission is omnidirectional: each message sent by p is effectively broadcast to all nodes in its neighborhood. Local broadcasts between sensor nodes have extremely small propagation delay. This differs from the message assumptions for satellite channels, ATM networks, or the Internet, where latency is a significant factor. Thus, the situation of having more than one message in flight from one node to another is impossible in a sensor network. We assume that communication capability is bidirectional: if p can transmit to neighbor q, then q can also transmit to p. Distribution of these nodes is such that the network is connected, meaning there exists at least one path of intersecting neighborhoods between any two nodes. Sensor nodes have fine-grained, real-time clock hardware.

Each node uses the same radio frequency (one frequency is shared spatially by all nodes in the network). Communication is half-duplex: node p cannot send one message and receive another message concurrently. In fact, while p is transmitting, p is unable to detect whether or not another node is also transmitting. Therefore, collisions are possible in this model. Nodes do not have collision detection hardware. If p's neighbors concurrently transmit, then p receives the superposition of their transmissions, but p cannot detect that the superposition is a result of collision, because noise can also be the source of corruption in messages. We assume that each message contains sufficient error detection codes so that the event of corruption or collision can be deduced, though the detection cannot distinguish between these two causes of corruption.

These facts preclude the use of CSMA/CD access to the radio medium, however some basic techniques of CSMA are applicable: if node p has a message ready to transmit, but is receiving some signal, then p does not begin transmission until it detects the absence of signal. The usual technique dealing with collisions in this type of system is CSMA/CA: before p transmits a message, it waits for some random period. We assume that nodes have such CSMA/CA capability (as implemented, for instance, in [1]). We assume that the implementation of CSMA/CA satisfies the following: there exists a constant $\tau > 0$ such that the probability of a frame transmission without collision is at least τ (this corresponds to typical assumptions for multiaccess channels; the independence of τ for different frame transmissions indicates our assumption of an underlying memoryless probability distribution in a Markov model). We similarly assume that the probability of noise corrupting a message is bounded above by some constant smaller than 1.

The description of sensor networks can include more realistic detail. For instance, it can be that communication is not always bidirectional; it could be that p is outside of radio range of q, yet when q is receiving a message, say from r, a transmission by p interferes with q's reception; it may be that nodes are mobile or other nondeterministic factors dynamically affect radio range. Some sensor networks allow access to globally synchronized time (perhaps by GPS), and others can sense their positional coordinates or distance to other nodes, which can be useful for distributed computation.

3 Processes, Communication Graph, Scheduling

With the exception of only a few papers that use a shared-memory (asynchronously PRAM) model, the literature of self-stabilization is based on a network model of communication. Let $G = (V, E)$ be an undirected connected graph, with $n = |V|$, where each vertex in V represents a process and the edge set E constraints communication between processes: p and q can communication only if $(p, q) \in E$, in which case we say p and q are neighbors. Let $\Delta = \max_{p \in V} |\{ q \mid (p, q) \in E \}|$ denote the maximum degree of any vertex.

The majority of self-stabilization research is based on a shared-variable abstraction rather than message-passing, even for this network-centric model. The reason is that distributed algorithms based on message-passing models are often event-driven (the arrival of a message triggers some action). The drawback of an event-driven, asynchronous system is that a transient fault could put the system into a state where all events have been "lost" and deadlock is the result. Self-stabilization must cope with all cases of transient faults, so researchers prefer models that do not rely on events.

The *state* of the system is, at any time, completely determined by the values of variables. Each variable can be written by exactly one process, the variable's *owner*, and the *local state* of a process is determined by the values of the variables it owns. (This model, unlike general shared-memory models, does not permit shared variables written by multiple processes.)

Algorithms are usually described by the following program notation. A program consists of a list of guarded assignment statements, with each statement having the form $G \to A$, where G is a predicate (boolean function of the state) and A is an assignment. For a given state σ, we say $G \to A$ is *enabled at* σ if guard G evaluates to *true*. The "$[\!]$" operator combines guarded assignment statements into a program; $[\!]$ is commutative and associative, so closed-form expressions can be used to express programs:

$$G_0 \to A_0 \;[\!]\; \cdots \;[\!]\; G_{n-1} \to A_{n-1} \quad \equiv \quad (\;[\!]\; i : 0 \le i < n : G_i \to A_i\;)$$

Each guarded assignment is associated with a process, and assignments and guards comply with the communication network: an assignment associated with p can assign only to variables of p and expressions evaluated in the guard and assignment can only refer to variables of p and its neighbors.

To explain program execution, various schedulers are introduced. A *schedule* is a sequence $S = \langle S_i \mid i \geq 0 \rangle$ of nonempty subsets of V. A schedule is the abstraction used to model interleaving semantics of execution (concurrency). Schedules are restricted to satisfy certain fairness and atomicity properties, but are otherwise undetermined. Typically researchers describe the selection of a schedule and restrictions on the selection by naming a *scheduler*, though of course this is just an abstraction rather than an implementation requirement. A scheduler is *fair* if for any schedule S it selects, for all $p \in V$, for infinitely many values of i, $p \in S_i$. A *central daemon* scheduler is one that satisfies ($\forall i :: |S_i| = 1$); it models the activation of one process at each step in an execution. A *synchronous* scheduler models fully parallel execution: ($\forall i :: S_i = V$). A *distributed daemon* scheduler satisfies ($\forall i :: 1 \leq |S_i| \leq n$), which subsumes both synchronous and central daemon schedulers.

The standard theoretical question for schedulers (and other model details) is whether one model is more powerful than another (able to solve more computational tasks) or improves complexity compared to another model. A constructive proof that one model is as powerful as another yields a *transform* (or compiler or simulator). If there is a transform from the distributed daemon model to the central daemon, then programs written for the central daemon can be automatically transformed into ones that work on the distributed daemon model.

An *execution* is a sequence of states $D = \langle \sigma_i \mid i \geq 0 \rangle$ obtained from a schedule and an initial state σ_0: state σ_i, $i > 0$, is obtained from σ_{i-1} by evaluating all the guards of the statements of processes selected by S_{i-1}, and then in parallel executing the assignments of each selected statement enabled at σ_{i-1}. We call the pair (σ_{i-1}, σ_i) a *transition* in the execution. Programs are typically written to satisfy the property that for central daemon scheduling, if $G \rightarrow A$ is the statement responsible for (σ_{i-1}, σ_i), then $G \rightarrow A$ is not enabled at σ_i. A state is a *fixed point* if no statement is enabled (sometimes called *deadlock*).

A *round* $\langle S_i \mid \ell \leq i \leq m \rangle$ is a segment of a schedule such that $V = \bigcup_{i=\ell}^{m} S_i$, and a round is minimal if it has no proper subsegment that is also a round. Round one (the first round) of a fair schedule is the minimal round that includes S_0, and round $k > 1$ is the minimal round beginning with the first state following round $k - 1$. Rounds better approximate the notion of concurrent time (where in one time unit, many processes can perform a statement) than counting transitions.

A program is *self-stabilizing* with respect to a model and a given state predicate $\mathcal{L}$ if two properties hold: (*Closure*) for any transition (σ, σ'), if $\mathcal{L}$ holds at σ, then $\mathcal{L}$ holds at σ'; and (*Convergence*) for any state σ, every execution containing σ also contains a state σ' satisfying $\mathcal{L}$. A state satisfying $\mathcal{L}$ is called a *legitimate* state. The *stabilization time* of an algorithm is the worst-case number of rounds, taken over all executions and all initial states, to reach a legitimate state.

4   Integrating the Sensornet Model

If self-stabilization is a viable technique for networks of sensor nodes, then we have to take into account the realities of communication hardware in the design of self-stabilizing algorithms and network protocols. Two extremes in this direction would be either to build or transform the hardware into an implementation of the abstract models used by researchers, or to directly program using the native hardware. The danger of the former direction is that significant performance could be lost; the danger of the latter direction is that we may lose access to valuable algorithms and methods developed by researchers. §5 charts an intermediate direction between the two extremes. This section adapts the programming model of §3 to the system described in §2.

The sensor network's communication primitive of neighborhood broadcast differs from any of the known basic operations in the model of §3. Our first result is to introduce a construction that transforms a sensor network to a central daemon model. The construction is initially described without considering faulty initial states, collisions and message corruption errors, and subsequently modified to handle these issues. The transform specifies a protocol for sensor nodes so that self-stabilizing programs in the notation of §3, written for processes scheduled by a central daemon, are correctly simulated in the sensor network.

Let each node $p \in V$ in the sensor network have a single variable v_p (the construction can be extended to support numerous variables of p). For each $(q,p) \in E$, let q have a variable $\boxtimes_q v_p$, which denotes a cached version of v_p. Atomically, whenever p assigns a new value to v_p, node p also broadcasts the new value to all its neighbors. Whenever a node q receives such a new value for v_p, it immediately (and atomically) updates $\boxtimes_q v_p$. Because sending and receiving operations are exclusive in the sensor nodes, we suppose that receiving a cache update message cannot interfere with concurrent assignment and broadcast by the receiving node. The programming notation from §3 can be used, subject to the restriction that a process statement only refers to its own variables and cache variables. Call this discipline of variable assignment, local broadcast, and restricted statements the *cached sensornet transform* (CST). The property $(\forall p, q : (p,q) \in E : \boxtimes_p v_q = v_p)$ is called *cache coherence*, and justifies the following lemma.

Lemma 1. If the initial state of an execution is cache coherent, then the CST execution of the program is equivalent to a central daemon model execution of the same program with all occurrences of $\boxtimes_q v_p$ replaced by v_p, for all $p, q \in V$.

Proof: It is easily verified that cache coherence is an invariant for any local broadcast without collision, since any change to a shared variable v_p atomically updates all $\boxtimes_q v_p$ copies. Formally showing equivalence to the central daemon model maps executions from each model to the other (omitted for brevity). □

Lemma 1 verifies the closure requirement of self-stabilization, where legitimacy for CST is taken to be cache coherence.

Lemma 2. In any CST execution, if there occurs a state σ such that every node p has locally broadcast the value of v_p at least once prior to σ, then σ is cache coherent.

Proof: After the first local broadcast containing the value of v_p, $v_p = \boxtimes_q v_p$ holds for all neighbors q, and this property is invariant for all subsequent assignments at p; thus $\mathcal{L}$ holds after all nodes update their shared variables.   ❐

Faulty Initial State. Lemma 2 does not prove self-stabilization of the cache update discipline, because there can occur an initial state where caches are not coherent, yet no statement is enabled — in other words, a deadlock. Therefore, we add to the CST the following: some new action at each node p perpetually broadcasts the value of v_p. (An implementation of this new action would presumably use timeout or some fairness mechanism so that the new actions have acceptable cost once cache coherence holds.) Call this new action *periodic retransmit*. With this modification, Lemma 2 applies to every execution, and we can use CST to support the programming model of choice for many researchers, the central daemon scheduler.

Message Corruption. To deal with the effects of noise and collisions, the cached sensornet transform is further modified: for each node p, introduce a boolean $b_p(q)$ for each neighbor q (the vector b_p can be a field of variable v_p). Each statement $G \rightarrow A$ of the program of p is changed to $(\forall q : (p,q) \in E : b_p(q)) \wedge G \rightarrow A$. Whenever p correctly receives a message from a neighbor q, p assigns $b_p(q) := true$, and whenever p receives a corrupt (or collided) message, p assigns *false* to $b_p(q)$ for every q. Thus p blocks program execution immediately upon receiving a corrupt message, and only allows program execution after it has correctly received messages from all neighbors. We call a state where $b_p(q) = true$ for all q an *unblocked state* for p.

One other modification of the transform specifies details about the frequency of cache updates and period retransmits. Periodic retransmit of v_p continues regardless of values of b_p, but the rate of these events in real time is controlled. We require that transmission is implemented so that no node p begins to transmit more than once in any time interval less than or equal to some known constant T time units; we also require that in any interval of T time, for any p, the independent probability that neighbor q transmits exactly once is at least $\gamma > 0$. Recall from §2 that each transmission is collision-free with nonzero probability; it follows that T must be large enough to allow all the processes in each neighborhood to transmit without collision (T is constrained by the maximum transmission time for a message and the maximum neighborhood size $1 + \Delta$).

Lemma 3. For any node p and any state σ in any execution, there occurs an unblocked state for p following σ, with probability 1.

Proof: To ensure an unblocked state for p it suffices to show there is a sequence of events so that p correctly receives a message from every neighbor without receiving any corrupt message. Within any interval of T time units, all neighbors of p transmit a message once with probability γ^{Δ}. If all these transmissions are received correctly by p in this interval, then p will be unblocked. The CSMA implementation has the property that any transmission is collision-free with probability of at least τ. A lower bound on a sequence of collision-free transmissions by the neighbors of p in such a time interval is at least $(\gamma\tau)^{\Delta}$. The limit probability that p remains forever blocked is at most $\lim_{i\to\infty}(1 - (\gamma\tau)^{\Delta})^i = 0$.

$\square$

There are likely much better ways to simulate the central daemon model than the transform suggested above, including message acknowledgments and TDMA scheduling of transmission. However all such schemes introduce overhead, which brings up the question: is it a good idea to have a sensor network as described in §2 directly simulate the central daemon model?

5   Unison with Collision

An *asynchonous unison* protocol is a program with the following behavior. The variable v_p of each node p is a positive integer called a *clock*. The legitimacy predicate for unison is

$$\mathcal{L} \equiv (\forall p, q : (p, q) \in E : |v_p - v_q| \leq 1)$$

In addition to maintaining $\mathcal{L}$, every node p should execute $v_p := 1 + v_p$ infinitely often in the execution (no other change to a variable is allowed). The problem of asynchronous unison is motivated by the need for phase synchronization in distributed systems. The simplest self-stabilizing program for this behavior in the central daemon model is:

$$(\fbox{}\ p ::\quad (\forall q : (p, q) \in E : v_q \geq v_p)\quad \to\quad v_p := 1 + v_p\quad)$$

The proposal of this section is to have sensor nodes execute the program

$$(\fbox{}\ p ::\quad (\forall q : (p, q) \in E : \boxtimes_p v_q \geq v_p)\quad \to\quad v_p := 1 + v_p\quad)$$

without using the message-corruption modification to CST, that is, without using the b_p vector to inhibit a node p's execution. Whenever a node p correctly receives a cache update or period retransmit, p atomically updates the cache. Corrupt messages or collisions are ignored. Thereby the program may execute without cache coherence. Let $\mathcal{C}$ be the following weakening of cache coherence:

$$\mathcal{C} \equiv (\forall p, q : (p, q) \in E : \boxtimes_p v_q \leq v_q)$$

Lemma 4. $\mathcal{L} \wedge \mathcal{C}$ is an invariant of any execution.

Proof: If $\boxtimes_p v_q$ is an underestimate of v_q, then the guard $(\forall q : (p, q) \in E : \boxtimes_p v_q \geq v_p)$ may block p's progress, however whenever the guard is enabled, the result of incrementing v_p satisfies $\mathcal{L}$. If the cache update message for such an increment is not correctly received by a neighbor q, then $\mathcal{C}$ still holds because v_p's increment satisfies $\boxtimes_q v_p \leq v_p$.

$\square$

Lemma 5. With probability 1, every execution eventually satisfies $\mathcal{L} \wedge \mathcal{C}$.

Proof: Probability is needed to ensure that eventually each node correctly receives messages from its neighbors, and this is given by the CSMA/CA implementation. Consider two cases for some node p, either (1) eventually p never increments v_p or (2) p increments v_p infinitely often. In case (1), with probability 1, for every neighbor q, cache coherence $\boxtimes_q v_p = v_p$ eventually holds. If case (1) holds for all $p \in V$, the system is eventually deadlocked and cache coherent, but this contradicts the self-stabilizing behavior of the basic algorithm. In case (2), in order for p to infinitely increment v_p, it must forever correctly receive messages from each neighbor q containing larger values of v_q. In turn, each neighbor q must correctly receive larger clock values from all its neighbors; by an inductive argument it follows that all nodes increment their clocks infinitely often. We deduce therefore from the inequality in the program's guard that $\mathcal{C}$ eventually holds and continues to hold after each clock increment. Once a state is reached where $\mathcal{C}$ holds, and subsequently every node has incremented its clock at least once, then $\mathcal{L}$ can be shown to also hold.   ❐

The specification of unison does not bound the value of a clock. The asynchronous unison algorithm [2] does use bounded clocks; we leave as an open problem how bounded clocks would be incorporated with the weakened cache coherence $\mathcal{C}$.

If we assume that clocks can be bounded, say represented by some fixed number of bits (which could be realistic if memory-corrupting transient faults do not occur), then we can take further advantage of the sensor model. The remainder of this section sketches an implementation of cache update and periodic retransmit for unison in a ring, that is, every node has exactly two neighbors. We assume the following about noise and radio signals: a node does not mistake silence for a transmitted bit. This allows noise to corrupt messages, but not to introduce messages (or pad messages).

Let each transmission of value v_p be a fixed-length message k bits, which includes the error detection code (CRC or better) needed to detect transmission faults and collisions. Call the two types of broadcast resend (periodic retransmit) and send (cache update message). A resend transmission is a message of k bits followed by a period of silence (no transmission) of a least $3k$ bit times. A send transmission is $3k$ bits formed by triplicating the value message, followed by a silent period of at least k bit times. Whenever a send or resend is correctly (without collision or error) received, the cache corresponding to the transmitted value is atomically updated by the receiving node.

We now consider all the cases of collision at a node p. There are two cases for collision, either (a) both of p's neighbors transmit concurrently while p is silent, or (b) p transmits while one or both neighbors are concurrently transmitting. CSMA dictates that nodes do not initiate transmission while they are receiving (even a garbled) message. For case (b) it follows that p and some neighbor(s) begin transmission at essentially the same instant. It is still possible that p can receive a value for case (b): if p starts a resend and one neighbor q starts a send, then p could correctly receive at least the final k bits of q's send; if this happens, then p updates $\boxtimes_p v_q$ from the correctly received k bits. For all other subcases of (b), node p does not change $\boxtimes_p v_q$ for any q.

Case (a) consists of node p being silent, followed by p receiving some sequence of bits composed from the concurrent transmissions of neighbors q and r. Two subcases of (a) are (i) that both q and r begin transmitting at the same instant, so that p will receive either k or $3k$ bits, or (ii) p receives between $k+1$ and $6k-1$ bits. If subcase (i) is the result of both q and r transmitting the same type of message (send or resend), then p detects an error and ignores the reception; else if (i) is the result of concurrent send and resend, node p updates either $\boxtimes_p v_q$ or $\boxtimes_p v_r$ from k correctly received bits. Subcase (ii) is the reception of m bits, where either $m \in [k+1, 2k-1]$ (two resends) or $m \in [3k+1, 6k-1]$ (two sends). If $m < 3k$, node p ignores the reception. If $m \geq 4k$, then node p will correctly obtain k bits from the send of q and also correctly obtain k bits from the send of r, and from these both $\boxtimes_p v_q$ and $\boxtimes_p v_r$ are updated. The final subcase is $m \in [3k+1, 4k-1]$, which overlaps two send transmissions so that none of the triplicated values can be recognized. In this case, p atomically executes

$$\boxtimes_p v_q := 1 + \boxtimes_p v_q \quad ; \quad \boxtimes_p v_r := 1 + \boxtimes_p v_r$$

(The idea is to deduce from the composite message length that both neighbors have incremented their clocks.)

The advantage of this implementation is that in almost all cases, a send does result in an update to neighbor caches. The only case where collision foils a send transmission is when two nodes simultaneously begin send operations. The CSMA/CA protocol inserts a random delay before any transmission, so the probability that two nodes simultaneously begin to transmit is small, perhaps comparable to the probability that a single cache update message is corrupted by noise (probability is also needed for convergence from bad initial states).

6   Maximal Independent Set

Issues of scheduling and complexity are nicely exposed by examining algorithms that self-stabilize to a solution of the Maximal Independent Set (MIS) problem. A set of processes $S \subseteq V$ is *independent* if $(\forall u, v \in S : (u,v) \notin E)$; set S is *maximal* if for any independent set T, $S \subseteq T : S = T$. For system applications the output of the MIS problem is a set S of "local leaders" so that each process is either a leader or has a neighbor that is a leader, with the constraint that no leaders are neighbors.

A self-stabilizing algorithm for the MIS problem has each process p maintain boolean v_p to indicate whether or not it is a leader; the independent set is $S = \{ p \mid v_p \}$ at any state. Let $L_p \equiv (\exists q : (q, p) \in E : v_q)$.

$$(\, [\!] \ p :: v_p \wedge L_p \ \rightarrow \ v_p := false \ \ [\!] \quad \neg(v_p \vee L_p) \ \rightarrow \ v_p := true \ )$$

This algorithm, presented in [3], has $O(1)$ stabilization time for central daemon scheduling. The same algorithm does not stabilize in all schedules of a distributed daemon, because a race condition may emerge due to symmetrical decisions by neighboring processes. A self-stabilizing algorithm for the distributed daemon in [4] breaks symmetry using process identifiers, however the stabilization time is $O(diam)$, where $diam$ denotes G's diameter ($O(n)$ in the worst case). Randomization could be a better method to solve MIS, and [3] shows that the algorithm stabilizes with probability 1 using a randomized distributed daemon. In the sensornet model, randomization is already present in the CSMA/CA implementation, suggesting the following implementation.

Whenever a node p receives a transmission from neighbor q with $v_q = true$, then p atomically assigns $v_p := false$ as well as updating $\boxtimes_p v_q$. Notice that this violates the CST protocol of only assigning to variables in an atomic transmission that transmits a cache update of v_p. In other respects, the CST protocol (removing of course the b_p waiting method) remains the same. Observe that if a node p broadcasts $v_p = true$ and this message is correctly received by all neighbors, then v_p remains $true$ throughout the execution. Since the probability of sending without collision and without error is bounded, this event occurs with probability 1 in any execution. Eventually, every leader is correctly recognized to be a leader by its neighbors, and execution reaches a fixed point with an MIS solution and cache coherence holding.

7   Conclusion

In [5] a randomized coloring algorithm is presented, also based on the model of cache updates subject to collision. The examples of unison, maximal independent set, and coloring illustrate how it can be useful to exploit communication network characteristics in sensor networks rather than directly simulate standard models such as the central daemon abstraction.

References

[1] A Woo, D Culler. A transmission control scheme for media access in sensor networks. In *Proceedings of the Seventh International Conference on Mobile Computing and Networking (Mobicom 2001)*, pp. 221-235, 2001.

[2] JM Couvreur, N Francez, MG Gouda. Asynchronous unison. In *Proceedings of the 12th International Conference on Distributed Computing Systems (ICDCS 1992)*, pp. 486-493, 1992.

[3] SK Shukla, DJ Rosenkrantz, SS Ravi. Observations on self-stabilizing graph algorithms for anonymous networks. In *Proceedings of the Second Workshop on Self-Stabilizing Systems (WSS 1995)*, pp. 7.1-7.15, 1995.

[4] M Ikeda, S Kamei, H Kakugawa. A space-optimal self-stabilizing algorithm for the maximal independent set problem. In *Proceedings of the Third International Conference on Parallel and Distributed Computing, Applications and Technologies (PDCAT 2002)*, pp. 70-74, 2002.

[5] T Herman, S Tixeuil. A distributed TDMA slot assignment algorithm for wireless sensor networks. Rapport de Recherche LRI 1370, Université Paris Sud, 27 pages, Septembre 2003.

Performance of a Collaborative Target Detection Scheme for Wireless Sensor Networks

Kai Li[1] and Asis Nasipuri[2]

[1] Department of Industrial Technology
East Carolina University, Greenville, NC 27858
lik@mail.ecu.edu
[2] Department of Electrical & Computer Engineering
The University of North Carolina at Charlotte, Charlotte, NC 28223-0001
anasipur@uncc.edu

Abstract. A sensor network is an ad hoc network of a large number of sensor nodes, each equipped with embedded processor, memory, and a wireless transceiver. The collective resources of such a system of networked sensors can be used to perform powerful distributed target detection and monitoring operations over a wide area. Key concerns for designing such networks are the energy efficiency and scalability of its communication and signal processing tasks. In addition, spatial variations of noise characteristics and differences in signal sensitivities amongst the sensors lead to challenging problems for designing a reliable distributed detection scheme. This paper addresses these issues and presents a collaborative signal processing technique that reduces the amount of traffic in the network and also maintains a constant false alarm rate under varying noise powers and sensor sensitivities.

1 Introduction

Recent advances in low-powered embedded processors and wireless technology have led to the development of small low-cost wireless sensors that are capable of forming an autonomous ad hoc network amongst themselves. Potentially hundreds or thousands of such wireless sensors can be used to form a *sensor network* for unmanned surveillance and tracking operations. Applications include environmental monitoring, intrusion detection, monitoring manufacturing plants, and target detection and tracking in tactical environments [4, 8, 3].

In order to derive maximum benefits from distributed sensor information, the sensors must be designed to collaborate to perform a common task. The main hurdles for achieving this are the limitations in computation and communication capabilities of the sensor nodes, which are due to their small size and limited battery power. It has been found that transmission and reception operations consume several times greater amount of electrical energy than that used for data processing [13]. Since distributed processing can be enacted only by the exchange of information amongst the sensor nodes over wireless data packets, a sensor network must combine optimum on-board processing with a controlled

S. R. Das, S. K. Das (Eds.): IWDC 2003, LNCS 2918, pp. 215–224, 2003.
© Springer-Verlag Berlin Heidelberg 2003

amount of data communication amongst the various sensors. The ideal solution to sensor collaboration would involve dynamic selection of the appropriate members for collaboration and optimizing the signal processing and communication tasks between them [11]. Several approaches to sensor collaboration for specific applications have been presented in recent times [18, 5]. The work in [18] presents information-driven approaches to sensor querying and data routing for applications involving target tracking and localizing a stationary target. Their approach is based on information measures for optimally selecting members in the collaborating group. Some algorithms for member selection and cluster formation for enumerating targets in a region and determining their approximate locations are presented in [5]. Other work related to sensor collaboration include directed diffusion routing [8] that provides an efficient mechanism for information propagation based on advertised data attributes. An outline of some of the major challenges for sensor collaboration is given in [11].

We propose a sensor collaboration scheme for target detection in which each sensor generates a multi-level statistic based on a set of observed signal samples to determine the significance of its observations for detecting a target. A sensor node informs its neighbors if it determines that the level of significance of its observations is sufficiently high. When a minimum number of neighbors have significant observations, a collective decision is made that the target is present. This decision is then transmitted to the user using multihop packet transmissions. The multi-level statistic is based on a *nonparametric* signal processing technique [9] that is guaranteed to have the same statistical properties (probability distribution function) at all sensors when a target is not present, irrespective of the amount of noise and its distribution at the sensor. Consequently, it has several properties that make it appropriate for its application in sensor networks. Firstly, the proposed scheme provides a constant false alarm rate (CFAR) for a wide range of noise densities. This implies that sensors located at different locations and experiencing different noise conditions will perform similarly when no target is present. Secondly, the proposed scheme is relatively unaffected by variations of the sensitivities of the sensors. Since a very large number of sensors may be involved, the effect of sensitivity variations is an important issue. Finally, the proposed scheme reduces unnecessary data packet transmissions in the network unless a target is detected collectively by a group of sensors, thereby conserving battery resources. We use computer simulations to evaluate the detection performance and transmission requirements of the proposed collaborative detection scheme using different degrees of collaboration.

2   Overview

A sensor network may be deployed by strategically placing sensor nodes (SN) at desired locations (such as in a uniform grid-like pattern) or distributing them randomly (such as when they are dropped from a flying airplane). The sensors take periodic observations to detect the presence of a target signal, such as light, temperature, chemical level, etc. The observations are noisy, which introduces possibilities of detection errors. There may be two types of detection errors. If noisy observations lead to the decision that a target is present in the absence of any target, the error is known as a "false alarm". On the other hand, if a target is present but the detector is not able to detect it, we get the second type of error known as a "missed detection". One of several different criteria based on these error probabilities may be used for designing the detector [17].

When a target is detected, a message is sent to a *gateway node* over a wireless data packet. Such data transmissions are performed using ad hoc networking protocols [6]. Accordingly, a SN can transmit a data packet directly to another SN that is located within its radio transmission range. Since the SN's have limited transmission ranges, in order to transmit to a node that is located outside its range, data packets are routed over a sequence of intermediate SN's using a store-and-forward multi-hop transmission technique. Each SN maintains multi-hop routes to the gateway node. For this work we use the communication protocols and the *SensorSim* simulation platform that are described in [12]. *SensorSim* is a *ns-2* based simulator [1] that includes protocols with special features to conserve battery power and also allow scalability for the network to perform efficiently with a large number of sensor nodes.

We assume that each observation of a sensor SN_j consists of a set of k observation samples, which we represent by $\mathbf{Y}_j = \{Y_{1j}, Y_{2j}, \cdots Y_{kj}\}$. Under the conditions that the target signal is absent or present, the observation samples Y_{ij}, $i = 1, 2, \cdots, k$ can be described by random variables (rv) as follows:

$$\begin{aligned} \text{Target is not present: } & Y_{ij} = N_{ij}, \\ \text{Target is present: } & Y_{ij} = N_{ij} + \theta_j \end{aligned} \tag{1}$$

Here, N_{ij} are noise rv's and θ_j is the strength of the target signal received at SN_j. We assume that the noise rv's at a sensor SN_j are independent and identically distributed (iid) having the probability density function (pdf) $f_j(y)$. This pdf as well as the signal strength θ_j may vary from one sensor to another, although we assume the noise rv's at all sensors to be statistically independent. We will use H_0 to represent the hypothesis that the target signal is not present, and H_1 to represent the hypothesis that the target signal is present.

The strength of a signal received at a sensor depends on the strength of the target signal at the source (which we represent by θ), its distance from the target, and the path loss model that depends on the type of signal involved (heat, light, etc.) and the physical environment. Accordingly, for a given target signal θ, the signal received at SN_j is modelled by

$$\theta_j = \frac{\theta}{|\bar{x} - \bar{x}_j|^\alpha} \tag{2}$$

where $\bar{x}$ and $\bar{x}_j$ are the locations of the target and SN_j, respectively, and α is the decay exponent for the signal path loss. We assume that $\theta_j > 0$ for all j.

Classical signal detection methods [17] may be applied to derive the optimum signal processing scheme at any sensor SN_j to determine whether the set of observations $\mathbf{Y_j}$ belongs to the hypothesis H_0 (signal absent) or H_1 (signal present). Moreover, if a specific set of collaborating sensors and their communication topology is chosen in advance, theories from distributed detection [15, 16, 2] may be applied to determine the local signal processing scheme at the sensors as well as the optimum combining scheme with the selected group of sensor observations. However, when the group of sensors and their communication links are unknown, optimum detection theoretical principles are harder to apply.

Typically, classical signal detection is based on comparing a function of the observations (test statistic) to a threshold. The choice of the optimum test statistic depends on the performance criterion and the characteristics of the signal and noise. In the following, we briefly comment on two tests statistics that are related to our work.

Linear Test Statistic: A commonly used technique for signal detection is the comparison of the *linear* test statistic of $\mathbf{Y}_j$ to a fixed threshold [14]. The linear test statistic, which we denote by $L(\mathbf{Y}_j)$, is obtained as:

$$L(\mathbf{Y}_j) = \frac{1}{k} \sum_{i=1}^{k} Y_{ij}. \tag{3}$$

It can be shown that in the presence of Gaussian noise, amongst all functions of $\mathbf{Y}_j$, a threshold test based on $L(\mathbf{Y}_j)$ will guarantee the minimum probability of decision errors [14].

However, a key concern with using a linear detector in sensors is that $L(\mathbf{Y}_j)$ is *parametric*, which implies that its characteristics are dependent on the parameters of the signal *as well as* the noise pdf. This may cause difficulties in standardizing the performance of the sensor network when the noise characteristics differ from one place in the network to another. Another concern with this test statistic is that its value depends on the sensitivity of the sensor. For instance, if the sensitivity of a particular sensor drops by a factor of two, all the observation samples will be scaled down by the same factor, and any decision based on the same test statistic will give completely different results.

Wilcoxon Test Statistic: Given a set of k observation samples $\mathbf{y}_j = \{y_{1j}, y_{2j}, \cdots, y_{kj}\}$, the Wilcoxon statistic [10] $W(\mathbf{y}_j)$ is obtained by first ranking the k observation samples in an increasing order of magnitude. Let r_{ij} be the rank of $|y_{ij}|$ among $|y_{1j}|, |y_{2j}|, \cdots, |y_{kj}|$. Then $W(\mathbf{y}_j)$ is computed as

$$W(\mathbf{y}_j) = \sum_{i=1}^{k} r_{ij} sgn(y_{ij}) \tag{4}$$

where we define the function $sgn(x) = 1$ if $x \geq 0$, and $sgn(x) = -1$ otherwise.

Observe that the Wilcoxon statistic is based on the relative order of the magnitudes of the samples rather than on its actual values. Hence, if two sensors receive observation samples that differ only by a scaling factor, which may happen due to different sensitivities, they will still generate the same value of the Wilcoxon statistic. Another property of the Wilcoxon statistic is that under the hypothesis H_0, the positive and negative ranked samples would be identically distributed as long as the noise pdf is symmetric, i.e. when $f(-y) = f(y)$, which is true for most types of noise. Hence, a threshold test based on $W(\mathbf{y}_j)$ will provide the same probability of false alarm for all noise densities that are symmetric.

The Wilcoxon statistic takes only integer values between $-k(k+1)/2$ (obtained when all samples are negative) and $k(k+1)/2$ (obtained when all samples are positive). When no signal is present, the samples have zero mean, and the average value of $W(\mathbf{Y}_j)$ is also 0. The average value of $W(\mathbf{Y}_j)$ is higher whenever a signal is present and increases monotonically with the strength of the signal. The probabilities of error of a threshold test based on $W(\mathbf{Y}_j)$ decreases with increasing number of samples in the observation. The Wilcoxon test is more than 95% as efficient[1] as the linear detector in Gaussian noise, asymptotically [7]. When the noise pdf has a "heavier tail", such as in the two-sided exponential distribution, the relative efficiency is even higher. A more interesting property of the Wilcoxon test is that its efficiency is never less than 86.4% (asymptotically) with respect to the linear detector for *any* noise pdf.

3   Proposed Collaborative Signal Detection Scheme

With these properties in mind, we present a   *collaborative Wilcoxon detector (CWD)*, whose details are as follows:

- Every time a node SN_j obtains a set of observations $\mathbf{y}_j$, it computes the Wilcoxon test statistic $W(\mathbf{y}_j)$ from the k samples in that set. This is obtained at intervals of T_s seconds. To reduce random fluctuations caused by noise, we consider that L successive $W(\mathbf{y}_j)$ values obtained over a period T_p are averaged at each node to obtain W_j at the end of the period T_p, where $L = T_p/T_s$.
- This average W_j is compared to a predetermined threshold τ_s to decide if an alerting signal is to be transmitted from SN_j.
  - If $W(\mathbf{y}_j) \geq \tau_s$, then SN_j alerts its neighbors by broadcasting a non-propagating *Type-0* packet over the wireless channel.
  - Else, SN_j discards its observations and continues to monitor future observations in the same way.
- All SN's monitor the channel for alerting signals from its neighbors. *Type-0* packets trigger a local combination of observations at all sensors that receive it. The fusion rule is given by a simple counting rule, as follows:
  - If SN_j receives alerting signals from C *or more* SN's and its own W_j also exceeds the threshold τ_s, then it decides that the target is present. SN_j immediately transmits this information using a *Type-1* packet to the

---

[1] The efficiency of a detector is determined by the number of observations k required to provide a given P_{FA} and P_{DET}. A smaller k implies a higher efficiency.

gateway node. This packet is routed to the *gateway node* using multihop transmissions.

- • If the above criterion is not met, then SN_j does not take any action.
– All SN's who overhear a *Type-1* packet that is transmitted by some other SN, and also have a *Type-1* packet of its own that it intends to transmit to the gateway node, will conserve their battery by not sending it.

The thresholds τ_s and C are predetermined, which are based on the desired probability of false alarm of the collaborative signal detection scheme. Since a higher C requires a larger number of sensors from the same neighborhood to simultaneously (and independently) believe that a target might be present, the corresponding threshold must be smaller to maintain the same error probabilities. However, a smaller τ_s increases the number of transmissions of *Type-0* packets, and hence the power consumption in the network.

4   Simulation Results

In order to evaluate the performance of this scheme, we simulate a 100-node network in *SensorSim*, and evaluate two performance measures: (a) the probability of detection, and (b) the total number of wireless packet transmissions in the network.

For all simulations, we assume that the SN's are deployed in a 10×10 uniform rectangular grid with a spacing of 14 meters between adjacent nodes. The radio transmission range is 15 meters for all nodes. We obtain results under Gaussian noise for three different values of the target signal θ: 0, 40.0 and 80.0. For each case, seven randomly chosen target locations are considered. In the first part of our simulations, we assume the same standard deviation (s.d.) of 1.0 for the noise pdf at all sensors. In the second part, we simulate a scenario where the sensors experience unequal amounts of environmental noise. In the third part of our simulations, we consider the same noise s.d. at all sensors, but some of the sensors have different sensitivities.

For comparison, we also present the performances of a non-collaborative linear and Wilcoxon detectors, in both of which each sensor obtains a hard decision by comparing the average linear/Wilcoxon test statistic obtained from its observations to predetermined thresholds. We represent the threshold for the linear detector by τ_L and that of the Wilcoxon detector by τ_W. Since the linear detector is optimum in Gaussian noise, it provides a scale of reference for comparing the performance of the proposed nonparametric methods. The comparison with the non-collaborative Wilcoxon detector illustrates the advantages achieved from collaboration. Two different values of C ($C=1$ and 2) are used for the proposed schemes with the same overall false alarm probability to depict its performance with different degrees of collaboration.

4.1   Uniform Noise Power and Equal Sensitivities

All detection schemes were simulated with $k = 15$ samples per observation, a sampling period of $T_s = 1$ second, and an averaging period of $T_p = 10$ seconds.

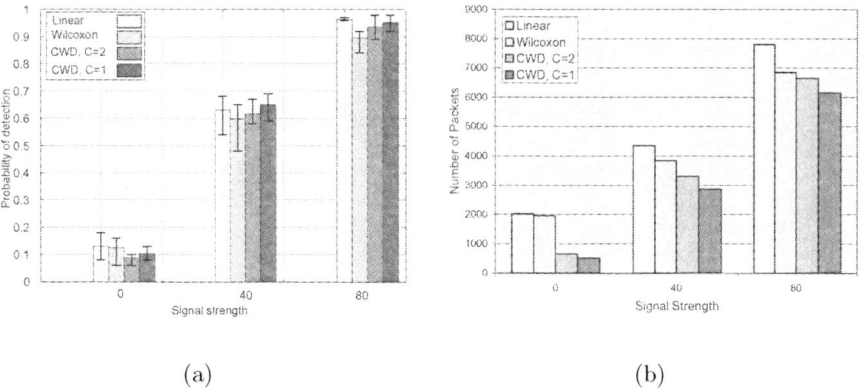

(a)   (b)

Fig. 1. Detection probabilities and packet counts under uniform noise conditions

The corresponding thresholds that give a false alarm probability of 0.1 were determined as $\tau_L = 6.43$ for the Linear detector, $\tau_W = 23$ for the non-collaborative Wilcoxon detector, $\tau_s = 9$ for CWD with $C = 2$, and $\tau_s = 13$ for CWD with $C = 1$.

Figure 1(a) shows the average detection probabilities as well as the extent of their variations obtained from seven different target locations. The average detection probabilities of the proposed scheme with $C = 1$ and 2 at intermediate and strong signal strengths (θ=40 and 80, respectively) are comparable to that of the (optimum) non-collaborative linear detector. The non-collaborative Wilcoxon detector has a lower detection probability, which reflects its lower relative efficiency with respect to the linear detector. The detection probability of CWD with $C = 1$ is higher than that with $C = 2$, which indicates that the performance improvement does not scale with the degree of collaboration. Another observation is that collaboration reduces the fluctuations of the false alarm rate (detection probabilities under zero signal strength).

The corresponding average number of packet transmissions over a simulation period of 1000 seconds are plotted in Figure 1(b). The proposed collaborative detection scheme generates fewer packets than the non-collaborative schemes. This is due to the difference in the number of multi-hop *Type-1* packets generated in the network. This difference is particularly significant at $\theta = 0$, where the proposed scheme generates less than 25% as many packets as the non-collaborative schemes. Since this represents the average packet transmissions in the network under idle operating conditions, i.e. when no target is present, it is largely indicative of the savings in the overall energy consumption in the network using the proposed scheme.

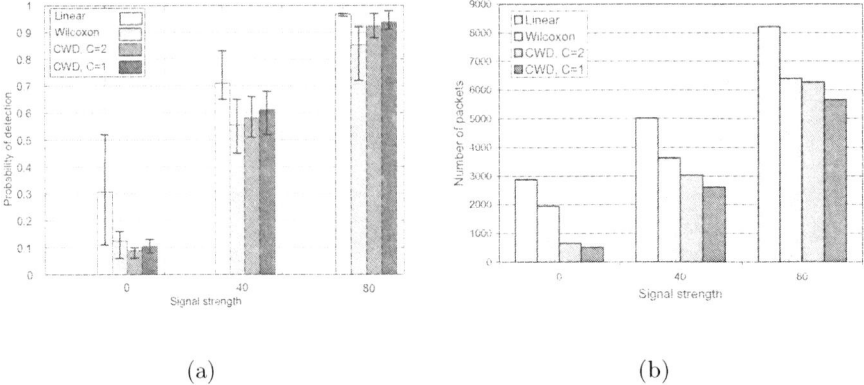

(a)   (b)

Fig. 2. Detection probabilities and packet counts under non-uniform noise conditions

4.2   Non-uniform Noise Power and Equal Sensitivities

To show how the proposed schemes perform under non-uniform noise powers, we simulate a scenario where the noise standard deviation in 15% of the nodes, chosen randomly, is set to 2.0, whereas the standard deviation at all other nodes are kept at 1.0. The thresholds τ_L, τ_W, and τ_s are kept unchanged. The detection probabilities obtained using the three schemes are shown in Figure 2(a). The non-collaborative linear scheme generates an average false alarm probability of 0.3, with the maximum value exceeding 0.5. This is a most undesirable effect and is the primary justification for using the more complex Wilcoxon statistic in the proposed collaborative scheme, which maintains the desired false alarm rate of 0.1 despite the non-uniform noise power distributions in the network.

Figure 2(b) shows that the number of packets generated in the schemes using Wilcoxon statistics do not change under noise power variations. The linear detection scheme, on the other hand generates much higher number of packets in this case. This is due to the fact that higher noise powers cause higher variations of the linear test statistic, thereby increasing the probability of exceeding the threshold in the linear detector. This causes a higher probability of target detection (either correctly or incorrectly) by the sensors, and consequently, a higher packet transmission rate in the network.

4.3   Uniform Noise Power and Unequal Sensitivities

Lastly, we consider a scenario where a random number of sensors have a sensitivity that is 50% lower than normal. This is simulated by scaling the observation samples Y_{ij}, $i = 1, 2, \cdots k$, by a factor of 0.5 in 15% of the nodes that are chosen randomly.

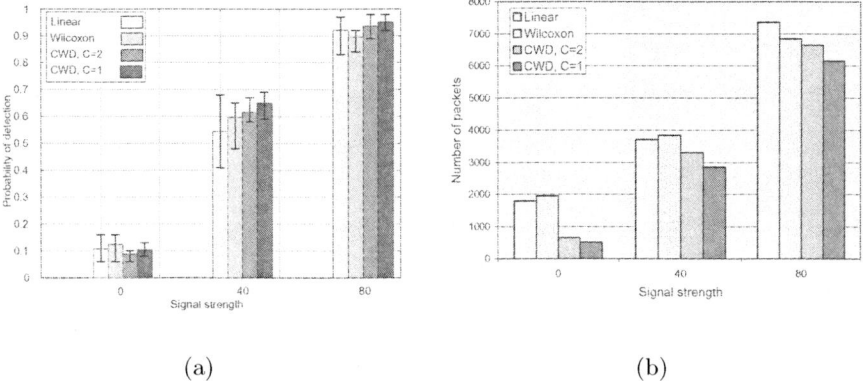

      (a)   (b)

Fig. 3. Detection probabilities and packet counts under non-uniform sensor sensitivities

A key characteristic of the results in this case is that the detection probabilities (Figure 3(a)) of the proposed collaborative schemes remain relatively unaffected by the sensitivity variations. For instance, while the detection probability of the linear scheme at $\theta = 40$ varies from 0.4 to 0.68, the proposed collaborative scheme provides detection probabilities in the range 0.57 to 0.68, for both values of C.

The corresponding packet counts, shown in Figure 3(b), show that the proposed schemes provide relatively higher energy efficiency under non-uniform sensitivities as well. In all simulations, the proposed scheme with $C = 1$ generates the smallest number of packet transmissions in the network.

5 Conclusions

We presented a distributed signal detection paradigm for wireless sensor networks. The proposed scheme utilizes the hardware present in the sensor nodes to perform some local processing that involves ranking and obtaining the weighted sum of the ranks. A group of sensors near a target exchange limited amounts of information to arrive at a reliable decision regarding the presence of the target, which is sent to the user. The proposed scheme generates a consistently low false alarm probability under variations of the physical environment such as noise and sensor sensitivities. The detection probabilities of the proposed schemes are comparable to a non-collaborative detection scheme using the optimum linear detection method, which, on the other hand, is seriously affected by variations of noise and signal sensitivities. Two different fusion rules are studied, which differ in the way that the information from the sensors are combined locally. While collaboration reduces detection errors, involvement of too many nodes does not always improve the detection performance. The proposed detection scheme also generates lower number of data packets in the network, thereby conserving battery power.

References

[1] ns-2 network simulator. http://www-mash.cs.berkeley.edu/ns/, 1998. 217
[2] R. Blum, S. Kassam, and H. V. Poor. Distributed detection with multiple sensors: Part ii - advanced topics. *Proceedings of the IEEE*, 85:64–79, 1997. 218
[3] R. T. Collins, A. J. Lipton, H. Fujiyoshi, and T. Kanade. Algorithms for cooperative multisensor surveillance. *Proceedings of the IEEE*, 89(10):1456–1477, 2001. 215
[4] D. Estrin, R. Govindan, J. Heidemann, and S. Kumar. Next century challenges: Scalable coordination in sensor networks. In *ACM/IEEE International Conference on Mobile Computing and Networking (MOBICOM)*, pages 263–270, August 1999. 215
[5] Q. Fang, F. Zhao, and L. Guibas. Lightweight sensing and communication protocols for target enumeration. In *ACM International Symposium on Mobile Ad Hoc Networking and Computing (MOBIHOC)*, pages 165–176, 2003. 216
[6] IETF MANET Working Group. http://www.ietf.org/html.charters/manet-charter.html. 217
[7] J. L. Hodges and E. L. Lehman. The efficiency of some nonparametric competitors of the t-test. *Ann. of Math. Stats.*, 27:324–335, 1956. 219
[8] C. Intanagonwiwat, R. Govindan, and D. Estrin. Directed diffusion: A scalable and robust communication paradigm for sensor networks. In *ACM/IEEE International Conference on Mobile Computing and Networking (MOBICOM)*, pages 56–67, August 2000. 215, 216
[9] S. A. Kassam. A bibliography on nonparametric detection. *IEEE Transactions on Information Theory*, IT-26:595–602, 1980. 216
[10] J. H. Klotz. Small sample power and efficiency of the one-sample wilcoxon and normal scores tests. *Annals of Mathematical Statistics*, 34:624–632. 218
[11] J. Li, J. Jannotti, D. S. J. DeCouto, D. R. Karger, and R. Morris. Collaborative signal and information processing in microsensor networks. *IEEE Signal Processing Magazine*, 19(2):13–14, March 2002. 216
[12] S. Park, A. Savvides, and M. B. Srivastava. Simulating networks of wireless sensors. In *Proceedings of Winter Simulation Conference*, 2001. 217
[13] S. Tilak, N. B. Abu-Ghazaleh, and W. Heinzelman. Infrastructure tradeoffs for sensor networks. In *Proceedings of WSNA'02*, 2002. 215
[14] H. L. Van Trees. In *Detection, Estimation, and Modulation Theory Part-I*. New York: Wiley, 1968. 218
[15] P. K. Varshney. *Distributed Detection and Data Fusion*. Spinger-Verlag, 1997. 218
[16] R. A. Viswanathan and P. K. Varshney. Distributed detection with multiple sensors: Part i - fundamentals. *Proceedings of the IEEE*, 85:54–63, 1997. 218
[17] A. D. Whalen. In *Detection of Signals in Noise*. Academic, 1975. 217
[18] F. Zhao, J. Shin, and J. Reich. Information-driven dynamic sensor collaboration for tracking applications. *IEEE Signal Processing Magazine*, 19(2):61–72, March 2002. 216

Improving End-to-End Delay through Load Balancing with Multipath Routing in Ad Hoc Wireless Networks Using Directional Antenna

Siuli Roy[1], Dola Saha[1], Somprakash Bandyopadhyay[1],
Tetsuro Ueda[2], and Shinsuke Tanaka[2]

[1] Indian Institute of Management Calcutta, Joka, Calcutta 700104, India
{siuli,dola,somprakash}@iimcal.ac.in
[2] ATR Adaptive Communications Research Laboratories
Kyoto 619-0288, Japan
{teueda,shinsuke}@atr.co.jp

Abstract. Multipath routing protocols are distinguished from single-path protocol by the fact that they use several paths to distribute traffic from a source to a destination instead of a single path. Multipath routing may improve system performance through load balancing and reduced end-to-end delay. However, two major issues that dictate the performance of multipath routing - *how many paths are needed* and *how to select these paths*. In this paper, we have addressed these two issues in the context of ad hoc wireless networks and shown that the success of multipath routing depends on the effects of route coupling during path selection. Route coupling, in wireless medium, occurs when two routes are located physically close enough to interfere with each other during data communication. Here, we have used a notion of *zone-disjoint routes* to minimize the effect of interference among routes in wireless medium. Moreover, the use of directional antenna in this context helps to decouple interfering routes easily compared to omni-directional antenna.

1   Introduction

Multipath routing protocols are distinguished from single-path routing by the fact that they look for and use several paths to distribute traffic from a source to a destination instead of routing all the traffic along a single path. Utilization of multiple paths to improve network performance, as compared to a single path communication, has been explored in the past [1, 2]. Classical multipath routing has focused on the use of multiple paths primarily for load balancing and fault tolerance. Load balancing overcomes the problem of capacity constraints of a single path by sending data traffic on multiple paths and reducing congestion by routing traffic through less congested paths. The application of multipath techniques in mobile ad hoc networks seems natural, as it

S. R. Das, S. K. Das (Eds.): IWDC 2003, LNCS 2918, pp. 225–234, 2003.
© Springer-Verlag Berlin Heidelberg 2003

may help to diminish the effect of unreliable wireless links, reduce end-to-end delay and perform load-balancing [2]. In addition, due to the power and bandwidth limitations, a routing protocol in ad hoc networks should fairly distribute the routing traffic among the mobile hosts. However, most of the current routing protocols in this context are single-path protocols and have not considered the load-balancing issue. An unbalanced assignment of data traffic will not only lead to congestion and higher end-to-end delay but also lead to power depletion in heavily loaded hosts. An on-demand multipath routing scheme is presented in [3], where alternate routes are maintained, so that they can be utilized when the primary one fails. However, the performance improvement of multipath routing on the network load balancing has not been studied extensively. The Split Multipath Routing (SMR), proposed in [6], focuses on building and maintaining maximally disjoint multiple paths.

Two key issues that dictate the performance of multipath routing are - *how many paths are needed* and *how to select these paths*. In this paper, we have addressed these two issues in the context of ad hoc wireless networks. It is shown that the performance of multipath routing through proper load balancing improves substantially, if we consider the effect of route coupling and use directional antenna instead of omni-directional antenna with each user terminal. In the context of ad hoc networks, the success of multipath routing depends on considering the effects of route coupling during path selection. In [5], the effect of route coupling on Alternate Path Routing (APR) in mobile ad hoc networks has been explored. It was argued that the network topology and channel characteristics (e.g., *route coupling*) can severely limit the gain offered by APR strategies. Route coupling is a phenomenon of wireless medium which occurs when two routes are located physically close enough to interfere with each other during data communication. As a result, the nodes in multiple routes constantly contend for access to the medium they share and can end up performing worse than a single path protocol. Thus, node-disjoint routes are not at all a sufficient condition for improved performance in this context.

In this paper, we use a notion of *zone-disjoint routes* in wireless medium where paths are said to be *zone-disjoint* when data communication over one path will not interfere with data communication along other paths. Our basic multipath selection criterion for load balancing depends on zone-disjointness. However, getting zone-disjoint or even partially zone-disjoint routes in ad hoc network with omni-directional antenna is difficult, since the communication zone formed by each transmitting node with omni-directional antenna covers all directions. Hence, one way to reduce this transmission zone of a node is to use directional antenna. It has been shown that the use of directional antenna can largely reduce radio interference, thereby improving the utilization of wireless medium and consequently the network throughput [7]. In our earlier work, we have developed the MAC and routing protocol using directional ESPAR antenna [7] and demonstrated the performance improvement. In this paper, we have investigated the effect of directional antenna on path selection criteria for multipath routing and obtained a substantial gain in routing performance through load balancing using multiple paths with directional antenna.

The paper is organized as follows. In section 2, we define the notion of zone disjointness and propose multipath selection criteria based on this notion. In section 3, we evaluate the performance of proposed mechanism in a simulated environment to show the effectiveness of our algorithm using directional antenna, followed by concluding remarks in section 4.

2   Selection of Paths for Multipath Routing

2.1   Zone Disjoint Routes with Omni-directional and Directional Antenna

Most of the earlier works on multipath routing in ad hoc networks try to find out multiple node-disjoint/ maximally node-disjoint paths between source s and destination d for effective routing with proper load balancing. [5,6,7]. Two (or multiple) paths between s and d are said to be node-disjoint, when they share no common node except s and d. However, because of route coupling in wireless environment, node-disjoint routes are not at all a sufficient condition for improved performance in this context. Suppose, two sources, s_1 and s_2 are trying to communicate data to destinations, d_1 and d_2 respectively. Let us assume that two node-disjoint paths are selected for communication- s_1-x_1-y_1-d_1 and s_2-x_2-y_2-d_2. Since the paths are node-disjoint, the end-to-end delay in each case should be independent of the other. However, if x_1 and x_2 and/or y_1 and y_2 are neighbors of each other, then two communications can not happen simultaneously (because RTS / CTS exchange during data communication will allow either x_1 or x_2 to transmit data packet at a time, and so on). So, the end-to-end delay between any source and destination does not depend only on the congestion characteristics of the nodes in that path. Pattern of communication in the neighborhood region will also contribute to this delay. This phenomenon of interference between communicating routes is known as *route coupling*. As a result, the nodes in multiple routes will constantly contend for access to the medium they share and can end up performing worse than a single path protocol.

In this paper, we use a notion of *zone-disjoint routes* in wireless medium where paths are said to be *zone-disjoint* when data communication over one path will not interfere with data communication along other paths. In other words, two (or multiple) paths between s and d are said to be zone-disjoint, when route-coupling between them is zero.

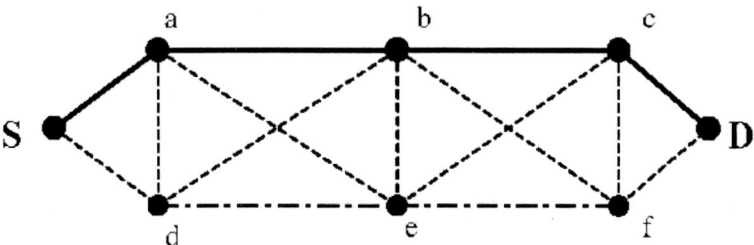

Fig. 1. Two node-disjoint path with $\eta = 9$

The effect of route coupling has been measured in [8] using a correlation factor η. In this paper, the correlation factor of a node n in a path P, $\eta^n (P)$, is defined as the number of *active neighbors* of n not belonging to path P, where *active neighbors* of n is defined as those nodes within the transmission zone of n that are actively partici-pating in any communication process at that instant of time. For example, in figure 1, S and D are communicating using two paths: S-a-b-c-D and S-d-e-f-D. So, all are active nodes in this context of communication. Now, the active neighbors of node a is {S, d,e,b}. So, correlation factor of node a in path {p= S-a-b-c-D}, η^a (p)= number of active neighbors not belonging to path p, i.e. 2.

The correlation factor η of path P, η (P), is defined as the sum of the correlation factor of all the nodes in path P. When η (P) =0, path P is said to be *zone-disjoint* with all other *active paths*, where active paths are those paths participating in communica-tion processes at that instant of time. Otherwise, the path P is η–related with other active paths.

Route coupling has a serious impact on path selection for load balancing via multi-ple path. Let us refer figure 1 and assume that source S is sending data traffic to desti-nation D along the path {S,a,b,c,D}. If S selects another path {S,d,e,f,D}, which is closely coupled with the first path (as shown), and tries to distribute traffic across both the path for load balancing, it may not result in performance improvement. In fact, it has been observed that larger the correlation factor, the larger will be the average end-to-end delay for both paths [8]. This is because two paths with larger correlation factor have more chances to interfere with each other's transmission due to the broadcast feature of radio propagation. In addition, larger the correlation factor, the larger will be the difference of end-to-end delay along multiple paths [8]. Based on this study, it can be concluded that the path selection criterion for multipath routing in ad hoc net-work needs to consider the correlation factor among multiple routes. In an environ-ment of multiple communication among several source-destination pairs, even if a path is less-loaded, that path may not be a good candidate for distributing traffic, if the route coupling of that path with respect to other active paths is high. One way to alle-viate this problem is to use zone-disjoint routes or maximally zone disjoint route for load balancing. However, it is difficult to get fully zone-disjoint routes using omni-directional antenna. As in figure 1, since both a and d are within omni-directional transmission range of S, a RTS from S to node a will also disable node d. Similarly, since both c and f are within omni-directional transmission range of D, a CTS from D will disable both c and f. So, even if {a,b,c} and {d.e.f} are zone-disjoint, the lowest possible η in case of omni-directional antenna with two multipath between s and d is 2. We call it minimal correlation factor η^{min}. With directional antenna, it is possible to de-couple these two routes, making them fully zone-disjoint. For example, if each of the nodes in figure 1 uses directional antenna where each node sets their transmission beam towards its target node only, then the communication between S-a-b-c-D will not affect the communication between S-d-e-f-D. Hence η^{min}(omni)=2 whereas η^{min}(dir)=0. This will be further illustrated in the next section.

2.2   Number of Paths in a Multipath Route

Even if we get multiple zone-disjoint routes with minimal correlation factor [η^{min}(omni)=2] using omni-directional antenna, the best-case packet arrival rate at the destination node will be 1 packet at every $2*t_p$, where t_p is the average delay per hop per packet of a traffic stream on the path p. The best-case assumption is, traffic stream in the network from S to D only with error-free transmission of packets. In contrast, if we use directional antenna, best-case packet arrival rate at destination will be one packet at every t_p. It was illustrated analytically in [10] that the destination D will receive packets in alternate time-tick with omni-directional antenna and *even if we increase the number of paths between s and d beyond 2, the situation will not improve.*

However, with directional antenna, when node a is transmitting a packet to node b, S can transmit a packet to node d simultaneously. Thus, destination D will receive a packet at every time-tick with two zone-disjoint paths using directional antenna. It is to be noted here that *two zone-disjoint paths with directional antenna are sufficient to achieve this best-case scenario* [10].

2.3   Selection of Paths Based on Correlation Factor η

Till now, we have considered communication over single s-d pair. However, situation will deteriorate, if we consider multiple s-d pairs, engaged in simultaneous communications. Let us assume that each s-d pair selects two paths between them with lowest possible η between them for effective load balancing. However, in the context of multiple s-d pairs, even if two multipaths between, say, s_1 and d_1 are zone disjoint, they may be coupled with other active routes between, say, s_2 and d_2. So, it is imperative to consider all active routes and to find out η for each of them with respect to other active routes in order to determine *maximally zone-disjoint multipath* between a s-d pair such that it is not only maximally zone-disjoint with respect to each other but also with respect to all active routes in the system.

However, it is a difficult task in the dynamic environment of ad hoc networks with changing topology and communication pattern. An approximate solution to alleviate this difficulty will be discussed in the next section. In this section, we will discuss the mechanism of finding maximally zone disjoint multipaths with multiple s-d pairs and, in the next section, will show the effectiveness of directional antenna over omni-directional antenna in this context. To do this, initially we have assumed a static scenario in our simulation environment. It has been assumed that each node is aware of the topology and the communication pattern in the network. We use the following algorithm to find out *maximally zone-disjoint path between s-d:*

Step I: Find out all node-disjoint paths between a s-d pair with number of hops H less than H_{max} (=5 in this experiment).

Step II: Find out η for each path between that s-d pair with respect to *other active paths.*

Step III: Discard the path with highest η.

Step IV: Repeat the process from Step II to step III with remaining paths between that s-d pair until number of paths between them is two. These two paths are *maximally zone-disjoint path between that s-d pair.*

2.4   Additional Criterion for Path Selection: Hop Count

However, zone disjointness alone is not sufficient for performance improvement. Path length is also another important factor in multipath routing. A longer path with more number of hops (H) will increase the end-to-end delay and waste more bandwidth. So, even if a longer bypass route between a s-d pair has a low η, it may not be very effective in reducing end-to-end delay. To deal with this problem, our route-selection criteria would be to minimize the product of η and H. Minimizing this factor will result in *maximally zone-disjoint shortest path*. We call this factor γ ($=\eta*H$). We use the following algorithm to find out *maximally zone-disjoint shortest path between s-d:*

Step I: Find out all node-disjoint paths between a s-d pair with number of hops H less than H_{max} (=5 in this experiment).

Step II: Find out η for each path between that s-d pair with respect to other active paths.

Step III: Find out γ for each path between that s-d pair.

Step IV: Discard the path with highest γ

Step V: Repeat the process from Step II to step IV with remaining paths between that s-d pair until number of paths between them is two. These two paths are *maximally zone-disjoint shortest path between that s-d pair.*

3   Multipath Routing Performance

The proposed mechanism has been evaluated on a simulated environment under a variety of conditions to estimate the basic performance. In the simulation, the environment is assumed to be a closed area of 1500 x 1000 square meters in which mobile nodes are distributed randomly. We present simulation results for networks with 40 mobile hosts, operating at a transmission range of 350 m. In order to evaluate the effect of changing topology due to mobility, several snap-shots of random topology with varying source-destination pairs are considered during our experiments. The effective width of directional beam in case of directional antenna is assumed to be 60°.

In order to implement any routing protocol using directional antenna, each node should know the best possible directions to communicate with its neighbors. So, each node periodically collects its neighborhood information and forms a Neighborhood-Link-State Table (NLST) at each node [7]. Through periodic exchange of this NLST with its neighbors each node becomes aware of the approximate global network topology and NLST at each node is upgraded to GLST (Global Link State Table). A directional MAC protocol, as discussed in our earlier work [7], has been implemented in our simulator using information kept in GLST. Implementation of omni-directional MAC follows the basic IEEE 802.11 scheme. A modified link-state routing protocol, based on our earlier work [7,9] has been implemented. In the context of directional antenna, GLST not only depicts the connectivity between any two nodes but also maintains the best possible directions to communicate with each other. Moreover, each node periodically propagates its knowledge about *active-node-list*, a list containing the node-ids of all nodes involved in any communication process at that instant of time. It is to be noted that the perception of each node about the network topology or number of active nodes in the network is only approximate. However, periodic re-computation of routes by each intermediate node on a path will adaptively adjust itself to the changing scenario.

Whenever a source s wants to communicate with a destination d, it computes multi-ple node-disjoint routes from s to d. From these multiple routes, it consults active node list and computes *maximally zone-disjoint multipath,* or, *maximally zone-disjoint shortest multipath* between s-d (as illustrated in previous section). However, due to mobility and slow information percolation, it may not be possible for a source node to perfectly compute maximally zone-disjoint multipath between s-d. To improve per-formance under this condition, each intermediate node periodically recomputes the same and adaptively modifies its routing decision.

We have compared the performance of (i) unipath routing with shortest path using omni-directional antenna, (ii) maximally zone disjoint multipath with directional and omni-directional antenna, and (iii) maximally zone disjoint *shortest* multipath with directional and omni-directional antenna. In order to evaluate the impact of of-fered load, we have experimented with 5, 10, 15 and 20 simultaneous communication. Observations are recorded for 20 snap-shots in each case and the average values of different parameters are computed. We evaluate the performance according to the following metrics:

Load balancing: To measure load balancing in each case, we observe the number of data packets forwarded at each node n [8]. If f(n) represents the number of data pack-ets forwarded at each node n, the load balancing factor is the ratio of standard devia-tion of f / mean of f, taken over all 40 nodes. Smaller the load balancing factor, better the load balancing [8].

Average end-to-end delay: Average end-to-end delay per packet between a set of selected s-d pairs is observed with increasing number of simultaneous communications and with omni-directional and directional antenna. The timing assumptions are the same as indicated in section 2.2.

Initially, to analyze average route-coupling among active routes, the experiment starts with finding *maximally zone-disjoint paths* between selected s-d pairs. In order to observe the impact of multiple simultaneous communications on route coupling factor η, number of simultaneous communications are taken as 5, 10, 15 and 20. In each case, we have found out average η, using omni-directional and directional an-tenna. The result (figure 2) shows that the increase of η is much sharper in case of omni-directional antenna. This implies that, as the number of s-d pair increases in the system, the route-coupling among active routes increase much more sharply in case of omni-directional antenna compared to that with directional antenna. This has an im-pact on end-to-end delay, as will be illustrated later.

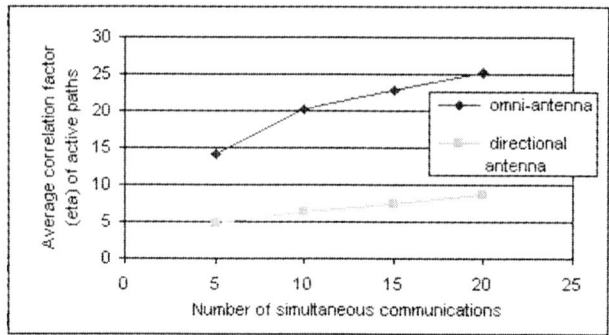

Fig. 2. Increase in route coupling with multiple multipath communications with omni-directional and directional antenna

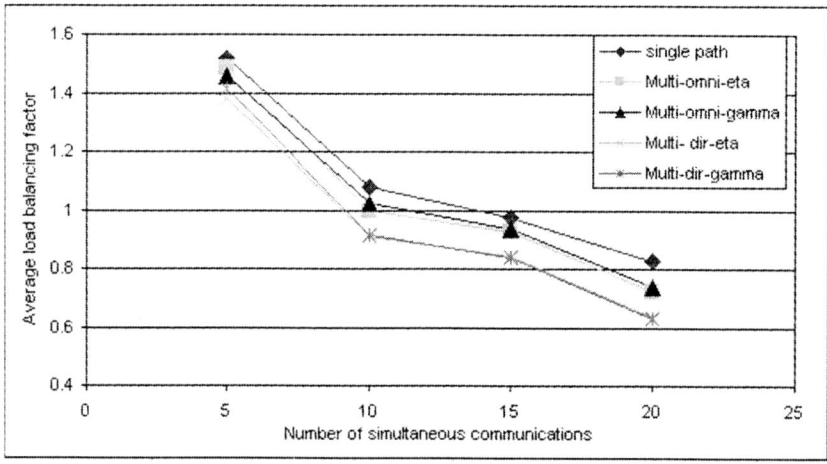

Fig. 3. Variation of load balancing factors with increasing number of simultaneous communications

As illustrated in figure 3, load balancing improves with increasing load with *maximally zone disjoint multipaths,* as compared to that with single shortest path. This improvement is more pronounced when we use directional antenna. It is to be noted that, smaller the load balancing factor, better the load balancing.

However, better load balancing does not imply better performance in this context. Because of the possibility of high route coupling with omni-directional antenna (as shown in figure 2), especially with increased number of simultaneous communication, average end-to-end delay using multipath with omni-directional antenna will not show any significant improvement as compared to that with single path. Since route coupling is far less with directional antenna, average end-to-end delay will be substantially less with directional antenna than that with omni-directional antenna. This is shown in figure 4. At the same time, path length is also another important factor in multipath routing. A longer path with more number of hops (H) will increase the end-to-end delay and waste more bandwidth. So, even if a longer bypass route between a s-d pair has a low η, it may not be very effective in reducing end-to-end delay. That is why, *maximally zone disjoint shortest path with directional antenna* will show best performance, so far as both end-to-end delay and load balancing are concerned (figure 4).

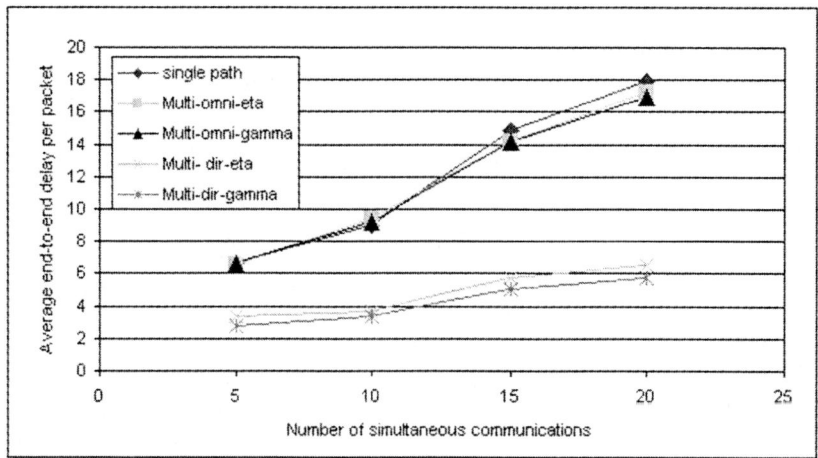

Fig. 4. Variation of Average End-to-End Delay per packet with increasing number of simultaneous communications

[*Multi-omni-eta*: Multipath communication with maximally zone-disjoint path with omni-directional Antenna

Multi-omni-gamma: Multipath communication with maximally zone-disjoint shortest path with omni-directional Antenna

Multi-dir-eta: Multipath communication with maximally zone-disjoint path with directional Antenna

Multi-dir-gamma: Multipath communication with maximally zone-disjoint shortest path with directional Antenna]

4   Conclusion

Multipath routing strategies in the context of ad hoc networks may improve load balancing, but may not improve system performance to the expected level through reduced end-to-end delay, unless we consider the effects of route coupling. However, high degree of route coupling among multiple routes between any source and destination pair is inevitable, if we use omni-directional antenna. The situation will worsen, if we consider multiple simultaneous communications with multipath routing scheme. This paper has analyzed the problem and proposed a mechanism to alleviate the problem of route coupling using directional antenna. The paper also considers the advantage of selecting maximally zone-disjoint as well as shorter route instead of longer by-pass routes for effective load balancing and better network performance. Thus, all active paths are maximally zone disjoint shortest paths. The final result shows that the routing performance using multiple paths improves substantially with directional antenna compared to that with omni-directional antenna.

Acknowledgement

This research was supported in part by the Telecommunications Advancement Organization of Japan.

References

[1]   Sajal K. Das, A. Mukherjee, Somprakash Bandyopadhyay, Krishna Paul, D. Saha, "Improving Quality-of-Service in Ad hoc Wireless Networks with Adaptive Multi-path Routing, Proc. Of the GLOBECOM 2000, San Francisco, California, Nov. 27-Dec 1, 2000.

[2]   Aristotelis Tsirigos Zygmunt J. Haas, Siamak S. Tabrizi , Multi-path Routing in mobile ad hoc networks or how to route in the presence of frequent topology changes , MILCOM 2001.

[3]   Nasipuri and S.R. Das, "On-Demand Multi-path Routing for Mobile Ad Hoc Networks," Proceedings of IEEE ICCCN'99, Boston, MA, Oct. 1999.

[4]   Johnson and D. A. Maltz, "Dynamic Source Routing in Ad Hoc Wireless Networks," T. Imielinski and H. Korth, editors, *Mobile Computing*, Kluwer, 1996.

[5]   M. R. Pearlman, Z. J. Haas, P. Sholander, and S. S. Tabrizi, On the Impact of Alternate Path Routing for Load Balancing in Mobile Ad Hoc Networks, MobiHOC 2000, p. 150, 3-10.

[6]   S.J. Lee and M. Gerla, Split Multi-path Routing with Maximally Disjoint Paths in Ad Hoc Networks, ICC 2001.

[7]   Siuli Roy, Dola Saha, S. Bandyopadhyay, T. Ueda, S. Tanaka. A Network-Aware MAC and Routing Protocol for Effective Load Balancing in Ad Hoc Wireless Networks with Directional Antenna. *ACM MobiHoc,2003,* Maryland, USA, 1-3 June 2003.

[8]   Kui Wu and Janelle Harms, On-Demand Multipath Routing for Mobile Ad Hoc Networks EPMCC 2001, Vienna, 20th – 22nd February 2001.

[9]   Romit RoyChouldhury, Somprakash Bandyopadhyay and Krishna Paul, "A Distributed Mechanism for Topology Discovery in Ad hoc Wireless Networks using Mobile Agents", Proc. of the First Annual Workshop On Mobile Ad Hoc Networking & Computing (MOBIHOC 2000), Boston, Massachusetts, USA, August 11, 2000.

[10]   Somprakash Bandyopadhyay, Siuli Roy, Tetsuro Ueda, Shinsuke Tanaka "Multipath Routing in Ad Hoc Wireless Networks with Omni Directional and Directional Antenna: A Comparative Study" Proc. of the IWDC, Calcutta, December 27-30,2003, (LNCS 2571), Springer Verlag , Dec 2002.

A Power-Efficient MAC Protocol
with Two-Level Transmit Power Control
in Ad Hoc Network Using Directional Antenna

Dola Saha[1], Siuli Roy[1], Somprakash Bandyopadhyay[1],
Tetsuro Ueda[2], Shinsuke Tanaka[2]

[1] Indian Institute of Management Calcutta
Diamond Harbour Road, Joka Calcutta 700104, India
{dola,siuli,somprakash}@iimcal.ac.in
[2] ATR Adaptive Communications Research Laboratories
2-2-2 Hikaridai, Keihanna Science City, Kyoto 619-0288, Japan
{teueda,shinsuke}@atr.co.jp

Abstract. The use of directional antenna in wireless ad hoc networks largely reduces radio interference, thereby improving the utilization of wireless medium and consequently the network throughput, as compared to omni-directional antenna, where nodes in the vicinity of communication are kept silent. In this context, researchers usually assume that the gain of directional antennas is equal to the gain of corresponding omni-directional antenna. However, for a given amount of input power, the range R with directional antenna will be much larger than that using omni-directional antenna. In this paper, we propose a two-level transmit power control mechanism in order to approximately equalize the transmission range R of an antenna operating at omni-directional and directional mode. This will not only improve medium utilization but also help to conserve the power of the transmitting node during directional transmission. The performance evaluation on QualNet network simulator clearly indicates the efficiency of our protocol.

1   Introduction

Usually, in ad hoc networks, all nodes are equipped with omni-directional antenna. However, ad hoc networks with omni-directional antenna uses RTS/CTS based floor reservation scheme that wastes a large portion of the network capacity by reserving the wireless media over a large area. Consequently, lot of nodes in the neighborhood of transmitter and receiver has to sit idle, waiting for the data communication between transmitter and receiver to finish. To alleviate this problem, researchers have proposed to use directional (fixed or adaptive) antennas that direct the transmitting and receiving beams toward the receiver and transmitter node only. This would largely reduce radio interference, thereby improving the utilization of wireless medium and conse-

S. R. Das, S. K. Das (Eds.): IWDC 2003, LNCS 2918, pp. 235–244, 2003.
© Springer-Verlag Berlin Heidelberg 2003

quently the network throughput [1-6]. As shown in Fig. 1, while node n is communicating with node m using omni-directional antenna, node p and r have to sit idle.
However, with directional beam forming, while node n is communicating with node
m, both node p and r can communicate with node q and s respectively, improving the
medium utilization or the SDMA (space division multiple access) efficiency. We can
even improve this SDMA efficiency by controlling power of directional transmission
to make the directional transmission range almost equal to omni-directional transmission range. Due to the high gain of the main lobe of directional antenna, the directional transmission range is much larger than that of omni-directional transmission
range. So, directional transmission creates unnecessary interference in the area beyond
the omni-directional transmission range and SDMA suffers. By almost equalizing the
omni- and directional transmission range, drastic improvement in medium utilization
and SDMA efficiency can be achieved resulting in improved average throughput. In
Fig. 1, the power controlled directional transmission range is shown in dotted lines.
With power controlled directional transmission of node n, nodes x, y or z can start
communication improving SDMA efficiency, which was not possible with full power
directional transmission of node n.

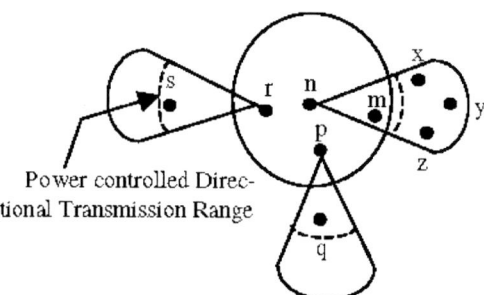

Fig. 1. Improving SDMA efficiency with Directional Antenna and power-controlled Directional Transmission

2   Related Work

In spite of the advantages of directional antennas, work on developing efficient MAC
protocol using directional antennas in the context of ad hoc networks is limited because of the inherent difficulty to cope up with mobility and de-centralized control in
ad hoc networks. Some researchers have tried to address this challenge in several
ways. In recent years, several MAC protocols that rely on RTS-CTS type handshaking
as in IEEE 802.11 have been suggested with directional antennas.

In [1], a directional MAC scheme has been proposed where directional or omni-
directional RTS is sent depending on the on-going communication in the vicinity. In
[2], a MAC protocol to achieve multihop efficiency has been proposed with multihop-
RTS-singlehop-CTS using directional antenna. In this mechanism, using larger range
of directional beam, a destination is reachable in less number of hops as compared to
that using omni directional antenna. In both the schemes [1-2], the mobile nodes are

assumed to know the physical locations of themselves and their neighbors using GPS. With the use of directional RTS and directional CTS, several issues like a *new hidden terminal problem* due to asymmetry in gain & due to unheard RTS/CTS, *deafness* and *higher directional interference,* as depicted in [2], remains unsolved. In [3], the proposed MAC protocol need not know the location information; the source and destination nodes identify each other's direction during omni-directional RTS-CTS exchange in an on-demand basis. It is assumed that all the neighbors of s and d, who hear this RTS-CTS dialog, will use this information to prevent interfering with the ongoing data transmission. However, because of omnidirectional transmission of RTS and CTS packets, this protocol provides no benefits in the spatial reuse of the wireless channel. However, it still improves the throughput over a MAC using omnidirectional antennas due to the reduced amount of interference caused by the directional data transmission. In our earlier work, we have developed a MAC protocol [4], where each node keeps certain neighborhood information dynamically through the maintenance of an Angle-SINR Table. In this method, in order to form AST, each node periodically sends a directional beacon in the form of a directional broadcast, sequentially in all direction at 30 degree interval, covering the entire 360 degree space. The nodes, which receive these signals at different angles, determine the best received signal strength and transmit the information back to the source node as data packet with RTS/CTS handshake. However, the overhead due to control packets is very high in this method [4].

In this paper, we will illustrate a receiver-oriented location tracking mechanism to reduce the control overhead and a simple MAC protocol for efficient medium utilization. On this directional MAC, we will show that power controlled directional transmission is a necessity and it improves network throughput by conserving power. We have done extensive performance evaluation using QualNet to demonstrate its effectiveness.

3   System Description

3.1   Antenna Model

There are basically two types of smart antennas used in the context of wireless networks: switched-beam or fixed beam antennas and steerable adaptive array antennas. A switched-beam antenna generates multiple pre-defined fixed directional beampatterns and applies one at a time when receiving a signal. In a steerable adaptive array antenna which is more advanced than a switched beam antenna, the beam structure adapts to Radio Frequency (RF) signal environment and directs beams towards the signal of interest to maximize the antenna gain, simultaneously depressing the antenna pattern (by setting nulls) in the direction of the interferers.

We have developed a wireless ad hoc network testbed using smart antenna [5] where each user terminal uses a small, low-cost smart antenna, known as ESPAR (Electronically Steerable Passive Array Radiator) antenna. The ESPAR antenna consists of one center element connected to the source (the main radiator) and several surrounded parasitic elements (typically four to six passive radiators) in a circle (Fig. 2).

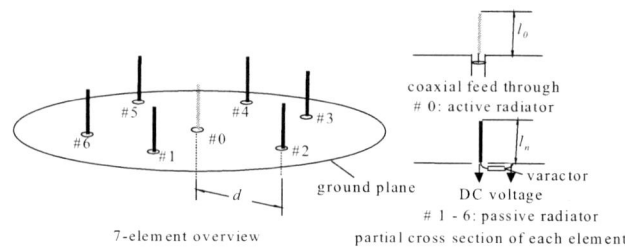

Fig. 2. Configuration of ESPAR antenna

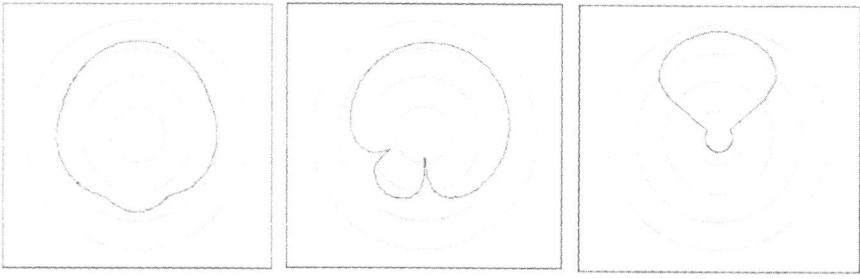

(a) ESAPR pattern at 0 degree   (b) ESAPR pattern at 30 degree   (c) Ideal Directional Antenna

Fig. 3. Different Directional Antenna Pattern Used in our Simulation

Developing suitable MAC protocols with adaptive antenna in ad hoc networks is a challenging task. That is why, most of the works in the context of ad hoc networks assume to use simpler switched beam antenna. In this work also, we are using smart ESPAR antenna as a switched beam antenna. ESPAR antenna can also be used as a generalized switched beam antenna or quasi-switched beam antenna [5], by selecting the value of reactance for one specific directional beam among multiple directional beam patterns, without using multiple receiver chains (frequency converters and analog-digital converters). Since ESPAR antenna would be a low-cost, low-power, small-sized antenna, it would help to reduce the power consumption of the user terminals in WACNet and would be able to deliver all the advantages of switched beam antenna.

The antenna pattern of ESPAR antenna with 60 degree beam width is shown in Fig. 3(a) and 3(b). Fig. 3(a) shows pattern at 0 degree: a beam pattern formed at each antenna element at an interval 0 to 60 degree, 60 to 120 degree and so on, thus forming 6 beams. Fig. 3(b) shows pattern at 30 degree: a beam pattern formed at each in-between antenna elements at an interval 30 degree to 90 degree, 90 degree to 150 degree and so on, thus forming 6 more pattern. Together they constitute 12 overlapping pattern at 30 degree intervals. Fig. 3(c) shows an ideal directional antenna with 45 degree beam-width with insignificant side-lobes. As will be demonstrated in performance evaluation, the performance of ideal directional antenna is the best (as expected); at the same time, ESPAR performance is also comparable to that of ideal directional antenna.

3.2   A Few Assumptions and the Rationales

When the antenna of a node operating in omni-directional mode, it is capable of transmitting and receiving signal from all direction with a gain, say, G^{omni}. While idle, a node operates in omni-directional receive mode.

When the antenna of a node operating in directional mode, a node can points its beam (main lobe) towards a specified direction with a beam width w and with a gain, say G^{dir} ($G^{dir} >> G^{omni}$).

Consequently, for a given amount of input power, the transmission range R^{dir} with directional antenna will be much larger than that with corresponding omni-directional antenna (R^{omni}).

4   A Directional MAC with Location Tracking Mechanism

In this work, our MAC protocol is basically a *Receiver-oriented, Rotational Sector Based Directional MAC protocol* which also serves as a Location Tracking mechanism. Here, each node waits in omni-directional-sensing-mode while idle. Whenever it senses some signal above a threshold, it enters into *rotational-sector-receive-mode*. In rotational sector receive mode, node n rotates its directional antenna sequentially in all direction at 30-degree interval, covering the entire 360-degree space in the form of the sequential directional receiving in each direction and senses the received signal at each direction. After one full rotation, it decides the best possible direction of receiving the signal with maximum received signal strength. Then it sets its beam to that direction and receives the signal.

However, in order to enable the receiver decoding the received signal, each control packet is transmitted with a preceding tone with a duration such that the time to rotate a receiver's rotational receive beam through 360 degree is little less than the duration of the tone (200 microseconds in our case). The purpose of this transmitted tone before any control packet is to enable the receiver to track the best possible direction of receiving the signal. Once it sets its beam to that direction, the purpose of tone signal is over and subsequently the control packet is transmitted.

In this proposed framework, we have used three types of omni-directional control packets: Beacon, RTS (Request to send) and CTS (clear to send) for medium access control. Another control packet ACK is directional control packet. Data is transmitted directionally after RTS/CTS handshaking is done. Beacon is a periodic signal, transmitted from each node at a pre-defined interval. As indicated earlier, beacon is transmitted with a preceding tone signal that helps the receivers to detect the best possible direction of receiving the signal. Then each receiver sets its beam to that direction and receives and decodes the packet. Since RTS is a broadcast packet and contains source address, nodes can decode that RTS also to form the Angle-Signal Table. So, we have used RTS as beacon. If an RTS is sent, beacon timer is reset. The use of RTS as beacon is advantageous at high traffic where overhead due to beacon is minimized. This is because, the transmitting nodes don't have to send an additional beacon to inform its neighbors of its presence.

Whenever node n wants to start data communication with, say j, it checks the medium and if it is free, n issues an omni-directional RTS. The target node j on receiving RTS, issues omni-directional CTS. The objective of RTS/CTS here is not to inhibit the neighbors of n and j from transmitting or receiving (as is the case with omni-directional antenna) but to inform the neighbors of j and n that j is receiving data from n. It also specifies the approximate duration of communication. All the neighboring nodes of n and j keep track of the communication between n and j by setting their Directional Network Allocation Vector (DNAV) towards n and j. Thus, nodes in the neighborhood of n and j can initiate communication in other directions *without disturbing the existing communication between n and j.* The source and destination nodes wait for Acknowledgement and Data respectively in directional receive mode.

5   Power Controlled Directional MAC Protocol

Most researchers [6-7] used power control schemes, which suitably vary transmit power to reduce energy consumption. But, this scheme has a shortcoming, which increases collisions and degrades network throughput, as pointed out in [7]. So, Jung et. al. [7] proposed to transmit each data packet at maximum power level periodically, for just enough time, so that nodes in carrier sensing zone can sense it. This work has been implemented using omni-directional antenna. But the scenario is completely changed when we use directional antenna to transmit and receive signals. In [6], RTS/CTS handshake at full power is used to decide transmission power for subsequent data and acknowledgement packets. But the marked difference between omni- and directional antenna gains has not been taken into account. So, the concept of controlling power, as suggested in [6] does not work in real scenario, which is illustrated below.

In order to demonstrate the advantage of using directional antenna in gaining SDMA efficiency, researchers usually assume that the gain of directional antennas is equal to the gain of corresponding omni-directional antenna. Under this assumption, it is easy to visualize improvement in SDMA efficiency: the coverage area or the area of the *transmission_zone_n* (α) of a node n forming a transmission beam with a beam-angle α ($\alpha << 360°$) and a transmission range R with respect to n is $\alpha R^2/2$ which is much low when $\alpha = 45$ degree(say) compared to that when $\alpha = 360$ degree (i.e. omni-directional). If the transmission zone of a node is small, it implies that the number of transmission zones that can be formed in a given area by a given number of nodes is high, giving rise to higher SDMA efficiency. However, in real situation, this assumption is invalid. For a given amount of input power, the range R of a user terminal using directional antenna will be much larger than that using omni-directional antenna. This implies that, narrower the beam-width, higher would be the gain of main lobe of the directional antenna and consequently the range would be larger. So, the transmission range R is not same in both the cases and is inversely proportional to α. Consequently, SDMA efficiency will not improve as much as it is expected. For example, in Fig. 1, nodes x, y and z are outside the omni-directional transmission range of node n, but in the directional transmission range of node n. So, even if they don't receive RTS from node n, they are captured by the directional transmission of node n and cannot start a communication in other directions. On the other hand, since node m is within omni-directional transmission range of node n, proper reception of signal at node m does not require higher directional transmission range of node n towards the direction of node m. If we reduce the power to reduce the directional transmission range, as shown in dotted line, nodes x, y and z can start communication in directions other than the direction of node m which is blocked by DNAV. Thus, SDMA efficiency is improved resulting in increased throughput.

In this paper, we study power control for the purpose of improving SDMA efficiency and as a result energy consumption is minimized. We propose a two-level transmit power control mechanism in order to approximately equalize the transmission range R of an antenna operating at omni-directional and directional mode. In other words, if P is the full power used during omni-directional transmission, a reduced power level p will be used during directional transmission so as to equalize the range of transmission approximately in both the cases. This will not only improve the SDMA efficiency but also help to conserve the power of the transmitting node during directional transmission of data. In this scheme, control packets like beacon, RTS and CTS are omni-directional and use full power P for their transmission. On the other hand, directional transmission of ACK (Acknowledgement) packets and data packets are done with reduced power p.

Since directional transmission range depends mainly on antenna pattern and the gain of its main lobe, the reduced power p will be different in different antenna patterns. If a node knows with which antenna it is equipped with, it can control its power accordingly during directional transmission to approximately equalize its directional transmission range with omni-directional transmission range.

6   Performance Evaluation

6.1   Simulation Environment

The simulations are conducted using QualNet 3.1 [8]. We have simulated ESPAR antenna in the form of a *quasi-switched beam antenna*, which is steered discretely at an angle of 30 degree, covering a span of 360 degree. We have simulated our MAC protocol with (i) Simulated ESPAR Antenna Pattern (ESPAR) and (ii) an Ideal directional antenna pattern without sidelobes (IDEAL) as described in Section 3.1.

We have used simple one-hop randomly chosen communication in order to avoid the effects of routing protocols to clearly illustrate the difference between 802.11 and our proposed MAC. Also, we have used static routes to stop all the packets generated by any routing protocol, whether it is proactive or reactive. In our simulation, we studied the performance of the proposed MAC protocol in comparison with the existing omnidirectional 802.11 MAC protocol by varying the Data Rate and number of simultaneous communications. In studying our MAC protocol, we have used different antenna patterns as described above to ensure the robustness of our proposed MAC protocol. In doing this, we have used ESPAR antenna as one of the antenna patterns, to evaluate the performance of the ESPAR antenna as well.

The set of parameters used are listed in Table I.

Table 1. Parameters used in Simulation

Parameters	Value
Area	1000 x 1000 m
Number of nodes	40
Transmission Power	15 dBm
Receiving Threshold	-81.0 dBm
Sensing Threshold	-91.0 dBm
Data Rate	2Mbps
Packet Size	512 bytes
Duration of Preceding Tone	200 microseconds
CBR Packet Injection Interval	2 ms to 50 ms
Number of simultaneous communication	4 to 12
Simulation Time	5 minutes

6.2   Results and Discussions

We have used the existing IEEE 802.11 MAC, which we caption as "802.11", as a benchmark to compare and evaluate the performance of our proposed MAC protocol with ESPAR antenna and an ideal antenna, which we caption as "ESPAR" and "IDEAL" respectively. The average Throughput and one-hop average End-to-End Delay is evaluated in random scenarios with increasing data rate (Fig. 4) and with increasing number of simultaneous communications (Fig. 5).

In Fig. 4, it is seen that with increasing data rate, average Throughput of our proposed MAC protocol (E-MAC) with any directional antenna pattern is much better than that of IEEE 802.11. Also, one hop average End-to-End Delay of E-MAC is nearly half of that obtained with IEEE 802.11 protocol. In Fig. 5, it is observed that with increasing number of simultaneous communication, average Throughput decreases in both E-MAC and 802.11, but E-MAC shows significant gain in average Throughput. Also, one hop average End-to-End Delay increases in both IEEE 802.11 and E-MAC, but the increase is much prominent in "802.11" than in E-MAC, irrespective of the directional antenna pattern used. This is because E-MAC does not inhibit neighboring nodes to transmit, but just informs neighbors of the ongoing communication and its direction, so that they can start communication in other directions. Thus, with directional transmission and directional reception, E-MAC performs much better in any random scenario with any directional antenna pattern.

By tuning the directional transmission power, it has been observed that if directional transmission is done with 55% (in case of ESPAR) and 20% (in case of IDEAL) of the full power, the directional transmission range nearly equals to omni-directional transmission range. The results in Fig. 6 show that by controlling power, less transmission energy is consumed than that with full transmission power and corresponding average throughput increases with both ESPAR and IDEAL directional antenna. This is mainly due to SDMA efficiency achieved by reducing the directional transmission power and almost equalizing directional and omni-directional transmission range.

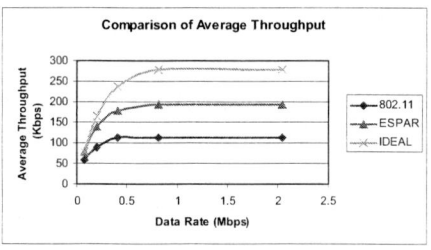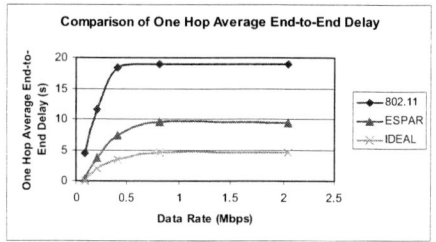

Fig. 4. Performance Evaluation of the proposed MAC protocol with directional antenna with increasing data rate

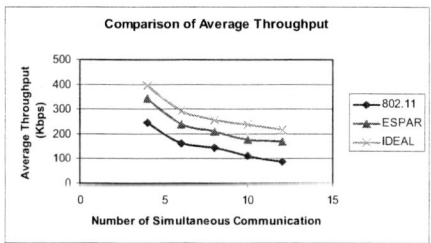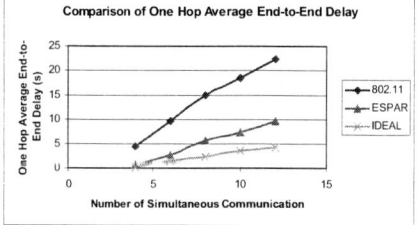

Fig. 5. Performance Evaluation of the proposed MAC protocol with directional antenna with increasing number of simultaneous communication

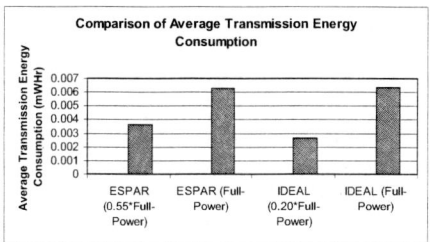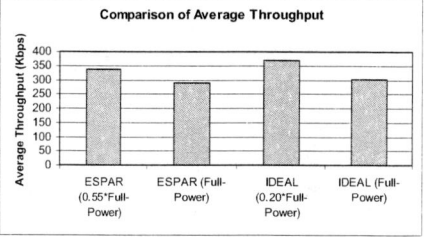

Fig. 6. Comparison of Average Transmission Energy Consumption and Average Throughput with and without controlling transmission power

7   Conclusion

Use of directional antenna in ad hoc wireless network can drastically improve system performance, if proper MAC protocol can be designed. The success of the MAC protocol highly depends on the directional antenna pattern. Currently, the ESPAR antenna is under development, where we are trying to modify the beam-pattern to get maximum gain from E-MAC. The location tracking mechanism as done in our proposed MAC protocol can be utilized in designing efficient Routing protocol. Presently, we are working on efficient controlling of transmission power dynamically to improve the proposed MAC performance.

Acknowledgements

This research was supported in part by the Telecommunications Advancement Organization of Japan.

References

[1]   Y.B. Ko, V. Shankarkumar and N. H. Vaidya, ``Medium access control protocols using directional antennas in ad hoc networks,'' Proc. Of the IEEE INFOCOM 2000, March 2000.

[2]   Romit Roy Choudhury, Xue Yang, Nitin H. Vaidya, Ram Ramanathan, "Using directional antennas for medium access control in ad hoc networks" Proceedings of the eighth annual international conference on Mobile computing and networking September 2002

[3]   Nasipuri, S. Ye, J. You and R.E. Hiromoto, "A MAC Protocol for Mobile Ad Hoc Networks Using Directional Antennas", Proc of the IEEE WCNC 2000.

[4]   S. Bandyopadhyay, K. Hasuike, S. Horisawa, S. Tawara, "An Adaptive MAC Protocol for Wireless Ad Hoc Community Network (WACNet) Using Electronically Steerable Passive Array Radiator Antenna", Proc of the GLOBECOM 2001, November 25-29, 2001, San Antonio, Texas, USA

[5]   T. Ueda, K. Masayama, S. Horisawa, M. Kosuga, K. Hasuike, "Evaluating the Performance of Wireless Ad Hoc Network Testbed With Smart Antenna", Fourth IEEE Conference on Mobile and Wireless Communication Networks (MWCN2002), September 2002

[6]   Asis Nasipuri, Kai Li, and Uma Reddy Sappidi, "Power Consumption and throughput in Mobile Ad Hoc Networks using Directional Antennas" in Proceedings of the *IEEE International Conference on Computer Communication and Networks (ICCCN2002)*, October 14-16, 2002, Miami, Florida.

[7]   Eun-Sun Jung and Nitin Vaidya, "A Power Control MAC Protocol for Ad Hoc Networks", ACM International Conference on Mobile Computing and Networking (MobiCom), September 2002.

[8]   *QualNet Simulator Version 3.1*, Scalable Network Technologies, www.scalable-networks.com

Achieving Routing Based Medium Access Fairness in Wireless Ad-Hoc Networks

Hend Koubaa[1] and Farid Naït-Abdesselam[2]*

[1] Mobile Distributed Information Systems Group
Fraunhofer IPSI, Darmstadt, Germany
`hend.koubaa@ipsi.fraunhofer.com`
[2] LIFL/IRCiCA Laboratory - INRIA POPS Project
University of Sciences and Technologies of Lille, France
`farid.nait@lifl.fr`

Abstract. Ad-hoc networks are dynamic multi-hop wireless networks that are established by a set of mobile nodes on shared wireless channel. In that way, each mobile node becomes potentially a router and it is possible to dynamically establish routes between itself and nodes to which a route exists. Because each node, in addition to its own packets, has to forward packets belonging to other nodes, selfish behavior may represent a significant advantage for a node to save its battery power and reserve more bandwidth for its own traffic. However, if a large number of nodes start to behave non cooperatively, the network may break down completely, depriving all users of services.
In order to avoid the misbehavior of certain nodes and alleviate the bandwidth share unfairness among all the mobile nodes, this paper describes and evaluates a routing-aware adaptive medium access control in IEEE 802.11 based wireless ad-hoc networks, which takes into account both node's own and routed data traffic, and try to give the same approximate bandwidth share to every mobile node to transmit its own traffic. Simulation results show that the proposed adaptive MAC scheme improves substantially the bandwidth share of the mobile nodes for their own traffic transmission while participating in the routing service.

1 Introduction

Recent advances in computer and wireless communication technologies have led to an increasing interest in wireless mobile ad-hoc networks [9]. In this new network paradigm, every mobile host plays the role of a router while being a terminal able to communicate with other wireless mobile nodes [4]. In fact, if a source and a destination nodes are not in the communication range of each other, data traffic is forwarded to the destination by relaying transmission through other mobile nodes which exist between the two communicating source and destination nodes. It is obvious that in this case, cooperation between involved nodes is very important in order for the network to emerge and operate. Since ad-hoc networks

---

* Authors are listed in alphabetical order

S. R. Das, S. K. Das (Eds.): IWDC 2003, LNCS 2918, pp. 245–254, 2003.

deploy multi-hop routing protocols here each of the nodes in addition to its own packets has to forward packets belonging to other nodes, selfish behavior may represent a significant advantage for a node, saving his battery power and reserving more bandwidth for its own traffic. However, if a large number of nodes start to behave non cooperatively, the network may break down completely, depriving all users of services.

To avoid misbehavior of the mobile nodes, compensation has to be made in order to encourage all the nodes in routing other nodes' packets without any degradation of their own data transmission.

While there has been lot of research work on improving fairness in the presence of hidden terminals [2, 5, 6] or high contention rate [3], to the best of our knowledge there is no research work focusing on the differentiation between the own and routed data traffic to achieve fairness improvement.

This paper describes an adaptive medium access in 802.11 based wireless ad-hoc networks that has been introduced in [1] and which takes into account both node's data traffic and routed data traffic of the other Neighboring nodes. It tends to give approximatively the same bandwidth share to each node for its own trafic sending, even though it is forwarding other nodes' packets. We describe our proposed adaptive MAC scheme and show by simulation the impact of certain parameters on its global performances.

In the reminder of this paper, we give in section 2 a short overview of the principle works that are proposed to enhance quality of service in wireless ad-hoc networks. In section 3, we describe our adaptive MAC mechanism in details. Section 4 describes the simulation configurations and gives some simulation results. Finally, section 5 gives the conclusions of this work and outlines future work extensions.

2   Related Work

In MACAW [2], the problem which has been addressed concerns the hidden terminal. In this scheme, additional control packets and a different backoff algorithm named Multiplicative Increase and Linear Decrease (MILD) with a backoff copy scheme were used to increase throughput and alleviate fairness problem. Another per-stream fairness mechanism was also introduced to treat equally each stream originated from the same node or other nodes and get equal share of the channel bandwidth. To do so, the per-stream fairness mechanism has to keep separate queue for each stream and runs backoff algorithms independently for each stream.

In [5], a probability based backoff algorithm has been proposed to address the fairness problem. A fairness index has been introduced to be the ratio of maximum link throughput to minimum link throughput. Each node calculates a link access probability for each of its links based on the number of connections from itself and its neighbors (connection based), or based on the average contention period of it and other nodes' individual links (time based) [6]. Whenever its backoff period ends, a node i will send RTS packet to j with probability p_{ij} or back off again with probability $1 - p_{ij}$.

In [6], the fairness index is introduced as in [5] to quantify fairness, and proposes a new estimation based backoff algorithm for the IEEE 802.11 MAC protocol [7]. The new algorithm can support the case when packets lengths are variable, which is a typical scenario of IEEE 802.11 compliant implementations that include both the basic CSMA/CA access method and the RTS/CTS access method. This work demonstrated that the fairness problem can be very severe with the original binary exponential backoff algorithm when packet length is variable and that their new backoff algorithm can achieve better fairness.

In [3], an adaptive service differentiation scheme for quality of service enhancement in IEEE 802.11e baesd wireless ad-hoc networks has been proposed. Their scheme is derived from the new Enhanced DCF Function introduced in the IEEE 802.11e standard. It aims to share the transmission channel efficiently. Relative priorities are provisioned by adjusting the size of the contention window (CW) of each traffic class taking into account both applications' requirements and network conditions in term of collision rate. The evaluation of the proposed scheme adaptive EDCF and its comparison with the native EDCF show a better behavior of AEDCF in presence of high traffic load.

3   Adaptive MAC Scheme Description

In this work, we are addressing the problem of efficient bandwidth sharing among all mobile nodes of a wireless ad-hoc network. We believe that when a node is participating in routing packets, it has to access more frequently the channel than another who is not participating in routing other nodes' packets. Moreover, depending on the amount of data to route, one node may access the channel more frequently than one who has less data to route. We believe also that by adopting this behavior, nodes will be more interested in building a connected and viable ad-hoc network and misbehaving nodes can be avoided or reduced.

In order to succeed in obtaining full nodes' cooperation and efficiently share the channel in 802.11 based wireless ad-hoc networks, we choose to dynamically change the contention window value CW after each unsuccessful transmission, which is caused by a contention or the end of a defer access period. We believe that by doing so, mobile nodes participating in routing other nodes' data traffic will not suffer from sending their own data traffic.

To facilitate the discussion, we introduce the following notations. For each mobile node i generating its own traffic and routing others' nodes traffic:

- $W_o(t)$: the amount of the *own* data traffic to send belonging to node i at the time t.
- $W_r(t)$: the amount of the *routed* data traffic to send belonging to other nodes at the time t.
- $\rho_i(t) = W_r(t)/(W_o(t) + W_r(t))$: the ratio of routed packets of node i at the time t over the total packets.

In the basic DCF scheme in IEEE 802.11 for ad-hoc networks, the contention window is reset to its minimum value (CW_{min}) after each successful transmission and doubled when collision occurs or the medium is sensed to be busy at the end of defer access period. In our scheme, we propose to change this mechanism and differentiate between nodes participating in routing packets from those who are using the routing service of the other nodes. To do so, we decide to change CW increase scheme by multiplying it by a multiplicative factor, smaller than the standard value 2, after each unsuccessful transmission. In this manner, we ensure that at each instant, the node who is sending its own traffic and participating in routing other nodes' traffic has the lowest contention window value so that it has the highest priority to access the channel.

In the next sub-sections, we explain in detail how the contention window of each mobile node is set after each successful transmission and after each collision or the end of defer access period.

3.1 CW Decrease Function

In the current version of IEEE 802.11 DCF function, after each successful transmission of packet by a mobile node, its CW is reset to its minimum contention window value (CW_{min}). In this work, we decide to keep this mechanism invariable since it helps the node to access the medium with a high probability.

3.2 CW Increase Function

After each unsuccessful transmission, caused by a contention with another transmitter or a busy medium sensed after a defer access period, the basic DCF function doubles the contention window CW, regardless of the type of packets being sent, while remaining less than the maximum contention window (CW_{max}):

$$CW_{new} = min(2 * CW_{old}, CW_{max}) \qquad (1)$$

In our adaptive MAC scheme, after each unsuccessful transmission, a mobile node i updates its contention window using a multiplicative factor MF_i, instead of the constant value 2. The MF_i value is calculated after each update period $\Delta(t)$ by taking into account the node's ratio of routed packets ρ_i over the total packets sent during the period $\Delta(t)$ (equation 2). By doing so, we allow to every node to access the channel more frequently when it has to forward other nodes' packets. The ratio ρ_i being in the range of 0 and 1 and in consequence MF_i in the range of 1 and 2, we allow more bandwidth to nodes having their MF_i approaching the value 1 and therefore their CW_{new} will not grow fast, and we allow less bandwidth to other nodes which are not involved extensively in routing packets by approaching the value 2 for their MF_i and in consequence will have their CW_{new} growing faster.

$$MF_i = 2 - \rho_i \qquad (2)$$

The new CW is then calculated following this equation:

$$CW_{new} = min(MF_i * CW_{old}, CW_{max}) \qquad (3)$$

In order to minimize the bias against transient routing period, we use an estimator of Exponentially Weighted Moving Average (EWMA) [3, 10] to smoothen the estimated values of the average routing ratio ρ. Note that ρ is calculated after each update period $\Delta(t)$, expressed in time-slots, following the equation 4:

$$\rho_i^{avg}(t) = (1 - \alpha) * \rho_i^{curr}(t) + \alpha * \rho_i^{avg}(t - \Delta(t)) \qquad (4)$$

where $\rho_i^{curr}(t)$ stands for the instantaneous routing ratio calculated at the time t, $\rho_i^{avg}(t - \Delta(t))$ being the previous average routing ratio calculated at the time $t - \Delta(t)$, $\Delta(t)$ represents the given update period, and α the *smoothing factor*. Note that the update period $\Delta(t)$ has to be chosen not too high and not too small to get the most accurate estimation and limit the complexity.

4   Simulation and Results

We have implemented our adaptive MAC scheme using glomosim simulator [11]. The main goals of our simulations is to study the capabilities of our protocol to enhance the channel access fairness, taking into account the willingness of a node to send its own packets at approximatively the same rate as other nodes' rates, while it is participating in routing other nodes' packets. For our simulations, we use two main scenarii to assess our statements (Figure 1). In the scenario 1, the three nodes A, B, and C are directly connected and no routing service is necessary. In this case, A sends packets to B, B sends packets to C, and C sends packets to A. In the scenario 2, the traffic flows are similar to scenario 1, but since C is hidden to node A, node C has to sollicitate the routing service of the node B to send its packets to node A.

4.1   Adaptive MAC Scheme Behavior

We have studied the impact of many parameters that characterize our adaptive MAC scheme. The smoothing factor α is varied from 0 to 1 and the update interval is varied from $10ms$ to $80ms$. To compare bandwidth share between

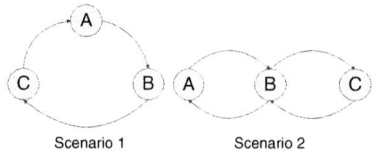

Fig. 1. Scenarii of different network configurations

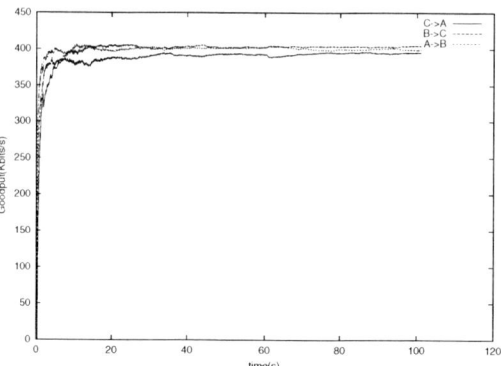

Fig. 2. Goodput as function of time in the scenario 1

nodes, we measured the goodput and the number of delivered packets for each node during the simulation.

Each node operates with the IEEE 802.11b standard. CW_{min} is set to 3 since higher values show no significant adaptation of our protocol. This is justified by the fact that with a 3 nodes topology, relative high values of CW do not ensure an access differentiation between the nodes. For this reason, we chose to set CW_{min} to 3 which helps better in studying our protocol behavior. The remaining of this section is organized as follows. In this section we show the unfairness problem when a node is routing other nodes' packets and we present a typical result of the medium access performances when the adaptive MAC scheme mechanism takes place. The section 4.2 presents the influence of the smoothing factor and the update interval on the performance of the adaptive MAC scheme.

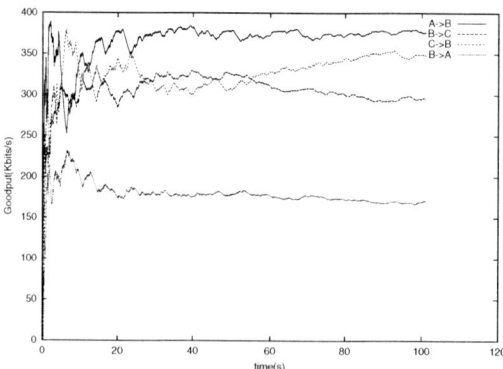

Fig. 3. Goodput as function of time in the scenario 2 using the native 802.11 MAC scheme

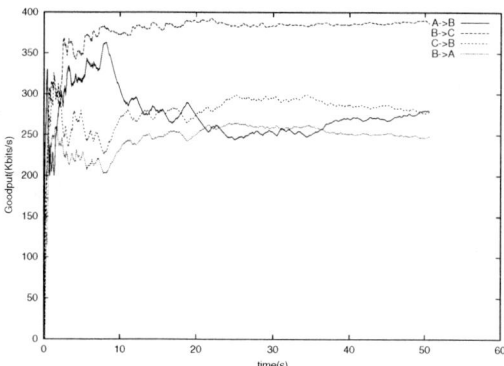

Fig. 4. Goodput as function of time in the scenario 2 using the adaptive 802.11 MAC scheme. ($\alpha = 0.9$, $\Delta(t) = 40ms$)

The figure 2 presents the goodput for each node in the case of a native 802.11 underlying MAC layer in the configuration of scenario 1. This figure shows that all nodes access the channel almost fairly as it is expected.

In the figure 3, we present the goodput for each node in the case of a native 802.11 underlying MAC layer in the configuration of scenario 2. In this figure, we can notice that the node B which is in charge of routing C node's packets to node A is penalized while comparing its goodput to the goodputs of the other nodes. In fact, its own traffic is sent with a throughput lower than the throughput of the other nodes and also the traffic it has to route is delivered with a lower throughput than the throughput of the own traffic of the other nodes. This behavior is a normal behavior if we do not ensure a node differentiation between

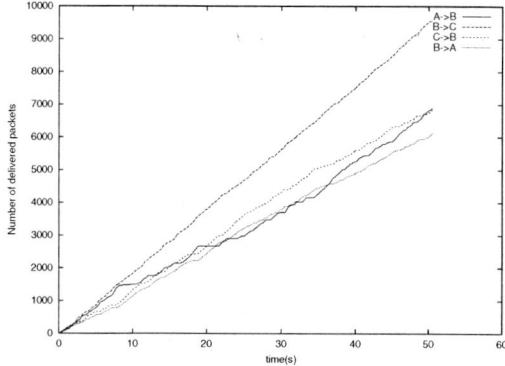

Fig. 5. Number of delivered packets as function of time in the scenario 2 using the adaptive 802.11 MAC scheme. ($\alpha = 0.9$, $\Delta(t) = 40ms$)

that aims to ensure a certain fairness between nodes while considering routed and own traffics.

With the adaptive MAC scheme, we have succeeded to obtain an improvement of the channel access performance. In fact, this is illustrated in the figure 4. First, we notice that with our protocol, we enhance the goodput corresponding to the routed traffic. Second, a relatively fair channel access is ensured between the three links $A \longrightarrow B$, $C \longrightarrow B$, and $B \longrightarrow A$ comparing to the native 802.11 behavior, where nodes which are not ensuring routing are more favored than the node ensuring routing. However, we can observe that with the adaptive MAC scheme, the node B, which was penalized in the native 802.11, is now too much favored. Indeed, as it is showed in the figure 4, the link $B \longrightarrow C$ has the higher goodput when compared to the other links. This is not a misbehavior of our protocol but just the result of differentiation missing at the IP layer which concerns the own and routed traffics. Since no differentiation is done to adapt channel access, to not favor routed packets over own packets and vis versa, the IP layer queue of the node B contains usually more own packets than packets to route. When the adaptive MAC scheme favors the node B, the own traffic benefits more than the routed traffic from the channel access share. Thus, our protocol improves partially the medium access while differentiating between nodes having different routing loads. We think that the adaptive MAC scheme can behave better by not favoring too much the own traffic of nodes if it is deployed with an IP layer which manages the IP differentiation in order to not favor own traffic over routed traffic. Therefore, this result is in accordance with the result presenting the number of delivered packets as function of time (see figure 5).

4.2   Impact of the Parameters α and $\Delta(t)$

In this section, we give simulation results showing the performance of the adaptive MAC scheme as function of the smoothing factor α and the update period $\Delta(t)$. The smoothing factor, as we described before, is used to improve the estimation of the routed packets ratio.

The figure 6 presents the goodput for different values of the smoothing factor α at a fixed update interval. Even if both parameters are correlated, we still be able to give some comments on α impact. With the current adaptive MAC scheme mechanism, the parameter α can not avoid the extra bandwidth consumption of the node B for its own traffic. This later is usually favored over the other traffics. Among the simulated update intervals, we noticed that the update interval of $40ms$ gives the less fluctuation of α impact on the goodput performances. We can also observe that when $\alpha = 0.5$ or 0.9, we have the best behavior of the adaptive MAC scheme. The goodput of the routed traffic is improved and we have more "fairness" of the bandwidth share. For this reason, we presented the adaptive MAC scheme enhancement in the section 4.1 only in the case where $\alpha = 0.9$. We also mentioned in the section 3 that our protocol adapts the backoff increase function within nodes ensuring routing. This adaptation is based on the estimated packets to route measured at each update interval $\Delta(t)$. The figure 6 presents the goodput for the update interval $\Delta(t) = 40ms$ which

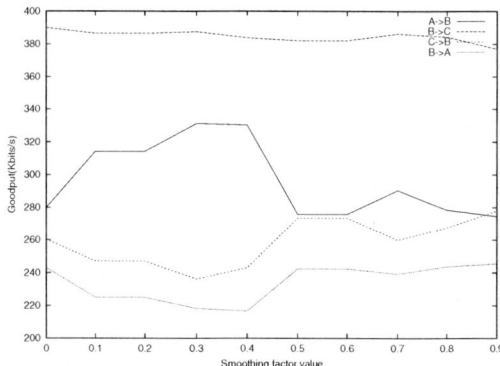

Fig. 6. Goodput as function of the smoothing factor ($\Delta(t) = 40ms$)

represents the most stable value leading to better performances when varying the smoothing factor α.

5   Concluding Remarks

In this paper, we propose a routing-aware adaptive MAC protocol for 802.11 based wireless ad-hoc networks. The characteristic of our adaptive MAC scheme consists in its use of a new backoff algorithm which aims to ensure a differentiation between nodes transmitting only their own packets and nodes transmitting their own packets and forwarding routing packets for the benefit of the other nodes. This new backoff algorithm supports thus the fact that nodes in the network have different loads of own and routed traffic. Simulation results show that our protocol can enhance the goodput corresponding to the routed traffic and can improve the fairness. However, the adaptive MAC scheme favors somehow the own traffic of the nodes having the charge of routing. This is not due to our protocol itself but to the fact that no differentiation between routing and own traffic is done at the upper layer (for instance IP layer). As a direct perspective of this work, it is worth to validate the behavior of our adaptive MAC scheme when deployed with an IP layer traffic differentiation to ensure a certain medium access fairness between the own and routed traffic.

Future works include also the tuning of the parameters that influence on the adaptive MAC scheme behavior, for instance the smoothing factor and the update period. This will be held in an adaptive MAC scheme to different network conditions such as node density, congestion, etc. In that way, the scalability of the adaptive MAC scheme represents one of our future efforts. Finally, we mention that we intend to study more in deep the increase and also the decrease functions corresponding to the backoff algorithm used in the adaptive MAC scheme. This study will be based on the distributed aspect of these functions and will be helped by the work related to ensuring the fairness in multi-hop wireless ad-hoc networks.

References

[1] Naït-Abdesselam, F. and Koubaa, H.: RAMAC: Routing-aware adaptive MAC in IEEE 802.11 Wireless Ad-Hoc Networks. The 8th International Conference on Cellular and Intelligent Communications (CIC), Seoul, Korea, October (2003) 246

[2] Bharghavan, V., Demers, A., Shenker, S. and Zhang, L.: MACAW: A Media Access Protocol for Wireless LANs. ACM SIGCOMM, London, England (1994) 246

[3] Romdhani, L., Ni, Q. and Turletti, T.: AEDCF: Enhanced Service Differentiation for IEEE 802.11 Wireless Ad-Hoc Networks. IEEE WCNC, New Orleans, Louisiana, USA, March (2003) 246, 247, 249

[4] MANet Working Group. Mobile Ad-hoc Networks (MANet) http://www.ietf.org/html.charters/manet-charter.html 245

[5] Ozugur, T., Naghshineh, M., Kermani, P. and Copeland, J.: Fair Media Access for Wireless LANs, IEEE Globecom, Rio de Janeiro (1999) 246, 247

[6] Bensaou, B., Wang, Y. and Chung Ko, C.: Fair Medium Access in 802.11 based Wireless Ad-Hoc Networks. Proceedings of the 1st ACM international symposium on Mobile ad hoc networking & computing, Boston, Massachusetts (2000) 246, 247

[7] IEEE: Wireless LAN Medium Access Control (MAC) and Physical Layer (PHY) Specifications. IEEE 802.11 standards, June (1999) 247

[8] Tobagi, F. A. and Kleinrock, L.: Packet Switching in Radio Channels: Part II - The Hidden Terminal Problem in Carrier Sense Multiple Access and the Busy Tone Solution. IEEE Transaction on Communications (1975)

[9] Shah, H., Chen. K. and Nahrstedt, K.: Dynamic Bandwidth Management for Single-hop Ad Hoc Wireless Networks. IEEE International Conference on Pervasive Computing and Communications (PerCom), Dallas-Fort Worth, Texas (2003) 245

[10] NIST/SEMATECH: e-Handbook of Statistical Methods, "http://www.itl.nist.gov/div898/handbook/" 249

[11] UCLA Parallel Comp Lab, Wireless Mobility Lab: GlomoSim, "http://pcl.cs.ucla.edu/projects/glomosim/" 249

Various Distributed Shortest Path Routing Strategies for Wireless Ad Hoc Networks

Subhankar Dhar[1], Michael Q. Rieck[2], Sukesh Pai[3], and Eun Jik Kim[1]

[1] San José State University, San José, CA 95192 USA
dhar_s@cob.sjsu.edu
eunkim@sjsu.edu
[2] Drake University, Des Moines, Iowa 50311 USA
michael.rieck@drake.edu
[3] Microsoft Corporation, Mountain View, CA 94043 USA
sukeshp@microsoft.com

Abstract. In this paper, we describe and compare several distributed greedy algorithms that produce sets of nodes that can be used to form a virtual backbone for an ad hoc wireless network. The backbone produced is always a d-hop dominating, d-hop connected set and has a desirable "shortest path property". The perfomance of these algorithms for various parameters are compared.

1 Introduction

Wireless ad hoc networks consist of a set of identical mobile devices (nodes) that communicate with each other via wireless links. The growing importance of ad hoc wireless networks can hardly be exaggerated, as portable wireless devices are now ubiquitous and continue to grow in popularity and in capabilities. In such networks, all of the nodes are mobile and so the infrastructure for message routing must be self-organizing and adaptive. Building an infrastructure for ad hoc network that guarantees reliable communication is an important problem. In recent years, there have been prolific research activities in this regard. However, there are quite a number of challenging problems yet to be solved in the area of ad hoc networks. Finding efficient and effective routing schemes is just one example, and the one that will be our focus here.

Ad hoc wireless networks are represented by a connected graph where all the links are bi-directional. Several researchers have used minimum connected dominating sets to do routing in ad hoc wireless networks [1], [2], [4], [10], [11]. The dominating set induces a virtual connected backbone. The CDS (Connected Dominating Set) problem is described as follows: Find a subset D of nodes, such that the subgraph induced by D is connected and D forms a dominating set *i.e.*, it is a set in which each node is either in the dominating set or adjacent to some node in the dominating set. It is well-known that finding a minimum connected dominating set is an NP-complete problem.

B. Liang and Z. J. Haas [8] use the greedy algorithm with redundancy elimination to obtain a d-hop dominating set (see next section), to serve as a small

S. R. Das, S. K. Das (Eds.): IWDC 2003, LNCS 2918, pp. 255–265, 2003.
© Springer-Verlag Berlin Heidelberg 2003

virtual backbone. The problem of finding such a set is reduced in a natural way to the classic NP-complete Set Covering Problem. Rather than seeking an optimal solution, they apply the greedy algorithm to settle for a reasonably small d-hop dominating set. Our approach in this paper is quite similar, except that we focus on a different covering problem to obtain our virtual backbone set.

To be more specific, in this paper, the methods of [3] and [10] are modified in a number of ways and compared. The notion of an "SPR-set" ("shortest path routing set") set is re-introduced and justified as being quite useful as a virtual backbone for an ad hoc wireless network. We look at different ways of computing such a set by mapping the problem to the Set Covering Problem. We propose a general strategy for reducing such sets through "greedy refinement". Finally, we compare the results from these algorithms to the results from previous work.

2   Shortest Path Routing

2.1 d-SPR Sets

Throughout this paper, a wireless ad hoc network will be represented by a connected graph G. In [3] and [10], the authors proposed an algorithm for constructing a d-hop connected, d-hop dominating set for G, where d is a fixed integer greater than one. A set of nodes is "d-hop dominating" (also called "d-dominating") if each node in the network is within d hops of a node in the set. "d-hop connected" means that given any two nodes u and v in the set, there is a path beginning with u and ending with v, such that the hop count between consecutive nodes along the path that belong to the set never exceeds d. In fact the sets obtained in [3] and [10] have the addition special property that there exists a path as just described between any two nodes u and v from the graph, and that this path is a shortest (possible) path connecting u to v, and that the hop counts from u to the first member of the set and from the last member of the set to v do not exceed d. We refer to this property as the 'shortest path property'. To construct the set, [3] introduces the notion of a d-SPR set, as follows.

Definition 1. *A set S of nodes of G will be called a d-SPR set if given any pair of nodes u and v of G such that $\delta(u,v) = d + 1$, there exists a $w \in S$ with $\delta(u,w) + \delta(w,v) = d + 1$ and $w \neq u, w \neq v$.*

Here $\delta(u,v)$ means the hop count from u to v, and d-SPR stands for 'd-shortest path routing'. Under reasonable assumptions, a d-SPR set can be shown to be a d-hop connected, d-hop dominating set that has the shortest path property. For details, please see [3]. Deciding whether or not a set is a d-SPR set is a local issue. To be more specific, it is possible to consider a certain test that can be conducted based solely on the induced subgraph of a given $d + 1$-hop neighborhood. Each node can conduct this test for its own $d + 1$-hop neighborhood. The set as a whole is a d-SPR set if and only if each node perceives it to be so locally. This follows easily from the definition and [3, Corollary 1], which is also reproduced below as Corollary 1 in this paper.

An alternative way to think about a d-hop connected d-hop dominating set is simply as a connected dominating set for the so-called *d-hop closure G_d* graph. This is the graph obtained from the original graph G by adding edges between any pair of nodes that are within d hops in G.

2.2   Set Covering Problem, Bipartite Graph

It is well-known that the Set Covering Problem is essentially a problem concerning bipartite graphs that can be stated as follows. Suppose that H is a bipartite graph, consisting of two sets of nodes A and B, where edges only connect nodes from A to nodes from B. Also assume that for every node in B, there is at least one edge connecting it to a node in A. The goal is to find a minimal (or at least small) subset C of A such that every node is B is "covered by" (*i.e.* adjacent to) some node in C.

The problem of finding a d-SPR set can be translated into the problem of finding a "covering set" C for the following bipartite graph. Let A be the set of all the nodes in the network. Let B be the set of all unordered pairs $\{x, y\}$ of nodes in the network satisfying $\delta(x, y) = d + 1$. Put an edge between a node v from A and a pair $\{x, y\}$ from B if v does not equal x or y, but v does lie along some shortest (possible) path connecting x and y. A subset C of A covers B if and only if it is a d-SPR set, as is straightforward to check. For more details, see [3]. The following definition will be useful, although it really is just another name for "node degree" in the context of the bipartite graph H.

Definition 2. *The* covering number *of each node w in A is the number of $\{x, y\}$ pairs in B that share an edge with w. Also, we say that w covers the pair $\{x, y\}$.*

2.3   Distributed Approaches to Obtaining d-SPR Sets

In order to produce small d-SPR sets in a distributed approach, we consider certain subgraphs as follows.

Definition 3. *The d-local view of a node v consists of all the d-hop neighbors of v, together with all edges between these, except for the edges that connect two nodes at a distance d from v.*

Each node maintains its d-local view by obtaining the necessary link-state information from its neighbors. This requires just d rounds of message passing, during which each node sends a message to its adjacent neighbors. We now state the following theorem that has been proved in [3].

Theorem 1. *For any node x, y, v in G, let $\delta(x, v) + \delta(y, v) = d + 1$, the distance $\delta(x, y)$ can be computed solely from a knowledge of the $(d + 1)$-local view of v. Moreover, all the shortest paths connecting x to y lie inside the $(d + 1)$-local view of v.*

The following corollary is a useful consequence of Theorem 1.

Corollary 1. *Given* $u, v \in A$ *(i.e. nodes in G), and a pair* $\{x, y\} \in B$ *such that u and v are both adjacent to $\{x, y\}$ in H, the four nodes u, v, x, y, as nodes of G, are within a distance $d + 1$ of each other.*

We will refer the $dCDS$ algorithm proposed by the authors in [10] as d-SPR-I here, as was done in [3]. Let H be the bipartite graph as discussed earlier. Each node in the original graph is randomly assigned a unique positive integer or ID. For each pair $\{x, y\}$ in B, a node v in $A_{\{x,y\}}$ is elected into the d-SPR set to cover this pair if v has the highest ID among the nodes in $A_{\{x,y\}}$. The node v is in a position to decide this question from its local information. The resulting set produced by the d-SPR-I algorithm is clearly a d-SPR set.

The d-SPR-C algorithm from [3] requires an additional $d + 1$ rounds of local broadcast. After the first $d+1$ rounds, each node is aware of its own $(d+1)$-local view, and is able to compute its own covering number. The subsequent $d + 1$ rounds of broadcast are used to allow each node to transmit its covering number to each of its $(d + 1)$-hop neighbors.

In the d-SPR-I algorithm, given a pair $\{x, y\}$ (from B), the node that covered $\{x, y\}$ with the highest ID was selected for inclusion into the d-SPR set. Here, instead we select the node having the highest *priority*. The priority of each node is defined to be the ordered pair of numbers (covering number, ID), lexicographically ordered. This is analogous to the approach taken in [6, Subsection 2.1].

2.4   An Example

Let us now consider an example graph to illustrate how we can construct the bipartite graph. In Fig. 1, the nodes with solid edges constitute the graph G which represents an ad hoc network consisting of 15 nodes. Let us choose d to be equal to 2 for this example. Let A be the set of all nodes in G, B be the set of all pairs of nodes in G that are separated by a distance $d+1$ i.e. 3, and H be the

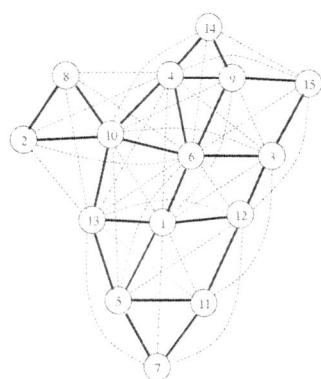

Fig. 1. Example graph G representing an ad hoc network

bipartite graph consisting of sets A and B where edges are defined as described earlier. $A = \{1, 2, 3, 4, 5, 6, 7, 8, 9, 10, 11, 12, 13, 14, 15\}$, and $B = \{\{1,2\}, \{1,15\},$ $\{1,14\}, \{1,8\}, \{2,9\}, \{2,5\}, \{2,14\}, \{2,3\}, \{3,13\}, \{3,5\}, \{3,14\}, \{3,7\}, \{3,8\},$ $\{4,5\}, \{4,12\}, \{5,9\}, \{5,8\}, \{6,7\}, \{6,11\}, \{7,10\}, \{8,9\}, \{8,14\}, \{9,13\}, \{9,12\},$ $\{10,12\}, \{10,11\}, \{10,15\}, \{11,15\}, \{13,14\}\}$.

Now we draw edges between A and B as shown in Fig. 2. For every pair of nodes $\{x, y\}$ in B, let $A_{\{x,y\}} = \{w \mid \delta(x, w) + \delta(y, w) = d + 1, w \neq x, w \neq y\}$ and we put an edge between node v and $\{x, y\}$ in H, for all nodes v in the set $A_{\{x,y\}}$. For example, $A_{\{1,2\}} = \{6,10,13\}$ and we add the following edges to the bipartite graph H: $(6,\{1,2\})$, $(10,\{1,2\})$, $(13,\{1,2\})$. Fig. 2 shows the bipartite graph H thus constructed. This example is fairly small, and it is possible to check by hand that $\{5, 6, 10, 12\}$ is a minimal d-SPR set.

3   Distributed Greedy Refinement of d-SPR Sets

In this section, we modify the greedy algorithm d-SPR-G discussed in [3] to reduce the size of any d-SPR subset A of all nodes in a given graph G. A might have been obtained by a method already discussed for example. The bipartite graph H is now modified to use this A in place of the set of all graph nodes. Since A is a d-SPR set, it covers the set B (with B exactly as before). Greedy refinement just amounts to applying the greedy algorithm in this situation to produce a subset C of A that also covers B, and hence is also a d-SPR set. We

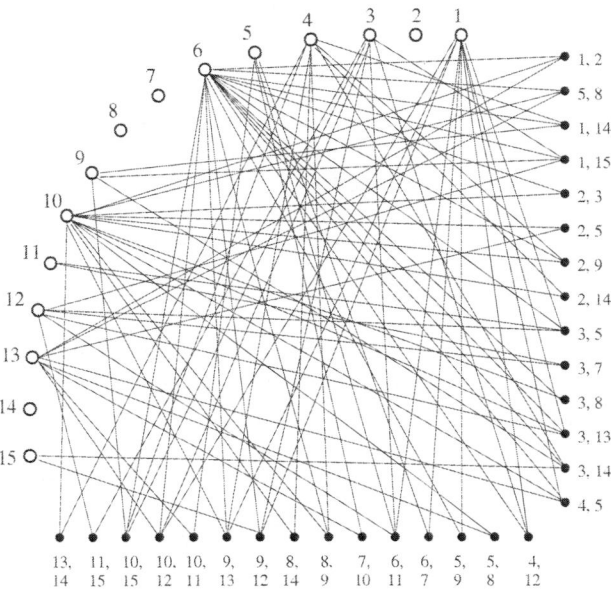

Fig. 2. Bipartite graph H generated from G in Fig. 1

refer to C as the *greedy refinement* of A. This set can be obtained in a distributed way as follows, where each node maintains information about its $(d+1)$-local view. In this case we know from Corollary 1 that a node v in the initial set A can "see" all the node pairs $\{x, y\}$ it covers in set B, and also the rest of the nodes in A that cover such pairs. The algorithm begins with each node is initially in an "undecided" state.

Step 1: Each node v in set A gathers information about its $(d+1)$-local view. This requires $d+1$ rounds of broadcast. Let B_v denote all the node pairs $\{x, y\}$ covered by v. Let C_v be the set of all nodes that also covers some node pair in B_v. So $v \in C_v$. v computes it covering number $|B_v|$.

Step 2: If v's covering number is zero, then v enters the "not selected" state.

Step 3: v sends covering number as well as its state information in a message to each node in C_v. If v decided to be in the "not selected" or "selected" state, it no longer participates in the algorithm.

Step 4: v waits until it receives the current covering numbers and the state information of each node in C_v.

Step 5: For each such node u that has become selected, v removes u from C_v, and removes any pairs from B_v that u covers.

Step 6: v recomputes its covering number and checks to see if its own priority is the highest among all the nodes of C_v. If so, then it enters the 'selected' state and jump back to step 3.

Note that we do not have a synchronization problem in the distributed algorithm even if each node does not wait as in step 4 before making its decision. Consider a node v that is told by some node u in v's $(d+1)$-hop neighborhood that it has become selected. When v learns about this, it is forced to update its covering number based on this information. After that, it would look at the covering numbers of all of its $(d+1)$-hop neighbors to decide if it needs to become selected. If v, after this computation finds that it has the highest covering number in its $(d+1)$ neighborhood, then it can safely elect itself to the set. This is true since the covering numbers could only reduce and never increase. Thus, even if it has stale information on the covering numbers of its neighbors, the decision would only be conservative and never wrong. After this computation, the node v necessarily has to propagate its new covering number to its $d+1$-neighborhood.

3.1 Distributed Greedy Refinement of d-SPR-I (d-SPR-IG), d-SPR-C (d-SPR-CG)

In general, with the exception of d-SPR-G, the greedy refinement algorithms have two phases: initial selection phase and optimization phase. In the initial selection phase, the initial set A is obtained by applying basic d-SPR algorithms such as d-SPR-I and d-SPR-C. In this phase, the nodes can be selected randomly - based on randomly assigned node ID as in d-SPR-I, or selected according to some desirable attributes of a node - based on covering number (*priority*) as in d-SPR-C. The second phase is the optimization phase, the purpose of which is to reduce the size of the set. In this phase, the distributed greedy refinement based on covering number on the bipartite graph is applied.

The distributed implementation of d-SPR-I and d-SPR-C involves every node sending control messages $(d + 1)$ times to all of its neighbors so that each node has a local view of its $(d + 1)$-hop neighborhood.

3.2   Distributed Greedy Refinement of d-SPR-I with Two Covering Nodes (d-SPR-C$_2$G)

This new algorithm is a variation made on d-SPR-C. In d-SPR-C, for every pair of nodes with distance $(d + 1)$, say $\{x, y\}$, a node in the set $A_{\{x,y\}}$ that has the highest priority is admitted to the d-SPR set. In the initial selection phase of d-SPR-C$_2$G, for each pair $\{x, y\}$ in set B, two such nodes are admitted to the initial set.

Obviously, the initial set A for d-SPR-C$_2$G is considerably bigger than the set obtained from d-SPR-C. Having a smaller initial however set limits the scope of optimization in the second phase. Thus, from the optimization point of view, it is more desirable to have a large initial set, although a large initial set requires more time and messages to process. d-SPR-C$_2$G was designed as a compromise between d-SPR-CG and d-SPR-G. The purpose of d-SPR-G is to produce a very small d-SPR set by using the largest initial set possible, i.e. the set of all nodes in the graph. However, the processing time is great and many messages need to be passed. d-SPR-CG takes far less time and involves far fewer messages, but results in a large set.

3.3   d-SPR Based on Weights Assigned by Node Pairs (d-SPR-PW)

Another interesting variant of d-SPR-I algorithm that produces a set somewhat better than the size produced by d-SPR-C is d-SPR-PW (for Pair Weighted). It is based on an algorithm of S. Rampone [9]. While this could be used to reduce any d-SPR set, just as greedy reduction for this purpose, we will only consider applying it to the set of all nodes in the network. It is worth mentioning, in the context of finding a d-SPR set in a distributed way, that no additional message passing is required when using Rampone's approach instead of the usual greedy approach. This is clear from Corollary 1. The d-SPR-PW method proceeds as follows. Each node v in A is selected into the d-SPR set based on the weights assigned to it by each of the node pairs in B_v. For $v \in A$ and $\{x, y\} \in B$, define the following:

$$I(v, \{x, y\}) = 1 \text{ if } v \text{ covers the pair } \{x, y\} \text{ and } 0 \text{ otherwise.}$$
$$PW(v, \{x, y\}) = I(v, \{x, y\}) \ / \ \Sigma_{k=1...n} I(v_k, \{x, y\})$$
$$PW(v) = \Sigma_{\{x,y\}} PW(v, \{x, y\})$$

where n is the number of nodes and v_k denotes each node.

The d-SPR-PW algorithm requires an additional $d+1$ rounds of local broadcast just like d-SPR-C. After the first $d+1$ rounds, each node is aware of its own

$(d+1)$-local view, and is able to compute its own pair weighted number. The subsequent $d+1$ rounds of broadcast are used to allow each node to transmit its pair weighted number to each of its $(d+1)$-hop neighbors. Then, we select the node having the highest *priority*. Here, the priority of each node is defined to be the ordered pair of numbers (pair weighted number, ID), lexicographically ordered.

3.4   Example

We consider again the graph in Fig. 1. We chose $d=2$. The graph G_2 is obtained by adding dashed edges as shown in the figure. The Wu-Li algorithm [11] was applied to G_2. We also computed the sets produced by d-SPR-I, d-SPR-C, d-SPR-G, d-SPR-IG, d-SPR-CG, d-SPR-C$_2$G and d-SPR-PW. Let $C_{algorithm}$ be the set produced by applying the *algorithm*. We summarize the resulting sets as follows.

$$C_{Wu-Li} = \{6, 10, 12, 13, 14, 15\}, C_{d-SPR-I} = \{5, 6, 9, 10, 12, 13, 15\}$$
$$C_{d-SPR-C} = \{1, 3, 6, 10, 12, 13\}, C_{d-SPR-G} = \{5, 6, 10, 12\}$$
$$C_{d-SPR-IG} = \{5, 6, 10, 12\}, C_{d-SPR-CG} = \{1, 6, 10, 12, 13\}$$
$$C_{dSPR-C_2G} = \{5, 6, 10, 12\}, C_{d-SPR-PW} = \{5, 6, 10, 12\}$$

Let us mention here that all the sets produced by the d-SPR algorithms have the shortest path property which the Wu-Li algorithm cannot guarantee. We also noticed that the d-SPR-G, d-SPR-IG, d-SPR-C$_2$G and d-SPR-PW produced sets which are smaller than the size of the set produced by the Wu-Li algorithm.

4   Performance Evaluation of the Algorithms

The main goal of our performance analysis is to find and compare the size of d-hop connected d-hop dominating sets produced by the Wu-Li algorithm and various versions of our d-SPR algorithms - d-SPR-I, d-SPR-C, d-SPR-G, d-SPR-IG, d-SPR-CG, d-SPR-C$_2$G and d-SPR-PW. Each algorithm is designed in a distributed manner where each node will gather its d-hop neighborhood information by exchanging messages with its direct neighbors. These algorithms are implemented and experiments are run on a single machine.

4.1   Experimentation

For each experiment, a random disk graph is used to model the topology of an ad hoc network. A disk graph is a graph in which a node is connected to all other nodes within a radius defined for the graph. A random disk graph is a disk graph in which nodes are positioned randomly. The radius of a disk graph represents the transmission range of a node in the ad hoc network assuming that all nodes in the network use the same transmission power. Also all links are assumed to be bi-directional.

In this experimentation, we tried to keep the density of the network constant and thus the degree of each node constant too, to some extent. Given a node density and the number of nodes in a network, the size of simulation area was computed and x and y coordinates of each node was randomly chosen within this boundary.

We ran experiments for each algorithm with different values of d and different number of nodes. The algorithms considered were Wu-Li algorithm with Rule 1 and 2, applied to a d-hop closure G_d of a random disk graph G; two d-SPR algorithms - d-SPR-I and d-SPR-C where the covering number in bipartite graph was used as the priority; and five greedy refinement algorithms - d-SPR-G, d-SPR-IG, d-SPR-CG, d-SPR-C$_2$G and d-SPR-PW.

4.2   Results

In terms of the set size, the Wu-Li algorithm performed slightly better than any of d-SPR algorithms considered, which is not surprising considering that d-SPR algorithms produce a d-hop connected d-hop dominating set with the shortest path property, an extra property that entails some overhead. Therefore, our aim is to produce a d-SPR set whose size is as small as that of the set produced by the Wu-Li algorithm.

Fig. 3 shows the average size of the sets produced by each algorithm when $d = 5$, while the number of nodes varies from 100 to 500. The first thing to notice is that the sets produced by basic d-SPR algorithms without the greedy refinement, i.e. d-SPR-I and d-SPR-C, are almost two times greater than the set produced by the Wu-Li algorithm. The greedy refinement produced a significant reduction in the set size; after applying the greedy refinement, d-SPR-I and d-SPR-C sets were reduced in size by $35 - 45\%$. Among various d-SPR algorithms, d-SPR-G and d-SPR-PW performed the best; the sets generated by d-SPR-G and d-SPR-PW are only 10% larger than the Wu-Li set, while retaining the shortest path property. Between d-SPR-G and d-SPR-PW, in most cases d-SPR-PW produced slightly smaller sets.

The algorithm, d-SPR-C$_2$G, is a variation on d-SPR-C and produced a set that is smaller than d-SPR-CG and very close to d-SPR-G. It should be noted that, unlike other greedy refinement algorithms, d-SPR-G does not have an initial selection process where a subset of nodes in the network with some desired attribute, e.g. high covering number, is selected before optimization algorithm is applied to the set.

Nodes	Wu-Li	d-SPR-I	d-SPR-C	d-SPR-G	d-SPR-IG	d-SPR-CG	d-SPR-C$_2$G	d-SPR-PW
100	7.1	19.9	14.0	8.0	10.8	8.4	8.1	7.8
200	17.3	45.7	38.3	21.4	26.6	22.1	21.8	20.7
300	25.1	68.3	56.4	31.4	36.8	32.5	31.1	30.1
400	37.3	101.9	80.3	45.6	54.8	47.7	45.5	44.7
500	43.7	123.8	97.7	55.7	67.9	58.6	57.2	54.8

Fig. 3. Size of the sets produced by various algorithms for $d = 5$

Both d-SPR-C$_2$G and d-SPR-PW, while producing a set with similar size to d-SPR-G, makes use of priority in their initial selection process. Making use of priority is significant if attributes other than the shortest path property (e.g. remaining battery power) need to be considered in the set selection process.

5   Conclusions

In this paper, we proposed three new distributed greedy algorithms that produce d-hop connected d-hop dominating sets. These sets can be used to create a virtual backbone of a wireless ad hoc network. In addition, these sets have a shortest path property which works efficiently in low mobility enviroments. This is the basis of our routing scheme.

In order to further reduce the size of the set produced by our algorithms, we are currently exploring some distributed probabilistic algorithms as well as hierarchical schemes that will preserve the shortest path property and produce d-hop connected d-hop dominating sets.

References

[1] K. M. Alzoubi, P. Wan, O. Frieder. New Distributed Algorithm for Connected Dominating Set in Wireless Ad Hoc Networks. *Proceedings of 35th Hawaii International Conference on System Sciences*, Hawaii 2002. 255

[2] Bevan Das and Vaduvur Bharghavan. Routing in Ad-Hoc Networks Using Minimum Connected Dominating Sets. *IEEE International Conference on Communications (ICC '97)*, (1) 1997: 376-380. 255

[3] Subhankar Dhar, Michael Q. Rieck and Sukesh Pai. On Shortest Path Routing Schemes for Wireless Ad-Hoc Networks. *10th International Conference on High Performance Computing (HiPC '03)*, December 2003. 256, 257, 258, 259

[4] S. Guha and S. Khuller. Approximation algorithms for connected dominating sets. *Algorithmica*, Vol 20, 1998. 255

[5] D. Johnson. Approximation Algorithms for Combinatorial Problems. *Journal of Computer and System Sciences*, 9:256-278, 1974.

[6] L. Jia, R Rajaraman, T. Suel. An Efficient Distributed Algorithm for Constructing Small Dominating Sets. *Proceedings of the Annual ACM Symposium on Principles of Distributed Computing*, pp 33-42, August 2001. 258

[7] L. Lovasz. On the Ratio of Optimal Integral and Fractional Covers. *Discrete Mathematics*, 13:383-390, 1975.

[8] B. Liang and Z. J. Haas. Virtual Backbone Generation and Maintenance in Ad Hoc Network Mobility Management. *Proc. 19th Ann. Joint Conf. IEEE Computer and Comm. Soc. INFOCOM, vol. 3, pp. 1293-1302, 2000*. 255

[9] Salvatore Rampone. Probability-driven Greedy Algorithms for Set Cover. *VIII SIGEF Congress "New Logics for the New Economy"*, Naples, September 2001. 261

[10] M. Q. Rieck, S. Pai, S. Dhar. Distributed Routing Algorithms for Wireless Ad Hoc Networks Using d-hop Connected d-hop Dominating Sets. *Proceedings of the 6th International Conference on High Performance Computing: Asia Pacific, (HPC Asia 2002)*, December 2002. 255, 256, 258

[11] Jie Wu and Hailian Li. A Dominating-Set-Based Routing Scheme in Ad Hoc Wireless Networks. *Special issue on Wireless Networks in the Telecommunication Systems Journal*, Vol. 3, 2001, 63-84.   255, 262

Secure Mobile Computing

Dharma P. Agrawal[1], Hongmei Deng[1], Rajani Poosarla[1], and Sugata Sanyal[2]

[1] Center for Distributed and Mobile Computing
University of Cincinnati, Cincinnati, OH 45221-0030
{dpa,hdeng,poosarrd}@ececs.uc.edu
[2] School of Technology and Computer Science
Tata Institute of Fundamental Research, Homi Bhabha Road, Mumbai 400005, India
sanyal@tifr.res.in

Abstract. As more and more people enjoy the various services brought by mobile computing, it is becoming a global trend in today's world. At the same time, securing mobile computing has been paid increasing attention. In this article, we discuss the security issues in mobile computing environment. We analyze the security risks confronted by mobile computing and present the existing security mechanisms.

1   Mobile Computing at a Glance

The last few years have seen a true revolution in the telecommunications world. Besides the three generations of wireless cellular systems, ubiquitous computing has been possible due to the advances in wireless communication technology and availability of many light-weight, compact, portable computing devices, like laptops, PDAs, cellular phones, and electronic organizers. The term of mobile computing is often used to describe this type of technology, combining wireless networking and computing. Various mobile computing paradigms are developed, and some of them are already in daily use for business work as well as for personal applications. Wireless personal area networks (WPANs), covering smaller areas (from a couple of centimeters to few meters) with low power transmission, can be used to exchange information between devices within the reach of a person. A WPAN can be easily formed by replacing cables between computers and their peripherals, helping people do their everyday chores or establish location aware services. One noteworthy technique of WPANs is a Bluetooth based network. However, WPANs are constrained by short communication range and cannot scale very well for a longer distance.

Wireless local area networks (WLANs) have gained enhanced usefulness and acceptability by providing a wider coverage range and an increased transfer rates. The most well-known representatives of WLANs are based on the standards IEEE 802.11 [1], HiperLAN and their variants. IEEE 802.11 has been the predominant standard for WLANs, which support two types of WLAN architectures by offering two modes of operation, ad-hoc mode and client-server mode. In ad-hoc (also known as peer-to-peer) mode (Figure 1(a)), connections between two or more devices are established in

S. R. Das, S. K. Das (Eds.): IWDC 2003, LNCS 2918, pp. 265–278, 2003.

an instantaneous manner without the support of a central controller. The client-server mode (Figure 1(b)) is chosen in architectures where individual network devices connect to the wired network via a dedicated infrastructure (known as access point), which serves as a bridge between the mobile devices and the wired network. This type of connection is comparable to a centralized LAN architecture with servers offering services and clients accessing them. A larger area can be covered by installing several access points, as with cellular structure having overlapped access areas.

The corresponding two architectures are commonly referred to as infrastructure-less and infrastructure-based network. Ad hoc network is a collection of wireless mobile hosts forming a temporary network without the aid of any centralized administration or standard support services regularly available on the wide area network [2]. Due to its inherent infrastructure-less and self-organizing properties, an ad hoc network provides an extremely flexible method for establishing communications in situations where geographical or terrestrial constraints demand totally distributed network system, such as military tracking, hazardous environment exploration, reconnaissance surveillance and instant conference. While we are enjoying the various services brought by mobile computing, we have to realize that it comes with a price: security vulnerabilities.

2   Why is Security an Issue?

Security is a prerequisite for every network, but mobile computing presents more security issues than traditional networks due to the additional constraints imposed by the characteristics of wireless transmission and the demand for mobility and portability. We address the security problems for both infrastructure-based WLANs and infrastructure-less ad hoc networks.

2.1   Security Risks of Infrastructure-Based WLANs

Because a wireless LAN signal is not limited to the physical boundary of a building, potential exists for unauthorized access to the network from personnel outside the intended coverage area. Most security concerns arise from this aspect of a WLANs and fall into the following basic categories:

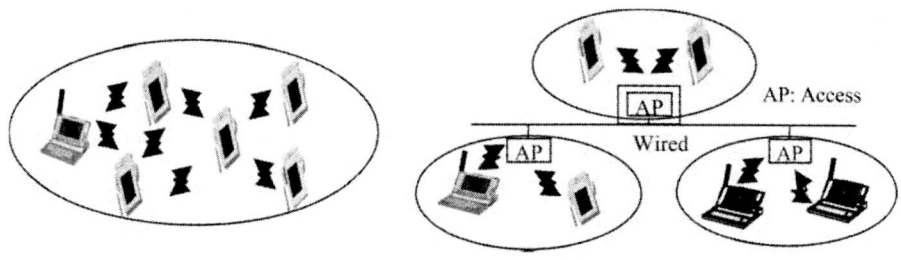

(a) Infrastructure-less   (b) Infrastructure-based Network

Fig. 1. WLAN Architectures

Limited Physical Security. Unlike traditional LANs, which require a wire to connect a user's computer to the network, a WLAN connects computers and other components to the network using an access point (AP) device. As shown in Figure 1 an access point communicates with devices equipped with wireless network adaptors and connects to a fixed network infrastructure. Since there is no physical link between the nodes of the wireless network and the access point, the users transmit information through the "air" and hence anyone within the radio range (approximately 300 feet for 802.11b) can easily intercept or eavesdrop on the communication channels. Further, an attacker can deploy unauthorized devices or create new wireless networks by plugging in unauthorized clients or setting up renegade access points.

Constrained Network Bandwidth. The use of wireless communication typically implies a lower bandwidth than that of traditional wired networks. This may limit the number and size of the message transmitted during protocol execution. An attacker with the proper equipment and tools can easily flood the 2.4 GHz frequency, corrupting the signal until the network ceases to function. Since the aim of this type of attack is to disable accessing network service from the legitimate network users, they are often named denial of service (DoS) attack. Denial of service can originate from outside the work area serviced by the access point, or can inadvertently arrive from other 802.11b devices installed in other work areas that degrade the overall signal.

Energy Constrained Mobile Hosts. To support mobility and portability, mobile devices generally obtain their energy through batteries or other exhaustive means, hence they are considered as energy constrained mobile hosts. Moreover, they are also resource-constraint relative to static elements in terms of storage memory, computational capability, weight and size. In WLANs, two wireless clients can talk directly to each other, bypassing the access point. A wireless device can create a new type of denial of service attack by flooding other wireless clients with bogus packets to consume its limited energy and resources.

2.2   More Vulnerabilities of Infrastructure-Less Ad Hoc Networks

In ad hoc networks, mobile hosts are not bound to any centralized control like base stations or access points. They are roaming independently and are able to move freely with an arbitrary speed and direction. Thus, the topology of the network may change randomly and frequently. In such a network, the information transfer is implemented in a multi-hop fashion, i.e., each node acts not only as a host, but also as a router, forwarding packets for those nodes that are not in direct transmission range with each other. By nature, an ad hoc network is a highly dynamic self-organizing network with scarce channels. Besides these security risks, ad hoc networks are prone to more security threats due to their difference from conventional infrastructure-based wireless networks.

The Lack of Pre-fixed Infrastructure means there is no centralized control for the network services. The network functions by cooperative participation of all nodes in a distributed fashion. The decentralized decision making is prone to the attacks that are designed to break the cooperative algorithms. A malicious user could simply block or modify the traffic traversing it by refusing to cooperate and break the cooperative algorithms. Moreover, since there are no trusted entities that can calculate and distribute the secure keys, the traditional key management scheme cannot be applied directly.

Dynamically Changing Topology aids the attackers to update routing information maliciously by pretending this to be legitimate topological change. In most routing protocols for ad hoc networks, nodes exchange information about the topology of the network so that the routes could be established between communicating nodes. Any intruder can maliciously give incorrect updating information. For instance, DoS attack can be easily launched if a malicious node floods the network with spurious routing messages. The other nodes may unknowingly propagate the messages.

Energy Consumption Attack is more serious as each mobile node also forwards packets for other nodes. An attacker can easily send some old messages to a node, aiming to overload the network and deplete the node's resources. More seriously, an attack can create a *rushing attack* by sending many routing request packets with high frequency, in an attempt to keep other nodes busy with the route discovery process, so the network service cannot be achieved by other legitimate nodes.

Node Selfishness is a specific security issue to ad hoc network. Since routing and network management are carried by all available nodes in ad hoc networks, some nodes may selfishly deny the routing request from other nodes to save their own resources (e.g., battery power, memory, CPU).

3   Security Countermeasures

Secure mobile computing is critical in the development of any application of wireless networks.

3.1   Security Requirements

Similar to traditional networks, the goals of securing mobile computing can be defined by the following attributes: availability, confidentiality, integrity, authenticity and non-repudiation.

Availability ensures that the intended network services are available to the intended parties when needed.

Confidentiality ensures that the transmitted information can only be accessed by the intended receivers and is never disclosed to unauthorized entities.

Authenticity allows a user to ensure the identity of the entity it is communicating with. Without authentication, an adversary can masquerade a legitimate user, thus gaining unauthorized access to resource and sensitive information and interfering with the operation of users.

Integrity guarantees that information is never corrupted during transmission. Only the authorized parties are able to modify it.

Non-repudiation ensures that an entity can prove the transmission or reception of information by another entity, i.e., a sender/receiver cannot falsely deny having received or sent certain data.

3.2   WLAN Basic Security Mechanisms

The IEEE 802.11b standard identifies several security services such as encryption and authentication to provide a secure operating environment and to make the wireless traffic as secure as wired traffic. In the IEEE 802.11b standard, these services are provided largely by the WEP (Wired Equivalent Privacy) protocol to protect link-level data during wireless transmission between clients and APs. That is, WEP does not provide any end-to-end security but only for the wireless portion of the connection. Apart from WEP, other well-known methods that are built into 802.11b networks are: Service Set Identifier (SSID), Media Access Control (MAC) address filtering, and open system or shared-key authentication.

SSID. Network access control can be implemented using an SSID associated with an AP or group of APs. Each AP is programmed with an SSID corresponding to a specific wireless LAN. To access this network, client computers must be configured with the correct SSID. Typically, a client computer can be configured with multiple SSIDs for users who require access to the network from a variety of different locations. Because a client computer must present the correct SSID to access the AP, the SSID acts as a simple password and, thus, provides a measure of security. However, this minimal security is compromised if the AP is configured to "broadcast" its SSID. When this broadcast feature is enabled, any client computer that is not configured with a specific SSID is allowed to receive the SSID and access the AP.

MAC Address Filtering. While an AP can be identified by an SSID, a client computer can be identified by a unique MAC address of its 802.11b network card. To increase the security of an 802.11b network, each AP can be programmed with a list of MAC addresses associated with the client computers allowed to access the AP. If a client's MAC address is not included in this list, the client is not allowed to associate with the AP. MAC address filtering (along with SSIDs) provides improved security, but is best suited to small networks where the MAC address list can be efficiently managed. Each AP must be manually programmed with a list of MAC addresses, and the list must be kept up-to-date.

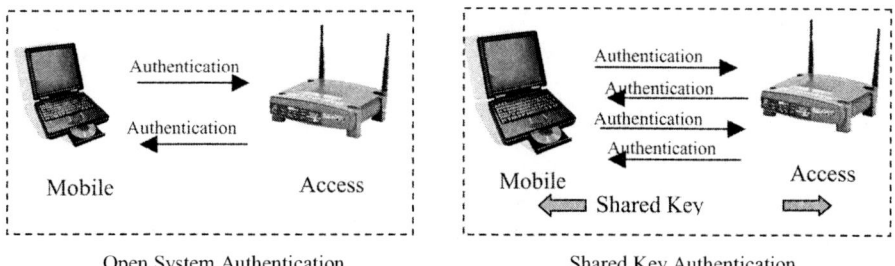

Open System Authentication   Shared Key Authentication

Fig. 2. IEEE 802.11 Authentication Modes

Authentication. In a WLAN, an AP must authenticate a client before the client can associate with the AP or communicate with the network. The IEEE 802.11b standard has defined two types of authentication methods: open system and shared Key. Open system authentication allows any device to join the network, assuming that the device SSID matches the access point SSID. Alternatively, the device can use the "ANY" SSID option to associate with any available AP within range, regardless of its SSID. With Shared Key authentication, only those PCs that possess the correct authentication key can join the network. When wireless devices are configured to operate in this mode, Wired Equivalent Privacy (WEP) data encryption is used and it requires that the station and the AP have the same WEP Key to authenticate, thus preventing the client from sending and receiving data from the AP, unless the client has the correct WEP key. Figure 2 illustrates the two authentication modes. By default, IEEE 802.11b wireless devices operate in an open system authentication mode. Both of these authentication modes are one-way authentication, i.e., the mobile clients can be authenticated by the APs, but the authenticity of APs is not authenticated. Thereby, a rogue node may masquerade as an AP and establish communication with the mobile nodes.

WEP-Based Security. WEP security protocol encrypts the communication between the client and an AP. It employs the symmetric key encryption algorithm, RC4 Pseudo Random Number Generator. Under WEP, all clients and APs on a wireless network typically use the same key to encrypt and decrypt data. The key resides in the client computer and in each AP on the network. The 802.11b standard does not specify a key-management protocol, so all WEP keys on a network usually must be managed manually and are static for a long period of time. This is a well-known security vulnerability. Support for WEP is standard on most current 802.11 cards and APs. WEP specifies the use of a 40-bit encryption key. The encryption key is concatenated with a 24-bit "initialization vector" (IV), resulting in a 64-bit key. This key is input into a pseudorandom number generator. The resulting sequence is used to encrypt the data to be transmitted. However, WEP encryption has been shown to be vulnerable to several cryptographic attacks that reveal the shared key used to encrypt and authenticate data, such as IV key reuse, keystream reuse, message injection, and so on [3][4]. Because of this, static WEP is only suitable for small, tightly managed networks with low-to-medium security requirements.

It is clear that these traditional WLAN security that relies on SSIDs, open system or shared key authentication, MAC address filtering, and static WEP keys is better than no security at all, but it is insufficient, and a new security solution is needed to secure mobile computing.

3.3   Advanced WLAN Security Mechanisms

WEP2. As an interim improved solution to the many flaws of WEP, the TGI Working Group of the IEEE proposed WEP2. Unfortunately, similar to major problems with WEP, WEP2 is not an ideal solution. The main improvement of WEP2 is to increase the IV key space to 128 bits, but it fails to prevent IV replay and still permits IV key reuse. The weakness of plaintext exploits and same IV replay are the same with that in WEP. In WEP2, the authentication is still a one-way authentication mode, and the problem of rogue AP is not solved.

Virtual Private Networking (VPN). To further address the concerns with WEP security, many organizations adopt the virtual private network (VPN) technology. The VPN approach has a number of advantages. Firstly, it is scalable to a large number of 802.11 clients and has low administration requirements for the IEEE 802.11 APs and clients. Secondly, the VPN servers can be centrally administered and the traffic to the internal network is isolated until VPN authentication is performed. Thirdly, if this approach is deployed then a WEP key and MAC address list management is not needed because of security measures created by the VPN channel itself. This is a good solution for networks, particularly with existing VPN infrastructure for remote access.

However, though the VPN approach enhances the air-interface security significantly, this approach does not completely address security on the enterprise network. For example, authentication and authorization to enterprise applications are not always addressed with this security solution. Some VPN devices can use user-specific policies to require authentication before accessing enterprise applications. Another drawback in the VPN solution is the lack of support for multicasting, which is a technique used to deliver data efficiently in real time from one source to many users over a network. Multicasting is useful for streaming audio and video applications such as press conferences and training classes. Also, a minor issue of VPNs is that roaming between wireless networks is not completely transparent. Users receive a logon dialog when roaming between VPN servers on a network or when the client system resumes from standby mode. Some VPN solutions address this issue by providing the ability to "auto-re-connect" to the VPN.

IEEE 802.11i Robust Security Network (RSN) Standard. To help overcome this security gap in wireless networks, the IEEE 802.11 working group instituted Task Group i (802.11i) has proposed significant modifications to the existing IEEE 802.11 standard as a long-term solution for security, called Robust Security Network (RSN). An interim draft of IEEE 802.11i is now available, known as Wi-Fi Protected Access (WPA). The draft of IEEE 802.11i standard consists of three major parts: Temporal Key Integrity Protocol (TKIP), counter mode cipher block chaining with message authentication codes (counter mode CBC-MAC) and IEEE 802.11x access control.

TKIP primarily addresses the shortcomings of WEP and fixes the well-known problems with WEP, including small initialization vector (IV) and short encryption keys. TKIP uses RC4, the same encryption algorithm as WEP to make it updateable from WEP, but it extends the IV from 24-bit to 48-bit in order to defend against the existing cryptographic attacks against WEP. Moreover, TKIP implements 128-bit encryption key to address the short-key problem of WEP. TKIP changes the way keys are derived and periodically rotates the broadcast keys to avoid the attack that is based on capturing large amount of data encrypted by the same key. It also adds a message-integrity-check function to prevent packet forgeries. TKIP is part of the existing WPA industry standard.

Counter mode CBC-MAC is designed to provide link layer data confidentiality and integrity. A new strong symmetric encryption standard, advanced encryption standard (AES) is deployed, in which a 128-bit encryption key and 48-bit IV are used. Different from TKIP, counter mode CBC-MAC has little resemblance to WEP, and it is set to be a part of the second generation WPA standard.

IEEE 802.11x is an authentication and key management protocol, which is designed for wired LANs, but has been extended to WLANs. IEEE 802.11x authentication occurs when a client first joins a network. Then authentication periodically recurs to verify the client has not been subverted or spoofed. The centralized, server-based 802.11x authentication process for WLANs is shown is Figure 3. A mobile client sends an authentication request to an associated access point. The access point forwards the authentication information to a back-end authentication server via Remote Authentication Dial-In User Service (RADIUS) for verification. Once the verification process completes, the authentication server sends a response message to the access point that the client has been authenticated and network access should be granted. In 802.11i, the response message should contain the cryptographic keys sent to the client. After that, the access point transfers the mobile client to authenticated state and allows the access of the mobile client.

IEEE 802.1X is not a single authentication method; rather it utilizes Extensible Authentication Protocol (EAP) as its authentication framework. This means that 802.1X-enabled switches and access points can support a wide variety of authentication methods, including certificate-based authentication, smartcards, token cards, one-time passwords, etc. However, the 802.1X specification itself does not specify or mandate any authentication methods. Since switches and access points act as a "pass through" for EAP, new authentication methods can be added without the need to upgrade the switch or access point, by adding software on the host and backend authentication server. Several common EAP methods have been defined in various IETF draft or other industry documents, such as EAP-MD5, EAP-TLS, etc. While TKIP and counter mode CBC-MAC are still unimplemented by most vendors, 802.11x support is already integrated into some operating systems.

Mobile Client   Access Point   Authentication Server

Fig. 3. IEEE 802.11x Authentication

In summary, TKIP/WPA provides enhanced security for existing infrastructure. Counter mode CBC-MAC protects the data integrity and confidentiality and 802.11x presents a fully extensible authentication mechanism. Combining these techniques, 802.11i RSN is significantly stronger than WEP. However, 802.11i has not yet been standardized. It requires changes to firmware and software drivers and may not be backward-compatible with some legacy devices and operating systems. Hence, not all users will be able to take advantage of it. A phased adoption process for this standard is anticipated because of the large amount of installed 802.11 devices.

3.4   Additional Security Requirements of Ad Hoc Networks

As ad hoc networking is somewhat different from the traditional approaches, designing an efficient security scheme to protect ad hoc networks is confronted with several new requirements.

First, the key management mechanism should be implemented in a distributed fashion[1]. Ad hoc network is a distributed network, in which network connectivity and network services, for example, routing, are maintained by the nodes themselves within the network. Each node has an equal functionality. There are no dedicated service nodes, which can work as a trusted authority to generate and distribute the network keys or provide certificates to the nodes, as the certificate authority (CA) does in the traditional public key infrastructure (PKI) supported approaches. Even if the service node can be defined, keeping the availability of the service node to all the nodes in such a dynamic network is not an easy task. Moreover, with limited physical protection, the service node is prone to a single point of failure, i.e., by only damaging the service node, the whole network would be paralyzed. Thus, distributed key generation and management approach is needed to secure ad hoc networks.

Secondly, light-weight authentication and encryption scheme with resource awareness are required. The low resource availability necessitates their efficient utilization and prevents the use of complex authentication and encryption algorithms. Public-key cryptography based authentication and encryption mechanisms are fully developed in securing traditional networks. Unfortunately, generation and verification of digital signatures are relatively expensive, which limits its acceptance to ad hoc networks. Symmetric cryptography is more efficient than public-key based asymmetric primitives due to its moderate resource consumption, but it requires both the sender and receiver to share a secret. In ad hoc networks, the problem is how to distribute the shared keys safely so that only the two parties (correct sender and receiver) would get it and not anyone else. It is thus challenging to define some new efficient cryptography algorithms for designing a light-weight authentication and encryption scheme.

Thirdly, combination of intrusion prevention and intrusion detection mechanisms is necessary. The work on securing wireless ad hoc networks can be classified into two types, intrusion prevention and intrusion detection [12] [13]. Intrusion prevention implies developing secured protocols or modifying the logic of existing protocols to

---

[1]   Here, we consider the ad hoc network working in a truly ad hoc mode. Depending on the network origin, an ad hoc network can be a planned network, in which some initial data structure such as pre-distributed public keys and shared keys can be assumed.

make them secure. Most of the key based security protocols belong to this type. The idea of intrusion detection is to characterize the user normal behavior within the network in terms of a set of relevant system features. Once the set of system features is selected, the classification model is built to detect the anomalies from its normal behavior. Currently, the research on intrusion prevention and intrusion detection is done separately, and intrusion prevention has been paid more attention. But actually, they are not independent of each other, and should work together to provide security services. For example, intrusion prevention approaches can efficiently deal with the attacks coming from the outsiders by constraining the network access control, but it has no way to handle the denial of service attacks performed by the compromised nodes who have all the keys to access the network. Indeed, some active attacks can be efficiently detected because of a large deviation of attackers' behavior from the normal user behavior. Therefore, a security scheme combining these two mechanisms is suitable to better secure ad hoc networks.

3.5   Security Schemes for Ad Hoc Networks

In the recent research of security in wireless ad hoc networks, several good security approaches have been proposed, and they generally fall into three categories, secure routing, trust and key management, and service availability protection.

Secure Routing

Establishing correct route between communicating nodes in ad hoc network is a prerequisite for guaranteeing the messages to be delivered in a timely manner. If routing is misdirected, the entire network can be paralyzed. The function of route discovery is performed by routing protocols, and hence securing routing protocols has been paid more attention. The routing protocols designed for ad hoc networks assume that all the nodes within the network behave properly according to the routing protocols and no malicious nodes exist in the network. Obviously this assumption is too strong to be practical. The use of *asymmetric* key cryptography have been proposed [5][6] to secure ad hoc network routing protocols. Dahill et al. [5] propose ARAN, in which every node forwarding a route request and route reply message must sign it. Although their approach could provide strong security, performing a digital signature on every routing packet could lead to performance bottleneck on both bandwidth and computation. In [6], Zapata proposed a secure extension of the Ad Hoc On-demand Distance Vector routing protocol, named SAODV. The basic idea of SAODV is to use RSA signature and one-way hash chain (i.e., the result of n consecutive hash calculations on a random number) to secure the AODV routing messages. The effectiveness of this approach is sensitive to the tunneling attacks. IP spoofing is still possible in SAODV routing protocol.

Using public-key cryptography imposes a high processing overhead. Some researchers have proposed the use of *symmetric* key cryptography for authenticating ad hoc routing protocols, based on the assumption that a security association (a shared key K_{SD}) between the source node S and the destination node D exists. In [7], a secure ad hoc network routing protocol based on the design of the Destination-Sequenced Distance-Vector routing protocol, called SEAD, has been proposed. In this approach, one-way hash function is employed to authenticate routing updates sent by a distance-vector protocol. Another approach, Ariadne [8], proposed by the same authors, uses one broadcast authentication scheme TESLA [9] for securing DSR routing protocol. Venkatraman and Agrawal [10] have proposed a scheme that prevents replay attack by authenticating route reply messages. The scheme implements *Message authentication code* (MAC) to ensure integrity of route request packets. Papadimitratos and Hass [11] also proposed a symmetric key based Securing Routing Protocol (SRP), which can be applied to several existing routing protocols. Symmetric encryption is more suitable for ad hoc networks due to its lower resource consumption. The problem is how to distribute the key in the first place.

Some efforts are also being made to use intrusion detection mechanism in protecting ad hoc networks. Zhang and Lee [12] present a distributed intrusion detection and response architecture, which provides an excellent guide on designing intrusion detection system in wireless ad hoc networks. Sergio Marti et al. [13] introduced Watchdog and Pathrater techniques that improve throughput in an ad hoc network by identifying misbehaving nodes that agree to forward the packets but never do so. The Watchdog can be considered as a simple version of intrusion detection agent to identify misbehaving nodes, and the Pathrater works as the response agent to help routing protocols avoid these nodes. However, the Watchdog can only detect the nodes who do not forward the packets, and the method only works on the source routing protocol since two-hop routing information is needed. In [14], two different detection models, distributed hierarchical model and completely distributed model, are proposed and the intrusion detection can be performed in a supervised or unsupervised way depending on the availability of attack data. The main problems of intrusion detection approach rely on two aspects: first, not all malicious behaviors are detectable, in particular, the dynamically changing topology in ad hoc networks makes detection more difficult; second, even if some attacks can be detected, a false alarm rate is still expected to be present. Therefore, intrusion detection usually works as a complementary approach to provide a second line of defense to the network.

Trust and Key Management

Most of the protocols discussed above make an assumption that efficient key distribution and management has been implemented by some kind of key distribution center, or by a certificate authority, which has super power to keep connecting to the network and can not be compromised, but how to maintain the server safely and keep it available when needed presents another major issue and can not be easily solved.

To mitigate this problem, the concept of threshold secret sharing is introduced and there are two proposed approaches. Zhou and Hass [15] use a partially distributed certificate authority scheme, in which a group of special nodes is capable of generating partial certificates using their shares of the certificate signing key. This work is the first to introduce the threshold scheme into security protocols in ad hoc networks and provides an excellent guide to the following work. The problem of this solution is that it still requires an administrative infrastructure available to distribute the shares to the special nodes and issue the public/private key pairs to all the nodes. How to keep the n special nodes available when needed and how the normal nodes know how to locate the server nodes make the system maintenance difficult. In [16], Kong et al. proposed another threshold cryptography scheme by distributing the RSA certificate signing key to all the nodes in the network. This scheme can be considered as having a fully distributed certificate authority, in which the capabilities of certificate authority are distributed to all nodes and any operations requiring the certificate authority's private key can only be performed by a coalition of k or more nodes. This solution is better in the sense that it is easier for a node to locate k neighbor nodes and request the certificate authority service since all nodes are part of the certificate authority service, but it requires a set of complex maintenance protocols.

Service Availability Protection

To protect the network from the problem of service unavailability due to the existence of selfish nodes, Buttyan and Hubaux proposed so-called Nuglets [17] that serve as a per-hop payment in every packet or counters to encourage forwarding. Both nuglets and counters reside in a secure module in each node, are incremented when nodes forward for others and decremented when they send packets as an originator. Another approach, the Collaborative Reputation Mechanism (CORE) [18] is proposed, in which node cooperation is stimulated by a collaborative monitoring and a reputation mechanism. Each network entity keeps track of other entities' collaboration using a technique called *reputation*. The reputation is calculated based on various types of information. Since there is no incentive for a node to maliciously spread negative information about other nodes, simple denial of service attacks using collaborative technique itself are prevented.

4   Conclusion

Mobile computing technology provides anytime and anywhere service to mobile users by combining wireless networking and mobility, which would engender various new applications and services. However, the inherent characteristics of wireless communication and the demand for mobility and portability make mobile computing more vulnerable to various threats than traditional networks. Securing mobile computing is critical to develop viable applications.

In this article, we discussed the security issues faced by mobile computing technology. We analyzed the various security threats and describe the existing current countermeasures. We have seen that many security solutions have been proposed to securing WLANs, but no one is able to claim that it solves all the security problems, or even most of them. In essence, secure mobile computing would be a long-term ongoing research topic.

Acknowledgement

This work has been supported by the Ohio Board of Regents, Doctoral Enhancement Funds and the National Science Foundation under Grant No. CCR-0113361.

Reference

[1]   "LAN Standards of the IEEE Computer Society. Wireless LAN medium access control (MAC) and physical layer (PHY) specification. IEEE Standard 802.11, 1999 Edition," 1999.

[2]   D. P. Agrawal and Q-A. Zeng, *Introduction to Wireless and Mobile Systems*, Brooks/Cole publisher, 2002.

[3]   J. Walker, "Overview of IEEE 802.11b Security", http://www.intel.com/technology/itj/q22000/pdf/art_5.pdf.

[4]   N. Borisov, I. Goldberg, and D. Wagner, "Intercepting Mobile Communications: the Insecurity of 802.11", http://www.isaac.cs.berkeley.edu/isaac/mobicom.pdf.

[5]   B. Dahill, B. N. Levine, E. Royer, and C. Shields, "A Secure Routing Protocol for Ad Hoc Networks," *Technical Report UM-CS-2001-037*, Electrical Engineering and Computer Science, University of Michigan, August 2001.

[6]   M. G. Zapata, "Secure Ad hoc On-Demand Distance Vector Routing," *ACM SIGMOBILE Mobile Computing and Communications Review*, Vol. 6 , No. 3, pp. 106-107, 2002.

[7]   Y. C. Hu and D. B. Johnson and A. Perrig, "SEAD: Secure Efficient Distance Vector Routing in Mobile Wireless Ad-Hoc Networks," *Proceedings of the 4th IEEE Workshop on Mobile Computing Systems and Applications (WMCSA'02)*, pp. 3-13, 2002.

[8]   Y. C. Hu, A. Perrig, and D. B. Johnson, "Ariadne: A Secure On-Demand Routing Protocol for Ad Hoc Networks," *Proceedings of the 8th ACM International Conference on Mobile Computing and Networking*, September, 2002.

[9]   Perrig, R. Canetti, B. Whillock, "TESLA: Multicast Source Authentication Transform Specification", *http://www.ietf.org/internet-drafts/draft-ietf-msec-tesla-spec-00.txt*, October 2002.

[10]   L. Venkatraman and D. P. Agrawal, "Startegies for Enhancing Routing Security in Protocols for Mobile Ad hoc Networks," *JPDC Special Issue on Mobile Ad Hoc Networking and Computing*, Vol. 63, No. 2, Feb. 2003, pp. 214-227.

[11]   P. Papadimitratos and Z. Haas, "Secure Routing for Mobile Ad Hoc Networks," Proceedings of the SCS Communication Networks and Distributed Systems Modeling and Simulation Conference, January 2002.

[12]   Y. Zhang and W. Lee, "Intrusion Detection in Wireless Ad-Hoc Networks," Proceedings of the *6th International Conference on Mobile Computing and Networking (MobiCom'2000)*, Aug 2000.

[13]   S. Marti, T. Giuli, K. Lai, and M. Baker, "Mitigating Routing Misbehavior in Mobile Ad Hoc Networks," *Proceedings of the 6th International Conference on Mobile Computing and Networking (MOBICOM'00)*, pp.255-265, August 2000.

[14]   H. Deng, Q-A. Zeng, and D. P. Agrawal, "SVM-based Intrusion Detection System for Wireless Ad Hoc Networks," *IEEE Vehicular Technology Conference*, Orlando, October 6-9, Fall, 2003.

[15]   L. Zhou and Z. J. Hass,"Securing Ad Hoc Networks," *IEEE Networks Special Issue on Network Security*, November/December, 1999.

[16] Kong, P. Zerfos, H. Luo, S. Lu and L. Zhang, "Providing Robust and Ubiquitous Security Support for Mobile Ad-Hoc Networks," *Proceedings of the IEEE 9th International Conference on Network Protocols (ICNP'01)*, 2001.

[17] Levente Buttyan and Jean-Pierre Hubaux, "Enforcing Service Availability in Mobile Ad-Hoc WANs," *Proceedings of the IEEE/ACM Workshop on Mobile Ad Hoc Networking and Computing (MobiHOC)*, Boston, MA, USA, August 2000.

[18] Pietro Michiardi, Refik Molva, "Core: A COllaborative REputation mechanism to enforce node cooperation in Mobile Ad Hoc Networks," Proceedings of the Conference on Communication and Multimedia Security, 2002.

Towards Privacy Preserving
Distributed Association Rule Mining

Mafruz Zaman Ashrafi, David Taniar, and Kate Smith

School of Business Systems, Monash University,
Clayton, VIC 3800, Australia
{Mafruz.Ashrafi,David.Taniar,Kate.Smith}@infotech.monash.edu.au

Abstract. Data mining is a process that analyzes voluminous digital data in order to discover hidden but useful patterns. However, discovery of such hidden patterns may disclose some sensitive information. As a result privacy becomes one of the prime concerns in data mining research. Since distributed association mining discovers global association rules by combining models from various distributed sites hence it breaches data privacy more often than it does in the centralized environments. In this work we present a methodology that generates global association rules without revealing confidential inputs of individual sites. One of the important outcomes of the proposed technique is that, it has an ability to minimize the collusion problem. Furthermore, the global model generated by this method is based on the exact global support of each itemsets, which is indeed a desirable property of distributed association rule mining.

1   Introduction

Data mining is a process of finding useful information from digital data collected by various business and government applications. The prime goal of data mining is to discover useful patterns that are hidden within the digital data. In order to discover useful patterns from such digital data we found a number of techniques such as association, classification, clustering etc. in the data mining literature [1].

Modern organizations are distributed in various geographical locations. Various business applications used by such organizations normally store their day-to-day data to corresponding site. Indeed, data of such organizations increase every day. Discovering useful patterns from such organizations using a centralized data mining approach is not always feasible due to a huge network communication cost which is imposed when merging all datasets into a central location. As a result distributed data mining emerges as a new sub-area of research in data mining domain [2].

Data mining algorithms analyze enormous digital data and discover hidden patterns within the dataset and hence impose a threat that such a discovery may breach the privacy of data. As a result preserving privacy appears as a prime concerns in the field of data mining. Moreover, distributed data mining algorithms discover patterns

S. R. Das, S. K. Das (Eds.): IWDC 2003, LNCS 2918, pp. 279-289, 2003.

beyond the organization boundary and combine various local patterns. Therefore, the resultant patterns may disclose private input of individual site and hence threaten the privacy.

One of the most important fields of distributed data-mining domain is association mining. It has attracted huge attention from numerous research communities. Many interesting and efficient mining of distributed/parallel association rule mining algorithms have been proposed in the data mining literature [3, 4]. Most of those algorithms overlook the privacy issue. However, we believe the privacy should be the main concern of distributed association mining; otherwise the resultant patterns may reveal sensitive information and hence participating sites may lose their business.

For example, a multinational company would like to mine its data to find global association rules, however, the laws of individual country may come as an impediment to share global data. Furthermore, privacy does not always means that disclosure of individuals or personal information rather it may disclosure of corporate information or their transaction details. Indeed, such disclosure may allow identifiable information or corporate plans, which may thread corporate business gain.

There are number of researches have been found in the data mining area to perform the mining task without compromising the privacy. Most of those works apply distortion or randomization techniques to the original dataset and generate association rules from that [5-7]. However, this distortion is not enough when we generate rules from various geographically distributed sites. Because distortion may still reveal individual support counts when we uniformly distort dataset of various sites.

In this work we present a methodology that generates global association rules without revealing confidential inputs of individual sites. One of the important outcomes of the proposed technique is that, it has the ability to minimize a collusion problem, which occurs when two sites on the chain collude to find the exact support of other site. We accomplish it without increasing the overall communication cost at a great extend. Furthermore, it diminishes the reconstruction problem, which is raised when we distort transactions of a dataset by using different randomization techniques.

The rest of the paper is organized as follows: In Section 2 we describe the background of distributed association mining and problem we encounter when considering privacy preserving rule mining. Related work of privacy preserving association mining is described in section 3. Section 4 describes our proposed privacy preserving distributed association rules generation methodology. Performance evaluation and comparison are described in section 5 and we conclude at section 6.

2   Distributed Data Mining: Background

Distributed Data Mining (DDM) intends to discover rules from different datasets that are distributed across multiple sites and interconnected by a communication network. It accepts that to combine those datasets in a centralized site require huge amounts of network communication. So, it offers a new technique to discover knowledge or patterns from such loosely coupled distributed datasets and produce global rule models by using minimal network communication.

DDM produces the global rules models in such a way that one would achieve the same model if the datasets from different sites were combined in a single site. It treats all distributed datasets as a single virtual table [9]. In general every distributed mining algorithm performs the following three tasks:

- Analyze local data and generate local models.
- Exchange local models with other sites.
- Generate global models by combining all local models.

2.1 Distributed Association Mining

Distributed association rule mining is subarea of distributed data mining. The prime motto behind it is to reduce communication cost in such a way that the overall cost will be less than the cost if we combined all datasets in a centralized site. For example, consider there are $1, 2, 3... n$ sites involved in the mining task and G_c be the overall communication cost for generating global frequent itemsets of various length and C be the total communication cost of combining n number of datasets in centralized site. Then, $G_c < C$ will be the prime and desirable property of the distributed association rule mining algorithms.

Distributed association rule mining can be described as follows: let DB^1, DB^2 and DB^3 be three datasets distributed geographically into three different sites such as O^1, O^2 and O^3 and their corresponding size is represented by D^1, D^2 and D^3. Let X^1, X^2 and X^3 three itemsets and $X^1.sup$, $X^2.sup$ and $X^3.sup$ their respective support counts generated from those three sites. For a given minimum global support S, let itemset $X = X^1.sup + X^2.sup + X^3.sup$ is the global support count. Itemset X will be globally frequent if $X > S$ and correspondingly $X^1.sup$ will be locally large if $X^1.sup > (S \times D^1)$ $/ (D^1 + D^2 + D^3)$. An association rule $X \Rightarrow Y$, $(\exists\ Y \subset X)$ is said to be a global rule if $X > S$ and $Y > S$.

2.2 Privacy Preserving Distributed Association Mining: Problem Definition

Consider $X = \{x_1, x_2, x_3 ... x_n\}$, $Y = \{y_1, y_2, y_3 ... y_n\}$ and $Z = \{z_1, z_2, z_3 ... z_n\}$ be support count of candidate *k-itemsets* geographically distributed over three different sites such as O_1, O_2 and O_3. Let S and C be the global support and confidence threshold values. In order to generate global frequent *k-itemsets* $F = \lambda\ (X + Y + Z)$ each site needs to send their respective support count of each candidate itemset λ to the other sites. Although broadcasting of support counts does not disclose any information about individual transaction, however it may disclose some valuable information about each site such as data size, exact support of each itemset etc., which may subsequently thread the privacy. Hence, the challenge is to find global frequent itemset F without revealing λ (i.e. support) of each site.

Furthermore, in a distributed association-mining context, messages exchange between different sites is considered as one of the main tasks. And this task becomes expensive when each site broadcasts large numbers of support counts. Indeed, the communication cost of all distributed association-mining algorithms will increase when privacy of each site input is considered as a prime objective. Hence, the problem can be defined as finding all association rules from various distributed sites without revealing the support counts of individual site and without increasing the overall communication cost.

3   Related Work

There are several numbers of frameworks that have been proposed for maintaining the privacy of association rules [5-8]. However, most of them were proposed for sequential or centralized association mining. The MASK [5] (Mining Association with Secrecy Konstraints) was proposed for centralized environment to maintain privacy and accuracy of resultant rules. This approach was based on simple probabilistic distortion of user data, employing random numbers generated from pre-defined distributed function. However, distortion process will employ system resources for long period when dataset has large number of transactions. Furthermore, if we consider this algorithm on distributed environment then we need uniform distortion among various sites in order to generate unambiguous rules. However this uniform distortion can disclose confidential inputs of individual site and may breach privacy of data (i.e. exact support); hence not suitable for distributed mining context.

Evfimievski et al. [6] provide a randomization technique in order to preserve privacy of association rules. The authors analyzed this technique in an environment where there are number of clients connected to a server. Each client sends a set of items to the server where association rules are generated. During the sending process client modifies that set (i.e. items) according to its own randomization policy, hence server is unable to find the exact information about a client. However, this assumption is not suitable for distributed association mining because it generates frequent itemsets by aggregating support counts of all clients (i.e. sites). If the randomization policy of each site differs with others then we will not be able to generate the exact support of an itemset. Subsequently, the resultant global frequent itemsets will be erroneous. Hence, we may not able to discover some useful rules. Furthermore, this technique individually disguises each attribute and data quality will degrade significantly when number of attributes in a dataset is large.

Vaidya et al. [7] proposed a technique to maintain privacy of association rules in vertically partitioned distributed across two data sources where each data site holds some attributes of each transaction. However, if the number of disjoints attributes among the site is high then this technique incurs huge communication costs. Furthermore, this technique worked only for two sites hence not scalable.

Another related work is [8], where they consider the problem of association rule mining for horizontally partitioned dataset, i.e. each site shares a common schema but has different records. The authors proposed two different protocols: secure union of locally large itemsets and testing support threshold without revealing support counts. The former protocol uses cryptography to encrypt local support count therefore it is not possible to find which itemset belongs to which site. However, it reveals the number of itemsets having a common support e.g. two itemsets are supported by two sites and rest by only one site. The latter protocol adds a random number with each support counts and finds excess support. Finally those excess supports are sent to the second site where it learns nothing about the first site actual dataset size or support. The second site adds its excess support and sends the value on until it reaches the last site. However, this protocol can raise a collusion problem. For example site i and $i+2$ in the chain can collude to find the exact excess support of site $i+1$. Furthermore, this protocol only discovers an itemset, which is globally large hence only able to discover the exact global support of an itemset when different datasets have equal number of transactions. As a result, when the number of transaction varies at different datasets then the inconsistence problem (i.e. same rule may varies in confidence at various site) may arise.

In this work, we propose an efficient technique that maintains privacy of distributed association rules according to secure multiparty computation definition [10]. The proposed technique accomplishes to find the exact support of each global frequent itemset without revealing the candidate support counts of individual site. The proposed technique has the ability to minimize the collusion problem without increasing the overall communication cost. Furthermore it diminishes the reconstruction problem, which is raised when we distort transactions by using different randomization techniques.

4   Proposed Method

In this section, we will describe a methodology that maintains privacy of distributed association rule mining. Before embarking on the details, let us find out the rationale, why each site in distributed association-mining share its support count of each itemset with all other sites. Firstly, in the context of distributed association mining each participating site needs to know whether an itemset is globally frequent or not in order to generate candidate itemsets for the next pass. Without that piece of information distributed association mining algorithms will not be able to generate global candidate itemsets. Secondly, if any site generates rules based on partial support count of an itemset then the inconsistency problem (i.e. same rule will have various confidence at various site) will arise.

Security and privacy in distributed computing has broader meaning. Maintaining privacy of distributed association mining is a multiparty computation problem [8]. For this reason we need secure multiparty computation solutions in order to maintain privacy of distributed association mining. The goal of secure multiparty computation is to compute a function where multiple parties hold their inputs and at the end of it all parties know about the result but nothing else. Details discussions about various approaches of the above mention problem can be found in [10,11].

Since, each participating site in distributed association mining should possess a minimum level of trust and for this reason let us consider all participating sites as *semi-honest*. However, sharing confidential inputs among the various semi-honest sites (i.e. support, number of transaction, etc.) are indeed vulnerable, because a semi-honest site possess following characteristics:

-   Follow multi-party computation protocols perfectly.
-   Keep record of all intermediate computation.
-   Be capable of deriving additional information using those records.

From the above discussion it is clear that a distributed association rule mining algorithm needs to share support counts global candidate itemsets with all other participating sites. And sharing confidential inputs with every participating semi-honest site is vulnerable. For this reason, the aim of this work is to discover global frequent itemsets with an exact support of each itemset without revealing support counts of each itemset of individual site. In to order to achieve this we made the following assumptions:

–   Support counts of each itemset are private and cannot be disclosed.
–   Number of records that each site's dataset has cannot be disclosed.
–   Number of participating sites is equal to n, where $n > 2$.
–   Global rules are generated using global support counts of each itemset.

4.1   Methodology

We maintain the privacy of each site input (i.e. support counts) by obfuscating the original support of each itemsets. In order to obfuscate a random number is added with each candidate itemset, hence each site that receives it, is unable to find the actual support of that itemset. For example, suppose we have a large number R, which is a sum of two numbers such as N_1 and N_2. If we consider the value of R is known and N_1 and N_2 are unknown then the value of N_1 or N_2 remains private and secure. This is because it is difficult to know either N_1 or N_2 without knowing the exact value of N_1 or N_2. In the proposed method we use this technique and consider N_1 as an exact support count of an itemset and N_2 as a random number.

The proposed method works in a round robin fashion. It has two distinct phases, namely *obfuscation* and *de-obfuscation*. In the *obfuscation phase* each support counts of candidate itemset is obfuscated and sent to the adjacent site. Each adjacent site obfuscates its own support counts and aggregates it with the receiving support counts. This sending process continues until it reaches the last site. Then those obfuscated support counts are sent to the first site where *de-obfuscation phase* is started by subtracting its own generated random number from each of the aggregated support counts. At the end of the de-obfuscation phase, each candidate itemset checks in order to prune away every non-frequent global itemset. Finally, those global frequent support counts are sent to all other sites. However, the proposed method does not introduce any itemsets generation algorithm, which means each site generates the local frequent itemset in same manner as it does each site of the DMA [3] algorithm.

One of the important outcomes of the proposed method is that it minimizes the collusion problem. It is because each site has its own random number and that random number is added and subtracted only by that site, hence it requires $n-1$ sites of the chain to collude in order to find out the exact support count of site n.

Example: Consider there are three sites such as O_1, O_2 and O_3 and there are three candidate itemsets such as $\{AB, AC, BC\}$. Let support counts of those candidate of site O_1 is equal to $\{5, 3, 4\}$ at S_2, is equal to $\{10, 9, 1\}$ and at O_3, is equal to $\{5, 5, 1\}$. Let site O_1, O_2 and O_3 generate a random number $R_1 = 100$, $R_2 = 200$ and $R_3 = 200$ and each site obfuscate its own support count of candidate itemsets by adding corresponding random number with each support counts. Then, consider site O_1 sends obfuscated support counts to site O_2. When O_2 receives that obfuscate support counts it add its own obfuscate support counts with that. Since each site share same candidate itemset hence this aggregation operation can be done on the fly. Upon performing those tasks site O_2 sends it to the site O_3 that perform the same task as it does for the site O_2 and finishes the obfuscation phase.

In next phase (i.e. de-obfuscation) site O_3 sends obfuscate support counts set {520, 517, 506} to site O_1 where it subtracts random number R_1 from each support counts. However, this subtraction does not reveal any information because exact support counts and random number of other sites is embedded with in it, hence site O_1 neither derive anything from that nor it knows the exact support counts of other sites. Finally site O_1 sends it to the site O_2 that subtract the random number R_2 and send support counts to the site O_3. After subtracting random number R_3, site O_3 find the aggregated support counts {20, 17, 6} of all sites and find out the global frequent itemsets from the aggregated support counts set. However, this global support counts is total of all candidate support counts of all sites and in this case it is not possible for site O_3 to distinguish the exact support counts of site O_1 and O_2. Upon deriving the global frequent itemsets, it broadcast that itemsets to the all other participating sites.

4.2   Implementation

A general distributed association mining algorithms has three distinct phases. First it finds out the total number of overall transactions then global support counts for various itemsets length and finally generating global rules. Our proposed method also works in same manner but obfuscate at every steps. At the beginning all sites generate local frequent 1-itemsets based on the simple DMA [4] algorithm. Upon generating frequent 1-itemset each site generate a random number and add that random number with the support counts and send it to the adjacent site as shown in the figure 1. This procedure continues for all different length of candidate itemsets and finishes when there are no more global candidate itemset.

```
Input:   Local support counts C,
         Obfuscate support counts C_R;
Output: Obfuscated candidate support counts C_O
R = generate_random_number();
for each element I of C {
   I_1 = I +R;
   add(I_1)
}
if(C_R == {})
   Send(C_O);
else{
   aggregate (C_O,C_R);
   send(C_O);
}
```

Fig. 1. Obfuscation procedure

The pseudo code of de-obfuscation procedure is shown is figure 2. Upon receiving the obfuscated support count each site subtract its own random number from it and send it to the adjacent site. At the end of the de-obfuscation (i.e. last site), discover the exact global support of all itemsets and generate the global frequent itemsets from that set. Finally those global frequent support counts are broadcast all other sites.

```
Input:  Obfuscate support counts C_OR, Random number R
Output: Partial candidate support counts C_p or
        Global frequent itemset Fg
for each element I of C_OR{
    I_1 = I - R;
    add(I_1)
}
if(current site is not end site)
    Send(C_p);
else{
    F_G = generate_global_frequent_itemset(C_p);
    broadcast(F_G);
}
```

Fig. 2. De-obfuscation procedure

5   Performance Evaluation

In distributed association mining, exchanging messages between different sites is considered as one of the main tasks. Due to this reason, message optimization becomes an integral part of distributed association mining algorithms. However, when we consider privacy as a main objective such optimization becomes very difficult to achieve. There are number of parallel or distributed association algorithms that had been found in the data mining literature [3, 4, 13-16]. Based on the type of design of each algorithm, we can divide those into two groups; data and task parallelism algorithm. The former relates to the case where datasets are partitioned among different sites, where as the latter relate to the case where each site performs tasks independently but need access to the entire dataset [13] and therefore it is not suitable for distributed environment where datasets are geographically distributed among various sites. Since the main objective of this work to maintain privacy hence in this section we will compare total number of message broadcast our proposed method (PP) and other well known distributed association rule mining algorithms such as Count Distribution [3] and DMA [4].

Table 1. Database Characterstics

Name	Avg. Transaction Size	Number of Items	Number of Records
Cover Type	55	120	581012
Connect-4	43	130	67557

We have established a socket based client/server environment. We have chosen two real datasets for evaluate the performance of our algorithm. Table 1 shows the characteristics of each datasets used in this evaluation. It shows number of items, average size of the each transaction and number of transactions of each dataset. Both datasets are taken from UC Irvine Machine Learning Dataset Repository [17]. The Cover Type dataset was considered as the largest dataset of the UCI machine-learning repository. It has 55 attributes, 10 of them have continuous value and rest of the attributes have boolean value. We divide each of the continuous valued attributes into

```
Input:  Obfuscate support counts C_OR, Random number R
Output: Partial candidate support counts C_p or
        Global frequent itemset Fg
for each element I of C_OR{
    I_1 = I - R;
    add(I_1)
}
if(current site is not end site)
    Send(C_p);
else{
    F_G = generate_global_frequent_itemset(C_p);
    broadcast(F_G);
}
```

Fig. 2. De-obfuscation procedure

5   Performance Evaluation

In distributed association mining, exchanging messages between different sites is considered as one of the main tasks. Due to this reason, message optimization becomes an integral part of distributed association mining algorithms. However, when we consider privacy as a main objective such optimization becomes very difficult to achieve. There are number of parallel or distributed association algorithms that had been found in the data mining literature [3, 4, 13-16]. Based on the type of design of each algorithm, we can divide those into two groups; data and task parallelism algorithm. The former relates to the case where datasets are partitioned among different sites, where as the latter relate to the case where each site performs tasks independently but need access to the entire dataset [13] and therefore it is not suitable for distributed environment where datasets are geographically distributed among various sites. Since the main objective of this work to maintain privacy hence in this section we will compare total number of message broadcast our proposed method (PP) and other well known distributed association rule mining algorithms such as Count Distribution [3] and DMA [4].

Table 1. Database Characterstics

Name	Avg. Transaction Size	Number of Items	Number of Records
Cover Type	55	120	581012
Connect-4	43	130	67557

We have established a socket based client/server environment. We have chosen two real datasets for evaluate the performance of our algorithm. Table 1 shows the characteristics of each datasets used in this evaluation. It shows number of items, average size of the each transaction and number of transactions of each dataset. Both datasets are taken from UC Irvine Machine Learning Dataset Repository [17]. The Cover Type dataset was considered as the largest dataset of the UCI machine-learning repository. It has 55 attributes, 10 of them have continuous value and rest of the attributes have boolean value. We divide each of the continuous valued attributes into require any further computational cost. Nevertheless, the resultant rule model achieved by this method is the same as if one generates it using some of the well-known distributed/parallel algorithms. The performance study shows that the overall communication cost incurred by the proposed method is less than the CD algorithm but increases slightly when we compare it with the DMA algorithm. However, it is worth to mention that the DMA is non-secure and does not maintain privacy of data according to the definition of multiparty computation.

References

[1]   U. Fayyed, G. Piatetsky-Shapiro, P. Smyth and R. Uthurusamy "*Advance in Knowledge Discovery and Data Mining*", The AAAI Press, 1996.
[2]   M. J. Zaki, "Parallel and Distributed Association Mining: A Survey", *IEEE Concurrency*, October-December 1999.
[3]   R. Agrawal and J. C. Shafer, "Parallel Mining of Association Rules", *IEEE Transactions on Knowledge and Data Engineering*, Vol. 8, No. 6, pages. 962-969, December 1996.
[4]   D. W. Cheung, V. T. Ng, A. W. Fu, and Y. Fu, "Efficient Mining of Association Rules in Distributed Databases", *IEEE Transactions on Knowledge and Data Engineering*, Vol. 8, No. 6, pages. 911-922, 1996.
[5]   S. J. Rizvi and J. R. Haritsa "Maintaining Data Privacy in Association Rule Mining", *In Proc. of* 20*th International Conference on Very Large Databases*, Hong Kong 2002.
[6]   Evfimievski, R. Srikant, R. Agrawal and J. Gehrke, "Privacy Preserving Mining Association Rules", *In Proc. of the SIGKDDD* 2002.
[7]   J. Vaidya and C. Clifton "Privacy Preserving Association Rule Mining in Vertically Partitioned Data", *In Proc. of ACM SIGKDD*, July 2002.
[8]   M. Kantercioglu and C. Clifton "Privacy Preserving Distributed Mining of Association Rules on Horizontal Partitioned Data", *In Proc. of ACM SIGMOD Workshop of Research Issues in Data Mining and Knowledge Discovery DMKD*, 2002.
[9]   R. Wirth, M. Borth, and J. Hipp, "When Distribution is Part of the Semantics: A New Problem Class for Distributed Knowledge Discovery", *In Proc. of Joint 12th European Conference on Machine Learning and 5th European Conference on Principles and Practice of Knowledge Discovery in Databases*, 2001.
[10]   O. Goldriech "Secure Multipart Computation" Working Draft Version 1.3, June 2001.
[11]   M. Naor and K. Nissim "Communication Preserving Protocols for Secure Function Evaluation", *In Proc. of the ACM Symposium on Theory of Computing* pages 590-599, 2001.
[12]   Park and H. Kargupta (2002). "Distributed Data Mining: Algorithms, Systems, and Applications", *To be published in the Data Mining Handbook*. Editor: Nong Ye.

[13] M. J. Zaki, M. Ogihara, S. Parthasarathy, and W. Li, "Parallel Data Mining for Association Rules on Shared-Memory Multiprocessors", *Technical Report TR 618,* University of Rochester, Computer Science Department, May 1996.

[14] A. Mueller, "Fast Sequential and Parallel Association Rule Mining Algorithms: Comparison", *Tech Report CS-TR 3515*, University of Maryland College Park MD, 1995.

[15] A. Schuster and R. Wolff, "Communication-Efficient Distributed Mining of Association Rules", *In Proc. ACM SIKMOD*, Santa Barbara.

[16] E. Han, G. Karypis and V. Kumar, "Scalable Parallel Data Mining for Association Rules", *In Proc. ACM SIGMOD,* pages 277—288, 1997.

[17] C. L. Blake and C. J. Merz. *UCI Repository of Machine Learning Databases,* University of California, Irvine, Dept. of Information and Computer Science, www.ics.uci.edu/~mlearn/MLRepository.html, 1998.

Performance Improvement of the Contract Net Protocol Using Instance Based Learning

Umesh Deshpande, Arobinda Gupta, and Anupam Basu

Department of Computer Science and Engineering
IIT, Kharagpur-721302, India
{uad,agupta,anupam}@cse.iitkgp.ernet.in

Abstract. The contract net protocol (CNP) is a widely used coordination mechanism in multiagent systems. It has been shown that its performance degrades drastically when the number of communicating agents and the number of tasks announced increases. Hence it has problems of scalability. In order to overcome this limitation, an Instance Based Learning (IBL) mechanism is incorporated with CNP that avoids the expensive bidding process and uses previously stored instances to select a target agent. The scheme is implemented in a simulated distributed hospital system where the CNP is used for resource sharing across hospitals. Experimental results demonstrate that with the incorporation of the IBL, the system performance improves significantly.

1 Introduction

In a distributed problem solving system, agents at different nodes (problem solvers) coordinate their individual problem solving activities to solve the complete problem. As pointed out in [ ], the coordination mechanisms should scale well along various dimensions of stress such as agent population, degree of interaction etc. A widely used coordination mechanism in a multiagent system is the well-known *Contract Net Protocol (CNP)* [ ]. According to this approach the process of task distribution consists of three activities: announcement of tasks by *managers* (i.e., agents that want to allocate tasks to other agents), submission of bids by potential *contractors* (i.e., agents that could execute announced tasks), and conclusion of contracts among managers and contractors. It is a simple and dynamic scheme but is communication intensive due to the broadcast of task announcements. It works well for small problem environments, but runs into problems as the problem size, the number of communicating agents and the number of tasks announced by them, increases [ ]. This limits its usability in a large scale multiagent system. Hence, in complex environments, mechanisms are needed to reduce the communication load.

In our previous work [ ], the CNP is used as a coordination mechanism for sharing resources across nodes. If a subtask T' of a task T submitted to a node N_i requires a resource r that is not present at N_i then T' has to be migrated to some other node of the system where r is present. The coordination mechanism chooses the appropriate target node where T' should be migrated depending on

S. R. Das, S. K. Das (Eds.): IWDC 2003, LNCS 2918, pp. 290–299, 2003.

the node loads, the qualities, and costs of the resources at the different nodes. Tasks in this system can have a deadline. A distributed hospital is modeled as a multiagent system and the coordination mechanism for resource sharing is used in it. It is observed that when the task arrival rate is high, many tasks can not finish within their deadlines and the average task waiting time of the system increases. At high arrival rate the usage of the CNP for resource sharing imposes an overhead due to the prohibitive communication time for acquiring bids.

In order to overcome this limitation, an Instance Based Learning (IBL) mechanism is designed in this paper that uses the history of subtask migrations in order to choose a target node for a new subtask. The history contains instances which consist of the system state and the target node selected by the coordination mechanism at that state. Whenever a task has less laxity (is close to the deadline), the target node is decided using the IBL mechanism. The k-Nearest Neighbor algorithm is used as the IBL technique which finds out the k instances in the history that are close to the current system state. Since this computation is local, the bidding process is avoided and thereby valuable time is saved. The communication load on the channels also reduces. The IBL mechanism is incorporated with the coordination mechanism for resource sharing in the simulated distributed hospital system. The results show that IBL improves the performance of the system appreciably. It is observed that more number of tasks finish within their deadlines, even at high loads, and the average task waiting time improves considerably. The system is better scalable with respect to the number of tasks.

In [ ], a method, called *addressee learning*, is proposed to reduce communication overhead of CNP. Agents acquire knowledge about other agents' task solving capabilities and then the tasks can be assigned more directly without the broadcast of task announcements. Case-based reasoning (CBR) is employed as an experience-based learning mechanism. Our work is similar to *addressee learning* in the sense that we also try to reduce broadcast of task announcements by using previous instances. But in our case, we cannot use CBR since the subtasks are atomic actions and do not have any attributes or values. In addition to this, we also handle real-time constraints. Ours is a dynamic system and hence learning has to be a continuous process which is not the case in [ ].

The rest of the paper is organized as follows. Section 2 gives an overview of the previously proposed coordination mechanism. The instance based learning mechanism is described in section 3. The performance evaluation of the learning mechanism is presented in section 4. The multiobjective case is briefly explained in section 5. Section 6 concludes the paper.

2   Background

In this section, a brief overview of the earlier work is given. We first present the system and the task model. The coordination mechanism for resource sharing across nodes is explained in section 2.1.

There are n nodes, $N_1, N_2, \ldots, N_n$ in a distributed system which communicate only through messages. Each node has a collection of agents controlling the resources present at that node. It is assumed that the communication delay for sending/receiving messages between any two nodes N_i and N_j can be estimated and is a constant Ψ_{ij}. A special unit called the *Liaison Unit (LU)*, with an agent in it, is identified at every node that is responsible for coordination across nodes for resource sharing. A task can be submitted to any node.

The tasks are described using the TAEMS representation language [ ]. TAEMS task structures are tree structures whose leaf nodes are *executable methods*. A TAEMS specification indicates the task deadline and the relationships between tasks, methods, or resources. A task (or subtask) represents a set of related subtasks or methods. Two of the relationships: *enables* and *mutex* are relevant to this work. If a subtask T *enables* a subtask T' then an agent should not execute T' before T is completed. In case of an *enables* relationship, the subtask at the source end of the relationship gives an *earliest start time* commitment to the subtask at the destination end indicating the time before which the destination subtask should not start execution [ ]. *Mutex* relationships for a resource R between a set of subtasks S_T indicate that at any point of time only one of the subtasks in S_T could use R.

2.1   The Coordination Mechanism

The coordination mechanism for resource sharing is based on CNP. When a subtask T at a node N requires a resource that is not present locally, then N asks for bids from a subset of the nodes present in the network. The remote nodes reply with a bid indicating how best they can solve T. The node with the best bid is chosen and T is migrated to it.

The Liaison Unit (LU) on each node N_i maintains the following two tables.

Local Resource Table (LRT) - This table keeps information about the resources that are locally present at N_i. Each entry is a five-tuple $< r_k, D, Q, C, S >$ and the fields are discussed below.

1. r_k - It is the name of the resource present locally at N_i.
2. $D, Q,$ and C - These are the duration of usage, the quality, and the cost of usage of r_k at N_i respectively.
3. S - This field is the surplus information. It is the ratio of the time r_k was not used at N_i in a past window of length WL_k. WL_k is chosen appropriately depending on the duration of usage for the resource r_k. The surplus is an indication of the load on a resource. Each agent of the node sends the surplus to the LU with a period of the window length for that resource.

Remote Resource Table (RRT) - This table keeps information about the resources at other nodes in the network as known by N_i. The entries for each resource r_k that is not present locally at N_i are kept together. Each such entry is a six-tuple $< N_j, D, Q, C, S, DE >$. The fields are as follows.

1. N_j - It is the identifier for the node where the resource r_k is present.
2. D, Q, and C - These are the duration of usage, the quality, and the cost of usage of r_k at N_j respectively.
3. S - It is the surplus that was most recently obtained from N_j. The entries are ordered in the non-increasing order of the surplus. The surplus information is sent by remote nodes periodically and is also piggybacked with message exchanges between nodes. RRT is updated then to reflect the changed values.
4. DE - For the node N_i, the delay estimate (DE), is computed for every node N_j where r_k is present at N_j. Let us say that N_j promised a bid for time t_0 but the actual finish time of the service of r_k was t_1. The difference $t_1 - t_0$ is referred to as the delay. This delay can be estimated for every remote node based on the previous history of the node. It is computed using exponential smoothening taking into account the previous requests serviced [ ].

The Bid Process When a resource r_k required by a subtask T' of a task T received by a node N_i is not available locally, the *Earliest Worst Case Start Time (EWCST)* of T' is calculated. T' must start execution before EWCST otherwise the deadline of T would be missed. Request for bids (RFBs) are sent to those nodes from the Remote Resource Table (RRT) for r_k such that each such node N_j satisfies the condition that $4 * \Psi_{ij} + DE + current\ time < EWCST$. The multiplication by 4 is used since there would be four communications between N_i and N_j (first for requesting the bid by N_i, second for sending the bid by N_j, third for sending the task by N_i, and the last for sending the results by N_j). The term DE is there in the condition since we give priority to those nodes whose estimated delay is less. It is expected that a node with a large delay is unlikely to service the request in a timely manner. RFBs are sent to the first ρ nodes of the RRT where ρ is a parameter used to reduce the communication overhead.

When a node, say N_h, receives a RFB message, it checks if it can *guarantee* the request (T' could be started before its EWCST). If the request can be guaranteed, it sends the *Earliest Possible Finish Time (EFT)*. N_i selects the best bidder which is the one that returns the least value of $2 * \Psi_{ih} + EFT$. The multiplication factor is 2 here since only the last two communications of the four mentioned above are required after a bid is obtained. N_i will wait for bids until it gets a favorable one or till $(EWCST - 2 * \Psi_{max})$ whichever is earlier. The worst case delay Ψ_{max} is the maximum time required to communicate between any two nodes. Still, if no favorable bids are available, the subtask T could be migrated randomly to any node.

The coordination mechanism requires various parameters like $EWCST, EFT$, and the guarantee routine which are provided by a heuristic real-time scheduler. The design of the real-time scheduler and the computation of these parameters are presented in detail in [ ].

3   Incorporating the Learning Mechanism

The coordination mechanism discussed in section 2.1 is communication intensive because of the broadcast for requesting the bids. In case of requests that have less laxity (the deadline is tight), it is not desirable to perform the complete bidding process for finding out the target node. A learning mechanism could be used to decide the target node for migration based on the previous history of requests. In this work, an instance based learning method is used.

Instance-based learning (IBL) methods such as nearest neighbor and locally weighted regression are simple approaches to approximating real-valued and discrete-valued target functions [ ]. Learning in these algorithms consists of simply storing the presented training data. When a new query instance is encountered, a set of similar related instances is retrieved from memory and used to classify the new query instance. In k-Nearest Neighbor (kNN) algorithm [ ], all instances correspond to points in n-dimensional space. The nearest neighbors of an instance are defined in terms of a distance function. The algorithm classifies a query x_q based on the most common of the k nearest instances.

3.1   Using kNN for Reducing the Communication Overhead of CNP

The basic idea of using kNN technique for CNP is as follows. When the complete bidding process of the CNP is used for choosing a target node for migration of a task, we store the system state and the target node as an instance in the history. History is maintained at every node for all the resources that are not present locally at that node. Subsequently, whenever a task has tight laxity, the current system state is used to find out the k-nearest neighboring instances from history. The target node that is most common among the k-nearest instances is chosen for the migration. Thus, we avoid the complete bidding process of the CNP whenever a task has tight laxity.

An instance I for a node N_n consists of a target node Γ found for a resource r when the state of the system was SS. The system state SS consists of the first ρ entries of the Remote Resource Table (RRT) of N_n for r when the decision for Γ was made. Let us recall that the RRT is sorted according to the decreasing values of the surplus (S) of the prospective destination nodes where r is present and that ρ is the system wide parameter that denotes the number of destination nodes to whom a request for bid (RFB) is sent. The other fields of RRT are the quality (Q) and the cost (C) of r at a destination node N_i. Let $RRT^r_{N_n}$ denote the RRT at a node N_n for a resource r. Thus,

$$I \equiv \langle SS, \Gamma \rangle \qquad \Gamma \in RRT^r_{N_n}$$
$$SS \equiv \{\langle N_i, S_i, Q_i, C_i \rangle | N_i \in RRT^r_{N_n}, \forall i \ 1 \leq i \leq \rho\}$$

The history H_{N_n} at a node N_n consists of a collection of such instances for every resource that is not present at N_n. The history $H^r_{N_n}$ for the resource r is organized as a circular queue of some predefined maximum size. The system under consideration is dynamic because the arrival rates vary and S changes continuously with the usage of the resources. We need to maintain the relevant history since the older instances may no longer be useful in such an evolving system. Hence the circular queue is required since it ensures that the old entries are automatically deleted from the history. An instance is considered for insertion into the history if it is good. The goodness is based on the delay observed for the service of the request corresponding to the instance. That means, if the delay observed for the target node Γ in an instance I, is less than some threshold Θ_H, then I is inserted into the history otherwise not. The intuition behind this is that if Γ selected by the coordination mechanism had a large delay, the choice for that was not proper since there was an appreciable difference between the actual service to the request and the bid promised.

When a task T submitted at a node N_n requires a resource r that is not present at N_n, the learning mechanism should be used rather than CNP if $EWCST \leq$ current time $+ 2 * \Psi_{max}$. The multiplication by 2 is used since the best case time for the bidding process is $2 * \Psi_{max}$. The current state of the system SS has to be matched with the entries in $H^r_{N_n}$ to find out the target node. Let us call Ω as the query which consists of the current system state, i.e. $\Omega \equiv \langle SS \rangle$. The k nearest neighbors for Ω from the instances in $H^r_{N_n}$ are found out where k is a system wide parameter. The learning mechanism is used only when the laxity is less even after we build the history for a large amount of time. If it is used indiscriminately, the insertion of the newly occurring instances into the history would not be done. The history would become stale and improper decisions would be made by matching the stale history with a new query. This hypothesis is also confirmed by the simulation experiments carried out which are explained in the next section.

For the nearest neighbor computation, distance between Ω and an instance I needs to be calculated. The k nearest neighboring instances are found out and the target node present in the majority of those is chosen. If there is a tie, i.e. there are more than one such target nodes, then one of them is randomly chosen. For computing the nearest neighbors, the distance (d_1) has to be computed based on the surplus (S). Let SS^I and SS^Ω denote the system states in the instance I and the query Ω respectively. Consider a node $N_i \in SS^I \cup SS^\Omega$. Note that there might be some nodes that are present only in SS^I and not in SS^Ω and vice-versa. Let $S^I_{N_i}$ and $S^\Omega_{N_i}$ be the surplus values of N_i in SS^I and SS^Ω respectively. If N_i is not present in either SS^I or SS^Ω then its corresponding surplus value is assumed to be zero. (This assumption is justified because if a node N_i is not present in a system state, its surplus is so low that it does not come in the first ρ entries of the RRT.) The distances for all such nodes are summed up to get the overall distance. Thus,

$$d_{1_{N_i}} = |S^I_{N_i} - S^\Omega_{N_i}| \qquad \text{and} \qquad d_1 = \sum_{\forall N_i \in SS^I \cup SS^\Omega} d_{1_{N_i}}$$

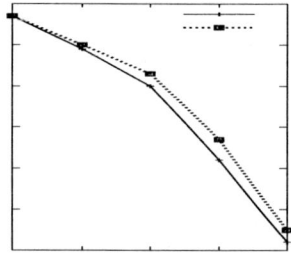 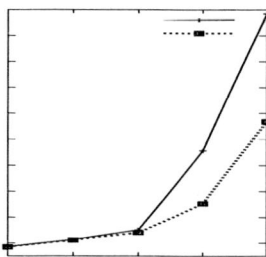

Fig. 1. Guarantee Ratios and Average Waiting Times of CNP with and without IBL

4 Performance Evaluation

A distributed hospital system is simulated which consists of a network of six nodes. The communication delay between any two nodes of the system and the duration of usage, the qualities, and the costs of the resources are randomly chosen. The task arrival is modeled as a Poisson process. Let X denote the *system wide arrival rate* which is the sum of the arrival rates at each of the nodes. Each of the experiments is executed for a total simulated time of 90 days. The arrival rate X is varied from 50 tasks per day to 500 tasks per day.

We first perform experiments to evaluate the system performance by measuring the *Guarantee Ratio (G = number of tasks that finish within their deadlines/total number of tasks)* and the *Average Task Waiting Time (W)* considering two scenarios - first, when the learning technique is not incorporated, and second, when the IBL is used with the coordination mechanism. Figure 1 shows the results. In the graphs, G_c denotes the guarantee ratio when the original CNP (without learning) is used and G_l is the guarantee ratio when the learning mechanism is incorporated. Similarly, W_c indicates the average waiting time without learning and W_l indicates the average waiting time with learning. The following conclusions are drawn.

When the arrival rate is low (below 200 tasks/day) there is not much performance improvement due to the learning technique. This is because the system is lightly loaded and the resources are scarcely used and hence the tasks will have more laxity. Most of the time the complete bidding process is used. Since, the laxity is high, there is no adverse effect on G even after going through the complete bidding process most of the time. Since, the learning mechanism is sparingly used, there is no significant gain in W, either.

At moderate and high arrival rate (above 300 tasks/day) there is significant performance gain due to the usage of the learning technique over CNP (without learning). The maximum gain in G is about 10% and in W about 45%. With higher arrival rate, the resources are heavily used and hence the probability of tasks having tight laxity is higher. The standard CNP mechanism spends a lot of valuable time in getting the bids by communicating with the remote nodes. With the instance based learning technique incorporated, the system adapts in view of the less laxity and uses the previously stored instances to decide a target node without going through the bidding process. A counter is maintained to find out the number of times the learning technique was used. It has been observed that when the arrival rate is 300 tasks/day, about 40% of the times the target node was decided by the learning technique and for about 70% of the times when the arrival rate is 450 tasks/day.

It is observed that after an initial learning period of about 30 days, the history has many (but not necessarily *sufficient*) instances and after this time, the k-nearest neighbor technique starts performing well. Having said this, we emphasize that the history should be used judiciously and only for cases when the task laxity is less. This is confirmed by using a mechanism which builds up history for 30 days and then blindly uses it for the next 60 days of the simulation for all the migration decisions. It has been found that this strategy pays off in the short-term but the overall performance *drops down* in the long-term. Initially, for some tasks this strategy seems to work well. Afterward, since the bidding is never used at all, the newer scenarios of the system state are never incorporated and the history becomes stale. This is because we are studying a continuously evolving system. The decisions made using the stale history become improper and the performance deteriorates.

The gain for W is more substantial over that of G. This can be attributed to the fact that the guarantee ratio is based on a binary decision - whether a task finishes within a deadline or not. Thus even though many tasks are not able to finish within their deadlines, the waiting time of tasks does reduce a lot and the total system waiting time improves. Also, for higher arrival rate, the laxity is less and there is a large impact on the guarantee ratio due to the presence of *enables* and *mutex* relationships of methods. For high arrival rate, the subtasks that enable a subtask T to be migrated hold up T due to the heavy usage of the resources. In order to confirm this conjecture, we experimented with different values of the probabilities for the presence of mutex resource (p_M) relationship for methods of a task and the presence of the enables (p_E) relationship between two methods of a task. We considered three cases - when the probabilities are *low* ($p_M = 0.3$, $p_E = 0.05$), *moderate* ($p_M = 0.5$, $p_E = 0.1$), and *high* ($p_M = 0.7$, $p_E = 0.15$). Due to lack of space, we are not displaying the results. It is observed that the gain in G and W decreases with the increase in the values of p_E and p_M. The maximum gain in G and W is about 7% and 38% respectively for high values of p_E and p_M. The corresponding values are 10% and 45% respectively for moderate and 12% and 51% respectively for low values of the p_E and p_M. For higher values of p_E and p_M, there are larger number of subtasks enabling a subtask T' that has to be migrated. Also, there are a larger number of subtasks sharing a mutually exclusive resource with T'. Due to the high usage of the resources at high arrival rate, these subtasks hold up the actual migration of T'. Hence there is a lesser impact of the early decision of the target node due to the incorporation of the learning mechanism.

To understand the impact of the average communication time between two nodes of the system, simulations are performed where the average communication

Table 1. Performance Evaluation Results for the Multiobjective Case

Update Rate	Arrival Rate	Max Gain in		
		G	W	P_l
Constant	All	13%	47%	45%
Low	All	11%	34%	40%
Moderate	Low (< 450 tasks/day)	7%	21%	33%
Moderate	High (> 450 tasks/day)	-3%	-2%	-15%
High	Low (< 400 tasks/day)	2%	4%	25%
High	High (> 400 tasks/day)	-7%	-17%	-25%

time is varied. We performed experiments for three different cases - when the communication time is *low* (two times the average method duration), *moderate* (three times), and *high* (five times). We are unable to show the results due to lack of space. For high average communication time, it is observed that the maximum gain in G and W is 16% and 56% respectively. The values for the moderate case are 13% and 51% respectively and for the low case, - 10% and 45% respectively. As expected, with higher communication times the performance gain is higher since the savings of the learning technique are higher.

5   Incorporating Multiple Objectives

In a real-life scenario, multiple criteria (or objectives) like minimizing cost and time required, maximizing quality, etc. could be present. Moreover, the qualities and costs of the resources at the nodes could evolve with time due to upgradation in quality or varying costs of the resources. A user can assign a preference rating, which could be specified subjectively using ordinal values, to each of the criteria along with an input task. The task assignment, which is important for the satisfaction of the criteria, is discussed in [ ]. The coordination mechanism of section 2.1 is also extended to handle multiple objectives in [ ]. The Instance Based Learning mechanism is proposed for the multiobjective case in the present work. The details are omitted due to lack of space.

In the simulation experiments the preference ratings of the objectives are randomly generated and are input with a task. When a task execution is over, a *performance measure* - which is the weighted average of the duration required, the quality accumulated, and the cost incurred for the task execution with the weights being the preference ratings - is calculated. An *average performance measure* (averaged over all tasks) is calculated. The experiments were performed for four different setups. First, the costs and the qualities were kept constant. Then the costs and the qualities of the resources at all the nodes were randomly updated at three different periodic update rates - *low* (every 30 days), *moderate* (every 15 days), and *high* (every 7 days). The additional parameter under investigation is the *Performance Measure Ratio* (P_l), which is the ratio of the *average performance measure* obtained for CNP with the learning technique to that obtained for CNP without the learning technique. The results, displayed in

Table 1, show a significant gain in the performance due to the incorporation of IBL. But when the update rate is from moderate to high and the arrival rate is high (rows 4 and 6 of Table 1), the learning technique performance degrades and it is worse than that of the technique without learning (i.e. the complete bidding process). The laxity is less in this condition and a very large number of decisions are based on the learning mechanism. The updating in history is performed very infrequently. The history quickly become stale and many improper decisions are made by the learning mechanism. The complete bidding process which uses the current state information for decisions performs better.

6   Conclusions

The contract net protocol (CNP) is a simple and dynamic scheme but is communication intensive due to the broadcast of task announcements. We demonstrate that the performance of a system that uses CNP can be improved by using the Instance Based Learning (IBL) Technique. The system is better scalable with respect to the number of tasks.

As part of the future work, we would be studying the system in more detail to further understand its dynamics. It would be particularly interesting to evaluate the performance with varying sizes of the history. Investigation of the evolution of the percentage of misclassified instances after updates of the parameters like surplus, quality, or cost needs to be carried out.

References

[1] D. W. Aha, D. Kibler, and M. K. Albert, "Instance Based Learning", *Machine Learning*, Vol. 6, pp. 37–66, 1991. 294

[2] R. Davis and R. G. Smith, "Negotiation as a metaphor for distributed problem solving", *Artificial Intelligence*, Vol. 20(1), pp. 63–109, Jan. 1983. 290

[3] K. S. Decker, "Environment Centered Analysis and Design of Coordination Mechanisms", *Ph.D. thesis*, University of Massachusetts, Amherst, 1995. 292

[4] U. Deshpande, A. Gupta, and A. Basu, "Adaptive Fault Tolerant Hospital Resource Scheduling", in *Proceedings of the 10th International Conference on Cooperative Information Systems (CooPIS - 2002), Irvine, USA*, LNCS-2519, pp. 503–520, Springer Verlag, Berlin, Nov. 2002. 290, 293

[5] U. Deshpande, A. Gupta, and A. Basu, "Coordinated Problem Solving through Resource Sharing in a Distributed Environment", *IEEE Transactions on Systems, Man, and Cybernetics B*, To Appear. 298

[6] R. Duda and P. Hart, *"Pattern Classification and Scene Analysis"*, Wiley, New York, 1973. 294

[7] E. H. Durfee, "Scaling up Coordination Strategies", *IEEE Computer*, Vol. 34(7), pp. 39–46, Jul. 2001. 290

[8] T. Ohko, K. Hiraki, and Y. Anzai, "Addressee Learning and Message Interception for Communication Load Reduction in Multiple Robot Environments", *In G. Weiss, ed., Distributed Artificial Intelligence meets Machine Learning, LNAI-1221*, pp. 242–258, Springer-Verlag, Berlin, 1997. 291

[9] S. Sen and G. Weiss, "Learning in Multiagent Systems", in *Multiagent Systems: A Modern Approach to Distributed Artificial Intelligence*, G. Weiss ed., MIT Press, Cambridge, Mass., 1999. 290

A Hybrid Model for Optimising Distributed Data Mining

Shonali Krishnaswamy, Arkady Zaslavsky, and Seng Wai Loke

School of Computer Science and Software Engineering
900 Dandenong Road, Monash University, Caulfield East, Victoria –3145, Australia
{Shonali.Krishnaswamy,Arkady.Zaslavsky,
Seng.Loke}@infotech.monash.edu.au

Abstract. This paper presents a hybrid model for improving the response time of distributed data mining (DDM). The hybrid DDM model uses cost formulae and prediction techniques to compute an estimate of the response time for a DDM process and applies a combination of client-server and mobile agent strategies based on the estimates to reduce the overall response time. Experimental results that establish the validity and demonstrate the improved response time of the hybrid model are presented.

1   Introduction

Distributed Data Mining (DDM) [3] aims to address specific issues associated with the application of data mining in distributed computing environments. Broadly, data mining environments consist of users, data, hardware and the mining software (this includes both the mining algorithms and any other associated programs). Distributed data mining addresses the impact of distribution of users, software and computational resources on the data mining process. There is general consensus that distributed data mining is the process of mining data that has been partitioned into one or more physically/geographically distributed subsets. In other words, it is the mining of a distributed.

Several researchers have presented the process of performing distributed data mining as follows [13, 5]:

- Performing traditional knowledge discovery at each distributed data site.
- Merging the results generated from the individual sites into a body of cohesive and unified knowledge.

There are predominantly two architectural frameworks for the development of distributed data mining systems – the client-server model and the agent-based model. The client-server model is characterised by the presence of one or more data mining servers. In order to perform mining, the user requests' are fed into the data mining server which collects data from different locations and brings them into the data mining server. The principle behind the mobile agent model for distributed data mining is that the mining process is performed via mobile code executing remotely at

S. R. Das, S. K. Das (Eds.): IWDC 2003, LNCS 2918, pp. 300-310, 2003.

the data sites and carrying results back to the user. The advantage of this model is that it overcomes the communication overhead associated with the client-server systems by moving the code instead of large amounts of data.

This paper presents a hybrid model for distributed data mining that integrates the client-server and mobile agent model to optimise the response time. A distinguishing feature of the optimisation in the hybrid model is the costing strategy. As stated by [1], *"optimisation is only as good as its cost estimates"* and the hybrid model uses accurate *a priori* estimates of the response time to support the optimisation process. Therefore, the ability to accurately determine the "cost" of alternate strategies is very important for any optimisation model.

IntelliMiner [10] is a client-server distributed data mining system, which focuses on scheduling tasks between distributed processors by computing the cost of executing the task on a given server and then selecting the server with the minimum cost. The cost is computed based on the resources (i.e. number of data sets) needed to perform the task. While this model takes into account the overhead of communication that is increased by having to transfer more datasets it ignores several other cost considerations in the DDM process such as processing cost and size of datasets. In [12], the optimisation strategy involves partitioning datasets and allocating them to distributed data servers using a linear programming algorithm. The partitioning is done using the linear programming approach to determine the optimal split in the datasets. The allocation is based on the cost of processing, which is assigned manually in terms dollars (e.g. $4 per GB of data) and the cost of transferring data, which is also assigned manually in terms of dollars (e.g. $2.5 per GB of data).

The paper is organised as follows. Section 2 presents the hybrid model and formalises its operation. Experimental evaluation and analysis of the performance of the hybrid model are discussed in section 3. Section 4 presents the conclusions and future directions of this research.

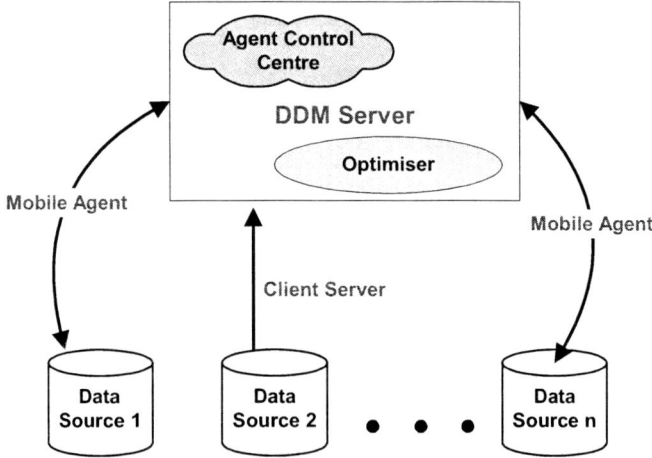

Fig. 1. Hybrid DDM Architectural Model

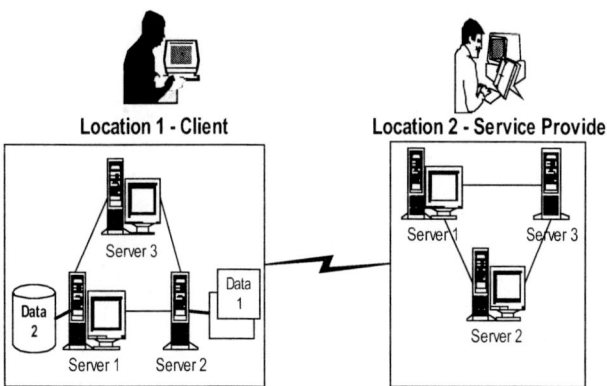

Fig. 2. A Scenario for Optimisation

2   Hybrid DDM Model

In this section, we propose a novel hybrid DDM, which integrates the traditional client-server and mobile agent architectural models for developing DDM systems. The hybrid DDM model, shown in figure 1 is driven by an optimiser that uses estimates of the DDM response time in order to choose a task allocation strategy that minimises the response time.

Thus by integrating the client-server and mobile agent models, the hybrid model endeavours to minimise the response time. The hybrid model by virtue of being able to support both the client-server and mobile agent models for distributed mining has the ability to use a combination of the two strategies for performing a task. This assumes added significance in view of the studies by [2, 11], which have shown that a mix of client-server and hybrid approaches can lead to improved communication times for certain distributed applications.

This operation of the hybrid DDM model is shown in the example scenario illustrated in figure 2. The scenario represents a distributed e-services environment and involves a client (based at location 1) who requires two datasets to be mined ("Data 1" and "Data 2") and has available two computational resources ("Server 1" and "Server 3") that it can make available for mining. One of the datasets (Data 1) is located on Server 2 that is not available for mining. A service provider (based at location 2) has the data mining software and has the ability to mine using its own computational resources (i.e. three servers) or dispatch mobile data mining agents to the client's site. The three servers may be geographically distributed, thereby making the communications costs variable between the three servers.

There are three possible options that the client can choose:

1.   The mining can be done locally using only the client's computational resources. Obviously, mobile agents will have to be employed in order to perform the task at the client's site since the client possesses no data mining software. However the optimisation issue will involve using the most appropriate allocation/transfer of the data to the locally available computational resources.

2. The mining should be done remotely using only the service provider's computational resources. In this case the client-server model will have to be used and the data has to be transferred to the service provider's servers and the optimisation issue will involve using these resources in the most suitable manner.

3. The client has no preference for the location and the service provider has the option of choosing the combination that is optimal for the given response time requirement. In this scenario, the optimisation question involves choosing the most appropriate combination of mobile agents and client-server models to achieve the optimal response time.

We have informally presented an overview of our hybrid DDM strategy to support minimising the response time. The hybrid model has the following principal characteristics: it combines the best aspects of the mobile agent model and the client-server approach by integrating an agent framework with dedicated computational resources. It brings with it the advantage of combining the concept of dedicated data mining resources (and thus alleviating the issues associated with lack of control over remote computational resources in the mobile agent model) and the ability to circumvent the communication overheads associated with the client-server approach. This gives the model the option of applying either approach to a particular DDM task. The hybrid model involves an optimisation component (i.e. the "optimiser"), that uses *a priori* estimates of the DDM response time for various alternative scenarios and choose the most cost-effective one. Thus, the optimiser provides the infrastructure necessary to build the cost-estimates and determine the strategy to be applied on a case-by-case basis. The key to the successful operation is the ability to estimate the DDM response time. The estimation in turn requires the identification of the cost components and formalisation of a cost model for DDM response time for different scenarios. We now formalise the cost components of the DDM response time and state the operation of the hybrid model in terms of the cost components.

2.1   Cost Components of the DDM Response Time

In this section we specify the different cost components of the response time in distributed data mining e-services. The response time of a task in a distributed data mining e-service broadly consists of three components: *communication, computation and knowledge integration*.

Communication: The communication time is largely dependent on the operational model. It varies depending on whether the task is performed using a client-server approach or using mobile agents. In the client-server model the communication time is principally the time taken to transfer data from distributed servers to a high performance machine where the data mining is performed. In the mobile agent model, the communication time revolves around the time taken to transfer mobile agents carrying data mining software to remote datasets and the time taken to transfer results from remote locations for integration.

Computation: This is the time taken to perform data mining on the data sets and is a core factor irrespective of the operational model.

Knowledge Integration: This is the time taken to integrate the results from the distributed datasets.

The response time for distributed data mining is as follows:

$$T = t_{dm} + t_{com} + t_{ki} \qquad (1)$$

In Eq. 1 above, T is the response time, t_{dm} is the time taken to perform data mining, t_{com} is the time involved in communication and t_{ki} is the time taken to perform knowledge integration. The modelling and estimation of the knowledge integration (t_{ki}) variable is dependent on the size and contents of the results obtained from the distributed datasets. Given that the primary objective of data mining is to discover hitherto unknown patterns in the data [3], we consider the *a priori* estimation of the time taken to perform knowledge integration to be outside the scope of this paper (since knowledge integration depends on the characteristics of the results of the data mining process). Having identified the cost components of the DDM response time, we now formalise the overall estimation cost for different models and scenarios.

2.2   Cost Matrix for Representing the Composite Response Time

We now present a cost matrix for computing the composite DDM response time estimates for different strategies. The response time for distributed data mining as presented in Eq. 1 consists of three components including communication (due to either transfer of mobile agents and/or transfer of data), computation (performing data mining) and knowledge integration. The following discussion focuses on the communication and computation components and does not consider the knowledge integration component. The strategy is to compute estimates for the individual cost components and then uses the estimates to determine the overall response times for different strategies. The cost matrix to calculate the composite response time for different DDM strategies is denoted by *CM* and is represented as a two-dimensional *m* x *n* matrix, where *m* is the number of available servers and *n* is the number of datasets. A fundamental feature of the cost matrix that makes it applicable for both the mobile agent and client-server models is that we incorporate location information on the datasets and servers.

The elements of the cost matrix represent the estimated response time and are defined as follows:

1. Let *m* be the number of servers.
2. Let $S = \{S_1, S_2, ..., S_m\}$ be the set of servers. A server S_j can either be located at the service provider's site or at the client's site. Therefore, let S^{SP} be the set of servers located at the service providers site and let S^C be the set of servers located at the client's site. The following properties are true for the sets S, S^{SP}, S^C.

      ❑ $S^{SP} \cup S^C = S$; the set of servers available at the client's site and the set of servers available at the service provider's site summarily constitute the total set of available servers. The obvious corollaries are $S^{SP} \subseteq S$ and $S^C \subseteq S$.

      ❑ $S^{SP} \cap S^C = \phi$; thus a server either belongs to the client or the service provider. It cannot belong to both.

❑ $S^{SP} = \phi$ is valid and indicates that the client is not willing to ship the data across and $S^{C} = \phi$ is also valid and indicates that the client's computational resources are unavailable or inadequate.

In order to specify the location of a server we use the following notation: $S_j \in S^{SP}$ and $S_l \in S^{C}$ where $l, j = 1, 2, ..., m$. This distinction is necessary for specification of how the response time has to be estimated in the cost matrix.

3.   Let n be the number of datasets and let $DS = \{ds(1), ds(2), ..., ds(n)\}$ represent the labelling of the datasets.

4.   Let $ds(i)_{S_j}$ represent the location of a dataset labelled $ds(i)$, $i=1, 2, ..., n$ at server S_j, where $ds(i) \in DS$ and $j= 1, 2, ..., m$.   Thus datasets are uniquely identified and multiple datasets at locations can be represented.

Let $cm_{ij} \in CM$ be the estimated response time for taking a dataset located at the server j and mining it at the server i, where $1 \le i \le m$ and $1 \le j \le n$. The value of cm_{ij} is computed as follows:

$$cm_{ij} = \begin{cases} MA_{S_X \to S_i} + TR_{S_j \to S_i}^{ds(k)_{S_j}} + W_{S_i} + DM_{S_i}^{ds(k)_{S_j}}, i \neq j, S_X \in S^{SP}, S_j \in S^{C}, S_i \in S^{C} \\ MA_{S_X \to S_i} + W_{S_i} + DM_{S_i}^{ds(k)_{S_j}}, i = j, S_X \in S^{SP}, S_j \in S^{C}, S_i \in S^{C} \\ TR_{S_j \to S_i}^{ds(k)_{S_j}} + W_{S_j} + DM_{S_i}^{ds(k)_{S_j}}, i \neq j, S_j \in S^{C}, S_i \in S^{SP} \end{cases}$$

In the above equation:

* $MA_{S_X \to S_i}$ is the time to transfer a mobile data mining agent from server S_X (which is a server of the service provider) to server S_i

* $TR_{S_j \to S_i}^{ds(k)_{S_j}}$ is the time to transfer a dataset $ds(k)_{S_j}$ located at server S_j to server S_i

* $DM_{S_i}^{ds(k)_{S_j}}$ is the time to mine dataset $ds(k)_{S_j}$ located originally at server S_j at server S_i

* W_{S_i} is the wait time at server S_i required for the completion of previous tasks and is generally more significant when $S_i \in S^{SP}$ (i.e. server S_i is located at the service provider's site).

As presented above there are three formulae for estimating the response time for different scenarios in the cost matrix. The first one is for the case where the server Si where the mining is to be performed is at the client's site but does not contain the dataset $ds(k)_{S_j}$ (which as indicated is located at the server Sj). Hence there is a need to transfer the data from its original location Sj to the server Si to perform the data mining. Further, the client's site would not have the data mining software and a mobile agent needs to be transferred from the service provider's site (represented as S_X). The second formula is for the case where the server where the mining is to be performed is at the client's site and the dataset is located on the same server (i.e. $i = j$). In this case, the mobile agent needs to be transferred but there is no need to transfer the data. The third formula is for the case where the server where the mining is to be performed is located at the service provider's site (i.e. $Si \in S^{SP}$) and therefore the data has to be shipped across from the client's site to perform mining. The three formulae map to the three scenarios outlined earlier, namely, that of mining at the client's site, mining at the service provider's site and using both sites.

We have modelled cost formulae and estimation algorithms to estimate the individual cost components in [6, 8]. The analysis of the cost matrices to determine the minimum response time reduces to the well-known Generalised Allocation Problem (GAP) [9]. We apply an approximate technique that has two phases to determine the task allocation strategy from the cost matrix. We first use the polynomial-time algorithm proposed by [9] to determine the initial allocation of tasks and then apply a Tabu search [4] to improve the initial allocation by exchanges. We now present experimental evaluation of the performance of the hybrid model.

3   Experimental Evaluating the Hybrid DDM Model

The objective of the experimental evaluation is to analyse the performance of the hybrid DDM model when compared with the current predominant models for constructing distributed data mining systems, namely, the client-server and mobile agent models. The comparison focuses on the overall response times obtained by using each approach to perform DDM tasks. We used our *a priori* estimation techniques to compute the cost for each model – mobile agent, client-server and hybrid. We note that we have the experimentally shown the accuracy of our *a priori* estimates [6, 8]. Therefore irrespective of the DDM model that is used, the estimates provide an accurate prediction of the response time. This facilitates a valid comparison of the different DDM models using the estimates as the costing strategy for task allocation. In order to experimentally evaluate the performance of the hybrid model we have a developed a DDM system – Distributed Agent-based Mining Environment (DAME). DAME is an implementation of the hybrid model presented in this paper and has been developed in Java. It has an optimiser and cost calculators to estimate the DDM response time for different scenarios. For details of the DAME system, readers are referred to [7].

The experiments were conducted using a network of four distributed machines located on two different campuses and by varying the following parameters:

1. We assigned different roles to the servers for different experimental runs. That is for a given run, certain servers were assigned the role of "Service Provider" and certain servers were assigned the role of "Client". As discussed in section 2, the cost matrix computes the composite response time for performing a given task at server based on its location at either the service provider's site or the client's site.
2. We varied the data mining tasks, the datasets and the number of datasets assigned per server.

3. We applied different "Wait Times" to different servers to simulate the unavailability of servers for a certain period that is typically brought about by the either the execution of previous tasks or by lack of computational resources.

For each experimental run, the task was allocated to servers for the following strategies:

1. *Mobile Agents Only.* In this case mobile agents are sent from one of the nominated "Service Provider" servers to the location of the datasets. The datasets are not moved and remained at the client's sites *in their original location* (i.e. a dataset originally located at the client's site c remains at c.

2. *Mobile Agents with Local Data Transfers.* In this case, mobile agents are dispatched as before, but the task allocation is used to perform local transfers of the datasets within the client's servers. That is, if there are two servers located at the client's site and one dataset is to be mined, the dataset may be mined at either of the two servers, depending on whichever provides the faster response time. The cost matrix and the task allocation strategy is used to determine the precise allocation.

3. *Client-Server.* In this case, the datasets are moved to the service provider's servers. However, if there are two or more servers nominated as service provider servers, then the server that can provide the faster response time is allocated the task using the cost matrix.

4. *Hybrid.* In this case, both mobile agent and client-server approaches are used to determine the task allocation strategy that results in the minimum response time.

The comparison between the response times of the client-server, mobile agent and hybrid DDM models is shown in figure 3. The figure clearly indicates that the response time achieved by the hybrid DDM model is always lower than or equal to the minimum response time of the other two models. The comparison between the hybrid DDM model and the mobile agent model with local data transfers is shown in figure 4. This model is the closest in performance to the hybrid DDM model. However, the hybrid DDM model outperforms even this model, which consists of mobile agents with local optimisation. It must be noted that the concept of local data transfers is also based on the model of task allocation based on *a priori* estimates to reduce the overall response time.

From our experiments we note that the mobile agent model performs better than the client-server model except where the wait times are considerably higher in the servers located at the client's site and the mobile agent model with local data transfers always performs better than or equal to the mobile agent model. In summary, the experimental results validate our hypotheses that *a priori* estimation is an effective basis for costing alternate DDM strategies (if the estimates are accurate) and that the hybrid DDM model lead to better overall response time than either the client-server or mobile agent models.

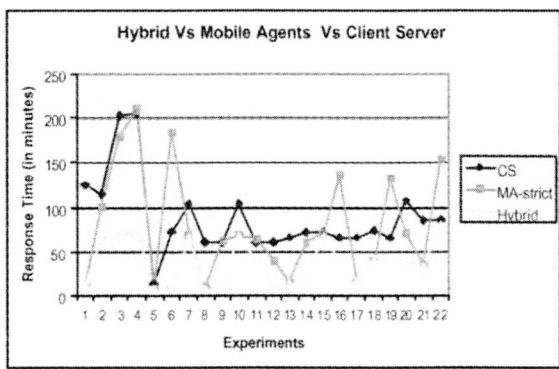

Fig. 3. Comparison: Hybrid Model with the Mobile Agent and Client-server Models

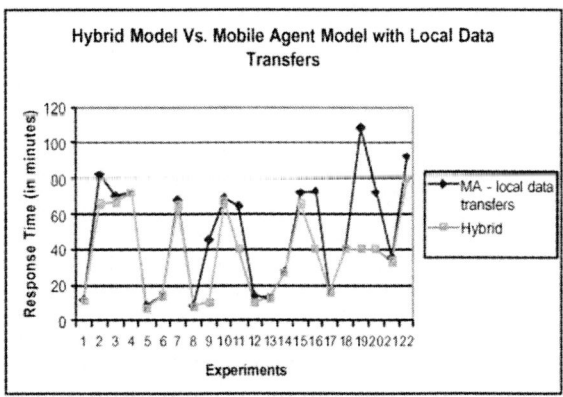

Fig. 4. Comparison: Hybrid Model with the Mobile Agent Model Including Local Data Transfers

4   Conclusions and Future Work

This paper has presented a hybrid DDM model to minimise the response time of the DDM process. The hybrid DDM model combines the best features of the mobile agent and client-server models by choosing the optimal combination of these techniques. Previous costing strategies in distributed data mining relied on manual estimates or were based on the number of datasets that required transferring. This paper has proposed the use of *a priori* estimates of the cost components of the DDM response time as a costing strategy to support the task allocation process for minimising the response time in the hybrid DDM model. Our model incorporates estimates of both communication and computation components. We have presented experimental results of performance of the hybrid DDM model that establish the minimum response time achieved using the hybrid model. The future directions of this work include applying this model in analysis and mining in mobile and ubiquitous environments.

References

[1]   Chaudhuri, S., (1998), *"An Overview of Query Optimization in Relational Systems"*, Proceedings of the Seventeenth ACM SIGACT-SIGMOD-SIGART Symposium on Principles of Database Systems, June 1-3, Seattle, Washington, ACM Press, ISBN 0-89791-996-3, pp. 34-43.

[2]   Chia, T., and Kannapan, S., (1997), *"Strategically Mobile Agents"*, First International Workshop on Mobile Agents (MA '97), Berlin, Germany, Lecture Notes in Computer Science (LNCS) 1219, Springer Verlag, pp. 149-161.

[3]   Fu, Y., (2001), *"Distributed Data Mining: An Overview"*, Newsletter of the IEEE Technical Committee on Distributed Processing, Spring 2001, pp.5-9.

[4]   4Glover, F., and Laguna, M., (1997), *"Tabu Search"*, Kluwer Academic Publishers, Massachusetts, USA.

[5]   Kargupta, H., Kamath, C., and Chan, P., (1999), *"Distributed and Parallel Data Mining: Emergence, Growth and Future Directions"*, (1999), Advances in Distributed Data Mining, (eds) Hillol Kargupta and Philip Chan, AAAI Press, pp. 407-416.

[6]   Krishnaswamy, S., Loke, S, W., and Zaslavsky, A., (2002), *"Application Run Time Estimation: A QoS Metric for Web-based Data Mining Service Providers"*, Proceedings of the Seventeenth ACM Symposium on Applied Computing (ACM SAC) 2002 in the Special Track on WWW and E-business Applications, Madrid, Spain, March 10-14, ACM Press, pp. 1153-1159.

[7]   Krishnaswamy, S., Zaslavsky, A., and Loke, S, W., (2000), *"An Architecture to Support Distributed Data Mining Services in E-Commerce Environments"*, Proceedings of the 2nd International Workshop on Advanced Issues of E-Commerce and Web-Based Information Systems (WECWIS2000), Milipitas, CA, USA, June 2000, IEEE Press, pp. 239-246.

[8]   Krishnaswamy, S., Zaslavsky, A., and Loke, S, W., (2002), *"Techniques for Estimating the Computation and Communication Costs of Distributed Data Mining"*, Proceedings of International Conference on Computational Science (ICCS2002) – Part I, Lecture Notes in Computer Science (LNCS) 2331, Springer Verlag. pp. 603-612.

[9]   Martello, S., and Toth, P., (1990), *"Knapsack Problems – Algorithms and Computer Implementations"*, John Wiley and Sons Ltd, England, UK.

[10]   Parthasarathy, S., and Subramonian, R., (2001), *"An Interactive Resource-Aware Framework for Distributed Data Mining"*, Newsletter of the IEEE Technical Committee on Distributed Processing, Spring 2001, pp.24-32.

[11]   Straßer, M., and Schwehm, M., (1997), *"A Performance Model for Mobile Agent Systems"*, in Proceedings of the International Conference on Parallel and Distributed Processing Techniques and Applications (PDPTA'97), (eds) H. Arabnia, Vol II, CSREA, pp. 1132-1140.

[12]   Turinsky, A., and Grossman, R., (2000), *"A Framework for Finding Distributed Data Mining Strategies that are Intermediate between Centralized Strategies and In-place Strategies"*, Workshop on Distributed and Parallel Knowledge Discovery at KDD-2000, Boston, pp.1-7.

[13]   Zaki, M, J., and Pan, Y., (2002), *"Introduction: Recent Developments in Parallel and Distributed Data Mining"*, Journal of Distributed and Parallel Databases, Vol. 11, No. 2, pp. 123-127.

Local Decision Procedures
for Avoiding the Tragedy of Commons

Sabyasachi Saha and Sandip Sen

Math & CS Dept, University of Tulsa
600 South College Avenue, Tulsa, OK 74104-3189
{sahasa,sandip}@ens.utulsa.edu

Abstract. The social sciences literature abound in problems of providing and maintaining a public good in a society composed of self-interested individuals [6]. Public goods are social benefits that can be accessed by individuals irrespective of their personal contributions. Such problems are also addressed in the domain of agent based systems [15]. In this paper we address the problem of the Tragedy of the Commons [9], a particularly common social dilemma that leads to inefficient usage of shared resources. We present a decision procedure following which rational agents can optimally use a shared resource using only local information. Our experimental results confirm that the tragedy of the commons is successfully avoided and the shared resource is utilized to its capacity when agents following our prescribed decision procedure.

1 Introduction

The viability of an individual in a society depends critically on the behavior of other members. Interesting computational problems in agent societies include paradoxes that involve reduction of system throughput when more resources are added to an existing system. Other social dilemmas arise when myopic, local-utility-maximizing decision-making by individual members of the society lead to a loss of utility for everyone. Such problematic scenarios appear frequently in natural and artificial societies.

In a society, the common infrastructures, goods and services are typically shared between members. For example, if we consider the problem of city traffic, we find that congestion problems arises out of self-interested drivers having to share common resources like roads, bridges etc. It often happens that the shared resource has a capacity and if the load is more than its capacity the resource performance or its perceived utility to the users decrease sharply. In a society of self-interested rational agents, or humans, each individual will try to maximize their utility from the shared resources. From the local perspective of a given agent, the more extensive use of a resource produces greater utility. If decision-making is predicated only on this local perspective, each user in the system can myopically try to maximize its load on the common resource. As a result, the combined load is likely to exceed the capacity of the common resource and adversely affect everyone and result in a decrease in everyone's utility from the

S. R. Das, S. K. Das (Eds.): IWDC 2003, LNCS 2918, pp. 311–321, 2003.

resource. This is the well-known *Social dilemma* problem of the *Tragedy of the commons*.

The examples of the *Tragedy of the commons* are now seen from the problem of global warming, congestion of traffic to the problem of sharing communication channel bandwidth.

An example of Tragedy of the commons lie in the example of network congestion if every packet is sent with highest possible priority. Suppose there are some routes of different quality. If everybody wants to route through the best possible route then it leads to a congestion which worsen every routing through that route.

More recently, attention has been drawn to the tragedy of the commons in the context of autonomous agent systems [15]. These and other problems arise in multiagent societies as multiple, distributed decision-makers try to maximize local utility by taking decisions based only on limited global knowledge. Correspondingly, multiagent system researchers have developed various approaches to resolve resource conflicts between distributed agents. For example, some researches have addressed the problem of effectively sharing common resources [2]. They proposed an agent as a planner who will make all resource allocation decisions. But this central planning approach requires nearly perfect global knowledge of all agents and the environment which is not very reasonable in complex, distributed and dynamic domains. Durfee and Lesser proposed a distributed partial-global planning [3] approach for coherent coordination between distributed problem solvers through the exchange of partial local plans. Approaches that emphasize economic mechanisms like contracting and auctions, allocate resources based on perceived utility [12]. While the economic approaches are interesting, we believe that they do not provide a satisfactory resolution to social dilemma problems without an adequate discussion of varying individual wealth and interpersonal utility comparisons. The *COIN* approach to solving social dilemmas allows distributed computation but requires an "omniscient" agent to set up the utility functions to be optimized locally [14].

Glance and Hogg [4] make the important observation that computational social dilemmas can produce situations where it is impossible to arrive at globally optimal system configurations based only on distributed, rational decision-making with local knwoledge. They contrast such computational problems with traditional complexity analysis in algorithm theory where solutions are hard, but not impossible to find.

The motivation of our work on computational social dilemma has been to investigate mechanisms to resolve conflicts while requiring minimal global knowledge or imposing minimal behavioral restrictions on the agents. For example, in [1] it is shown that a genetic algorithm based optimization framework can solve a well-known social dilemma problem, the Braess' Paradox [8]. The GA-based function optimization approach is a centralized mechanism. Munde *et. al.* used a more decentralized, adaptive systems approach using GAs, to address both the Braess' paradox and the Tragedy of the Commons [11]. Though decision making is decentralized in this approach, the survival of individuals, as determined by fitness-proportionate selection scheme, is a centralized procedure. Though the latter procedure can be approximated in a decentralized manner, a further criticism of the approach, the somewhat altruistic decision-procedure used by the distributed agents, is difficult to address.

In this paper we concentrate on *Tragedy of the commons* problem with the goal of designing defensible decentralized procedures relying on only minimal local information that can still solve this dilemma. In the following we first review the problem of social dilemmas and discuss the tragedy of the commons in more detail. Then we present a local decision procedure for addressing the tragedy of the commons. We assume that all agents in the system use our suggested decision procedure. We then experimenatlly demonstrate that our suggested local decision procedure produces optimal global utilization of the shared resource.

2   Social Dilemmas

A social dilemma arises when agents have to decide between contributing or not contributing towards a public good without the enforcement mechanism of a central authority [5]. Individual agents have to tradeoff local and global interests while choosing their actions. A selfish individual will prefer not to contribute towards the public good, but utilize the benefits once the service is in place. If a sufficient number of agents make the selfish choice, the public good may not survive, and then everybody suffers. In general, social laws, taxes, etc. are enforced to guarantee the preservation of necessary public goods. Consider a scenario where a public good is to be initiated provided enough contribution is received from the populace. Let us assume that the public good, $\mathcal{G}$, costs C, and the benefit received by individual members of the populace is B. Let us also assume that in a society of N agents, $P < N$ individuals decided to contribute to the public good. Assuming that the cost is uniformly shared by the contributors, each contributing agent incurs a personal cost of $\frac{C}{P}$. If enough agents contribute, we can have $\frac{C}{P} < B$, that is even the contributors will benefit from the public good. Since we do not preclude non-contributors from enjoying the public good in this model, the non-contributors will benefit more than the contributors. If we introduce a ceiling, M, on the cost that any individual can bear, then the public good will not be offered if $\frac{C}{P} > M$. In this case, everybody is denied the benefit from the public good.

Similarly in a resource sharing problem, where the cost of utilizing a resource increases with the number of agents sharing it (for example, congestion on traffic lanes). Assume that initially the agents are randomly assigned to one of two identical resources. Now, if every agent opts for the resource with the least current usage, the overall system cost (cost incurred per person) increases [7]. So, the dilemma for each agent is whether or not to make the greedy choice.

2.1   Tragedy of the Commons

In his book, *The Wealth of Nations* (1776), Adam Smith conjectured that an individual for his own gain is prompted by an "invisible hand" to benefit the group [13]. As a rebuttal to this theory, William Forster Lloyd presented the *tragedy of the commons* scenario in 1833 [9]. Lloyd's scenario consisted of a pasture shared by a number of herdsmen for grazing cattles. This pasture has a capacity, say C, i.e., each time a cattle added by a herdsman result in a gain as long as the total number of cattles in the pasture, x, is less than or equal to C. When $x > C$, each addition of a cattle result in a decrease in the quality of grazing for all. Lloyd showed that when the utilization of the pasture gets close to its capacity, overgrazing is guaranteed to doom the pastureland. For each herdsman, the incentive is to add more cattles to his herd as he receives the full proceeds from the sale of additional cattle, but shares the cost of overgrazing with all herdsmen. Whereas the common resource could have been reasonably shared by the herdsman exhibiting some restraint, greedy local choices made by the herdsmen quickly leads to overgrazing and destruction of the pasture. The question the herdsman will face is "What is the utility of adding one more animal to my herd?" [6]. He observes that "Freedom in a commons brings ruin to all." and convincingly argues that enforced laws, and not appeals to conscience, is necessary to avoid the *tragedy of the commons*.

Muhsam [10] has shown that if some or all other herdsmen add cattle when $x > C$, an individual must add a head if he or she wishes to reduce the loss suffered as a result. A rational, utility-maximizing agent will have no choice but to add to the herd, and hence, to the overall deterioration of the resource performance. This means that it is only possible to reach a *co-operative equillibrium*.

In our paper, we now define an abstract version of the Tragedy of the Commons problem, to be used in the rest of the paper, as follows: a shared resource can effectively support C units of load, but if the jointly applied load, x, exceeds C, the quality of the service received from the resource deteriorates. We call C the *critical load* of the resource. The above constraint is modeled by a utility per unit load function as follows:

$$U(x) = K, \quad \text{when } x \le C,$$
$$= K * \frac{C}{x}, \quad \text{otherwise,} \tag{1}$$

where K is a constant and $U(x)$ denotes the utility per unit load when a total of x units of load is applied on the system. It can be shown here that for any rational, self interested, utility maximizing agents it is always a better option to add more loads when the other agents are adding more load to the system. And also it is clear that when every agent will go on adding load the utilization of the system will be deteriorating as a result of decreased per unit utility. In such a situation, intelligent agents will try to reach a co-operative equillibrium to optimize the resource utilization. In this paper, we have presented such a mechanism.

3   A Local Decision Procedure for the Tragedy of the Commons Problem

In this section, we will provide an probabilistic distributed algorithm to solve the problem of the *Tragedy of the commons* and also discuss the convergence of the algorithm.

3.1   Algorithm

We assume the following. $A[i], i = 1..N$ are the agents in the society. C is the critical load of the shared resource, and x is the current combined load on the shared resource. $upper_i$ and p_i^u are private fields for agent i.

Step 1 Each agent apply a random load, l_i^0, on the shared resource. (We assume that the initial combined load is less than the capacity of the shared resource, i.e., $\sum_i l_i^0 < C$). Then the following steps are independently followed by each agent.

Step 2 Each agent i increments its load, l_i, on the shared resource. Note that, every time an agent increases (or decreases) its load implies increases (or decreases) it's load to the system by one unit.

Step 3 An agent i recieve its utility from the resource based both on the load it applied and the total load on the resource(i.e. x). The resource derives each agent's per unit utility from the equation 1 and send it to each agent. If the per unit utility received by this agent is not less than the best per unit utility it has received in the past, go to Step 2.

Step 4 Agent i decrements its load and sets its increasing probability, $p_i^u = 1$, and $upper_i$ to *false*.

Step 5 If $upper_i$ is *false*, agent i increments its load by one on the shared resource with probability p_i^u. Otherwise, agent i maintains its previously applied load.

Step 6 Agent i receives its updated utility based on current system load x and load applied by this agent. If an agent had increased its load in Step 5 and the new per unit utility is worse than the best per unit utility that agent has ever received, it decrements its load by one and sets p_i^u to half of it's previous value. Otherwise, an agent i who has increased its load in Step 5 sets $upper_i$ to *true*.

Step 7 If $upper_i$ is *false* and $p_i^u > p_t$ (where p_t is a small thresold probability) go to Step 5.

Step 8 Agent i maintains its current load l_i. Repeat Step 8.

3.2   Convergence to Equilibrium

The system reaches equilibrium when all agents reach Step 8 of the algorithm. At this state, each agent feels that any increment/decrement of its load will reduce its utility. Hence the load on the system do not change.

After an agent has passed through Step 3 it realizes that increasing its load may decrease its per unit utility which in turn decreases the utility obtained from the resource. So, every agent at Step 4 removes one load it added last time. Observe that every agent will execute Step 4 in the same iteration. Here we assume that in *Step*1 after all the agents add a random load, the total load administered to the resource did not exceed the resource capacity, C^1. So, after Step 4 there may be three possibilites from the perspective of each autonomous agent:

- It is the only agent who is using this resource and adding one more load will decrease the per unit utility.
- More than one agent is using this resource, and the resource has reached it's critical load *i.e.* if any of the agents adds one more load everybody's per unit utility will be decreased.
- More than one agent is using this resource and some but not all of the agents may add one more load without crossing the critical load.

A rational agent can reason after Step 4 that to prevent over-utilization of the resource it should add one more load. For the first two possibilities above, it should not increment load. But as it is not sure which of the possibilities correspond to the current situation, it can use a probabilistic exploration scheme outlined in Steps 5 through 7 to reach its optimum load. It starts with initial increment probabilty of 1. At Step 5 it adds one load with its *increment probabilty*. If the load increment produces increased per unit utility then it does not change it's load any more. Otherwise, if the addition of one more load reduces it's per unit utility, it halves it's *increment probability* and tries later to add one more load with this reduced *increment probability*. It keeps on probing in this manner until it's probability falls below a threshold. The motivation behind this halving exploration process is similar to exponential backoff used for conflict resolution in shared communication channels like a token ring. The realization is that there must be other agents in the system who are trying to increase their loads as well, and unless every agent back off a little from their eagerness to increase system load, no one can benefit. Such exponentially decaying probabilities make it more likely for the system to converge to an equilibrium.

In our algorithm, an agent uses only local feedback to determine the load it applies on the system. Global information about individual loads used by other agents is never used. Our claim is that when *equilibrium* is reached, the combined load on the resource is exactly the *critical load* or capacity of the resource, i.e., the agents are using the resource optimally. They have reached this optimality through a distributed decision procedure using only local knowledge and without the directive of any central authority.

Now, we present some arguments for our decision procedure producing convergence to the resource capacity, C. Suppose after Step 4 the load reaches L

---

[1] If this assumption is violated a minor modification to the algorithm is required to reduce the loads and bring the total load near C. Our implementation includes this modification.

and the increment probability to 1. So, it is clear that $L \leq C < L + N$ has to be satisfied, i.e., there must be some y for which $L + y = C$, where $0 \leq y < N$. After Step 5 of the first iteration of the loop Step 5 through Step 7 the expected number of load added to the system is N. So, the total load on the system exceeds critical load and every agent reaches to Step 5 with system load L and increment probability 0.5. Now in the second iteration of the loop, the expected load after Step 5 is $L + \lfloor \frac{N}{2} \rfloor$. C may be greater than, equal to or less than this load. In the first two possibilities the load on the resource is expected to increase to $L + \lfloor \frac{N}{2} \rfloor$ and the number of agents who will be further willing to increase their load by one is remaining $N - \lfloor \frac{N}{2} \rfloor$ with their increment probability set to 0.5 in the next iteration of the loop. When $C < L + \lfloor \frac{N}{2} \rfloor$, all of the N agents will withdraw their added load resulting in the load on the resource be still at L and each agent's increment probability reduced to 0.25 in the next iteration of the loop. So, after each subsequent loop through Steps 5 through 7, the system progresses towards convergence because either the number of agents willing to increment their load is reduced to half compared to that of the last iteration of the loop with the same individual increment probability, or the number of agents willing to increment their load remains the same with the corresponding increment probability reduced to half compared to that of the last iteration of the loop.

So, after $\lfloor \log_2 N \rfloor + 1$ iterations of the loop Step 5 through Step 7, the load on the resource will be C and $C - L$ agents will feel content with their maximum possible load and $N - C + L$ number of agents will be still trying to add one more load with increment probability lies between 0.5 to $2^{-\log_2 N}$.

Now, we can say that after sufficient steps (with a loose upper bound of $\lceil \frac{(-\log_2(p_t))^2}{2} \rceil$, where p_t is the small thresold probability) of the loop Step 5 through Step 7 these $N - C + L$ agents satisfy the criterion of equillibrium as its increment probability will be less than a threshold. So, in any of the three possibilities of perception, every agent will not change its current load. This is also the optimal procedure for reaching the *critical load* of the resource as can be inferred from the theory of binary search.

4   Experimental Section

We have experimented with different scenarios considering different number of agents in the system and different critical loads. We now show results from some of the experiments with 10 and 100 agents. Here we take the value of K as 1. We set the critical load to different values with each agent following the decision procedure outlined above. We have chosen resource capacities such that equal distribution of load will not produce optimal resource utilization. These scenarios provide more difficult challenges compared to the case whre system capacity can be uniformly shared by the agents.

In Figure 1 (left), the load capacity is set to 62 and there are 10 agents in the society. The figure shows how the autonomous agents reaches equilibrium with the total load applied equalling the resource capacity. In this figure

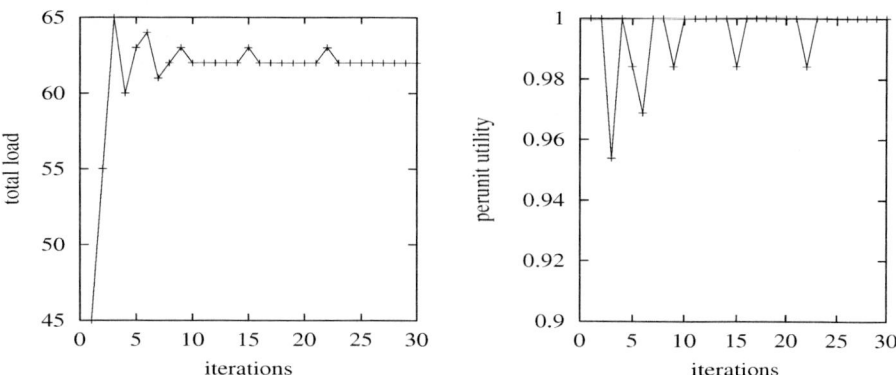

Fig. 1. (left)Variation of total load and (right) variation in average per unit utility of an agent: in the system with 10 agents and a load capacity of 62

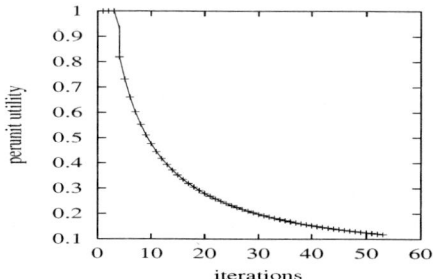

Fig. 2. Variation of average per unit utility of an agent in a system with 10 agents and a load capacity of 62 where the agents are not using local decision procedure and go on adding load in each iteration

we have shown that after initial random allocation of the load to the resource, agents steadily increment the load and then when the load exceeds the resource capacity, agents decrease their loads to reach equilibrium. The convergence phenomena is similar to the overshooting and undershooting typically observed in control systems where the control variable overshoots and undershoots the desired set point before settling. In Figure 1 (right), we present the variation of the average per unit utility of an agent over the course of a run. We can observe that initially there are a lot of deviations in the average per unit utility per agent as the system overshoots the critical load. Finally, however, optimal capacity is used at equilibrium and average per unit utility of an agent is reached to the maximum(which is 1 here as the value K) .

In this framework of the experiment, in Figure 2 we show how the average per unit utility of an agent goes down, if all agents go on adding load instead of using this decision mechanism.

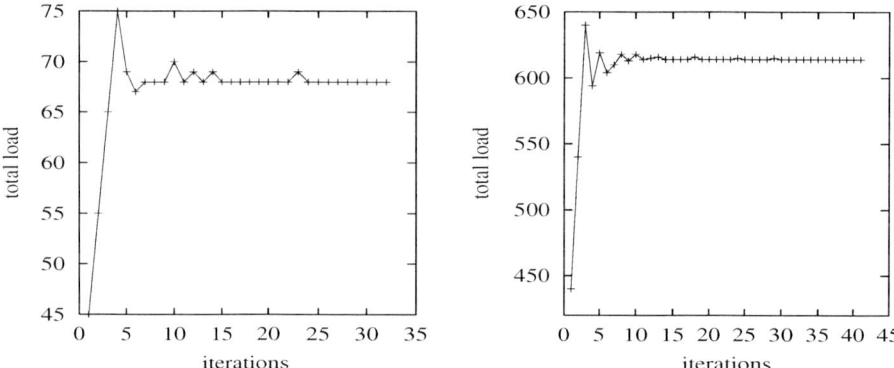

Fig. 3. (Left) Variation of total load on a system with 10 agents and a load capacity of 68 and (Right) Variation of total load on a system with 100 agents and a load capacity of 614

In Figure 3 (Left), the critical load is changed to 68. This to ensure that the algorithm works for all types of critical loads. Here we have shown what happens after Step 4 of the algorithm. In the right figure of Figure 3, we use a larger society of autonomous agents where the size of the society is 100. Here the critical load is set to 614.

We have noted that the system reaches equilibrium with combined load equal to resource's critical capacity in all the scenarios we have experimented with. We also verify this claim with running each experiment for 100 times and observe no deviation from the convergence.

5   Conclusions

In this paper we have presented an algorithm to avoid the problem of Tragedy of the commons in a society of rational agents based only on local feedback in the form of utility received for the current load applied on a shared resource. We have shown that our proposed procedure results in the shared resource is used at its capacity load starting from arbitrary initial loads. As the tragedy of the commons is an important, and common problem which can lead to inefficiencies in the usage of shared resources, our procedure can have wide applicability.

Our procedure results in an equilibrium where some agents have higher utility than others even when everyone starts with the same load. This is because the equilibrium nature is static in the sense that no agent change their load after reaching equillibrium. We plan to work towards a more "fair", dynamic eu-qilibrium where agents increase/decrease their load around so that the capacity load is maintained while individuals with higher than average utility change over time. We also plan to augment our procedure such that equilibrium is reached in fewer iterations.

One of the drawback of this approach is that this considers only integral load. It will be interesting to study how it can be improved to work with real-valued loads.

Acknowledgements

This work has been supported in part by NSF grants IIS-9724672 and IIS-0209208.

References

[1] ARORA, N., AND SEN, S.: Resolving social dilemmas using genetic algorithms: Initial results. In *Proceedings of the 7th International Conference on Genetic Algorithms*, pages 689–695, San Mateo, CA, 1997. Morgan Kaufman. 312

[2] CAMMARATA, S., MCARTHUR, D., AND STEEB, R.: Strategies of cooperation in distributed problem solving. In *Proceedings of the Eighth International Joint Conference on Artificial Intelligence*, pages 767–770, Karlsruhe, Federal Republic of Germany, August 1983. 312

[3] DURFEE, E. H., AND LESSER, V. R.: Using partial global plans to coordinate distributed problem solvers. In *Proceedings of the Tenth International Joint Conference on Artificial Intelligence*, pages 875–883, Milan, Italy, August 1987. 312

[4] GLANCE, N. S., AND HOGG, T.: Dilemmas in computational societies. In *First International Conference on Multiagent Systems*, pages 117–124, Menlo Park, CA, 1995. AAAI Press/MIT Press. 312

[5] GLANCE, N. S., AND HUBERMAN, B. A.: The dynamics of social dilemmas. *Scientific American*, 270(3): pages 76–81, March 1994. 313

[6] HARDIN G.: The tragedy of the commons. *Science*, 162: pages 1243–1248, 1968. 311, 314

[7] HOGG, T., AND HUBERMAN, B. A.: Controlling chaos in distributed systems. *IEEE Transactions on Systems, Man, and Cybernetics*, 21(6): pages 1325–1332, December 1991. (Special Issue on Distributed AI). 313

[8] IRVINE A. D.: How Braess' paradox solves Newcomb's problem. *International Studies in the Philosophy of Science*, 7(2): pages 141–160, 1993. 312

[9] LLOYD. W. F.: *Two Lectures on the Checks to Population*. Oxford University Press, Oxford, England, 1833. 311, 314

[10] MUHSAM H. V.: A world population policy for the World Population Year. In *Jouranl of Peace Research*, 1(2): pages 97–99, 1973. 314

[11] MUNDHE, M. AND SEN, S.: Evolving agent societies that avoid social dilemmas. In *Proceedings of the Genetic and Evolutionary Computation Conference, GECCO-2000*, pages 809–816, 2000. 312

[12] SANDHOLM, T. W. AND LESSER, V. R.: Equilibrium analysis of the possibilities of unenforced exchange in multiagent systems. In *14th International Joint Conference on Artificial Intelligence*, pages 694–701, San Francisco, CA, 1995. Morgan Kaufman. 312

[13] SMITH A.: *The Wealth of Nations*. A. Strahan, Printer-stree; for T. Cadell Jun. and W. Davies, in the Strand, Boston, MA, 10 edition, 1802. 314

[14] TUMER, K. AND WOLPERT, D. H.: Collective intelligence and Braess' paradox. In *Proceedings of the Seventeenth National Conference on Artificial Intelligence*, pages 104–109, Menlo Park, CA, 2000. AAAI Press. 312

[15] TURNER R. M.: The tragedy of the commons and distributed ai systems. In *Working Papers of the 12th International Workshop on Distributed Artificial Intelligence*, pages 379–390, May 1993. 311, 312

packets get queued for service, which slows down network traffic [6]. Such delay can translate to loss of productivity or negative utility to users. The second issue relates to firewall quality, which determines the incidence of misdetection and false alarm. In this paper we assume that a single firewall is being used, and perform an economic analysis by deriving a mathematical expression for the net benefit that accrues to the organization. By adjusting the firewall quality settings, we can then maximize the net benefit.

Section 2 is a brief general introduction to firewall technology. Section 3 describes the general problem to be solved, and section 4 discusses the single firewall case. The concluding section summarizes the paper and lists problems that merit further investigation.

2   Firewalls

Firewall technology [3, 9] plays a most important role in protecting informational assets. It helps to divide networks, and it checks for the legitimacy of the data that flows between networks. It is often considered to be the first line of defense, because it separates external networks from internal networks by disallowing potentially harmful messages to transfer in and out. It also allows layer security and segregates internal networks in order to offer better granularity in system protection.

Firewalls typically rely on configured rule sets to determine which messages are allowed to enter or leave the network. A *rule set* [9] is just a set of binary (Yes/No) rules for decision making. A firewall employs a *chokepoint strategy, i.e.*, it forces all traffic to pass through it, including potentially harmful message packets from intruders. Firewalls can be classified into two main types. The first type is a *packet filter firewall* (simple firewall, SF), and the second type an *application gateway proxy firewall* (complex firewall, CF). A packet filter examines the validity of information at the Data Link, Network and Transport Layers. Most of the information examined by it is found in the packet header, including the source and destination IP addresses, the port number, and the protocol being used. It finds its main use at the network perimeter where it provides fast inspection of the incoming traffic. An application gateway proxy firewall checks the information at the Application Layer and prevents direct connection between hosts. It scans packet *contents* upon receiving a request from a host, and then transmits the data using its own identity once the data has been verified. Thus it implements Network Address Translation (NAT), which allows the identity of the internal machines to be hidden from outsiders. In addition, it can provide user authentication and can thus implement a *least privilege strategy, i.e.*, users can only access information and perform tasks for which they have the requisite authorization. As a result, a complex firewall is able to inspect and log a greater amount of information about a request than a simple firewall. Because its operations are more extensive, a complex firewall has a significantly higher service time than a simple firewall [5]. A number of firewalls can be installed at different network checkpoints to create security layers. This *defense in depth strategy* allows networks to be physically separated into layers based on security sensitivity.

The firewall architecture shown in Fig 1 is typical of many implementations. For variants of this architecture, see [9]. The internal network generally has a number of servers, such as a database server, a web server and a mail server. All message packets from the external world first pass through a simple firewall, which is assumed to have routing and load balancing capabilities. The simple firewall forwards those packets it identifies as *benign* (*i.e.*, desirable or harmless) to an appropriate complex firewall connected to the server as determined by the protocol of the incoming message. Packets identified as *intrusive* (*i.e.*, undesirable or harmful) are blocked. Complex firewalls are needed because a simple firewall is unable to identify all intrusive packets. As stated before, there may be multiple types of servers in a network, and in practice a server might receive message packets from a number of complex firewalls. Fig 1 shows only one type of server.

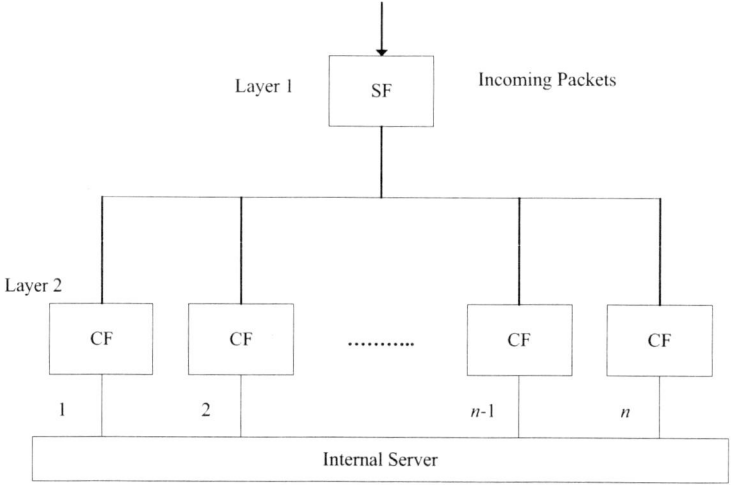

Fig. 1. Firewall Architecture

3   Problem Description

Our objective here is to undertake a preliminary analysis of the cost tradeoffs among the different cost factors related to firewalls. We consider only the incoming traffic and ignore the outgoing traffic. In the general case depicted in Fig 1, each packet goes through a simple firewall and then through a complex firewall before reaching its destination server. Here, for simplicity, we restrict ourselves to the case in which the incoming packets pass through only one firewall. The firewall has a buffer in which the packets can be queued. The size of the buffer depends on the available memory in the machine, but for simplicity we assume it is unlimited. Message packets from the external world, which are assumed to be of the same length, arrive at the firewall at the rate of λ packets per unit time. The inter-arrival time is assumed to have an exponential distribution, *i.e.*, the arrival process is Poisson. The firewall services the packets at the rate of μ packets per unit time, where $\mu > \lambda$. (If $\mu \leq \lambda$, there can be an indefinite buildup of packets in the queue.) Thus the queue behaves as an M/G/1 queue [4]. Since the firewall employs a rule set, with the special cases at the beginning of the set and the default rule at the end, it is unreasonable to suppose that the service time has an exponential distribution. For simplicity, we assume that the service time is constant and does not vary with the packet. Under this assumption, the firewall functions like an M/D/1 queue.

The major task of a firewall is to detect intrusive packets and block them. When a firewall is not perfect in its job, two types of errors may arise: i) *False Alarm*: A benign packet gets blocked; ii) *Misdetection*: An intrusive packet gets through. Let q_1 and q_2 be the fractions of benign and intrusive packets that get blocked. These are called the *quality parameters* of the firewall. Ideally, we should have $q_1 = 0$ and $q_2 = 1$, but in practice q_1 and q_2 are functionally related. This relationship can be depicted using the Receiver Operating Characteristics (ROC) curve. The ROC curve is a commonly used evaluation technique in clinical medicine and shows the relationship between an illness condition and a no illness condition [7, 8]. Axelsson [1] has made use of the ROC curve in the analysis of the effectiveness of intrusion detection systems.

Fig 2 shows how the two quality parameters are related. The feasible operating points of an imperfect firewall lie on the ROC curve, which has a steadily decreasing slope and lies above the unit slope line.

When the detection rate q_2 is high, the false alarm rate q_1 is also high and a significant portion of benign traffic is blocked. To achieve a low false alarm rate the rules in the rule set have to be relaxed, which results in more intrusive traffic getting through. At point a, no packets are blocked, corresponding to the situation when there is no firewall. At point c, all traffic is blocked. The situation is ideal at point b, because there are no false alarms and all intrusive packets are correctly identified and blocked. But an imperfect firewall cannot operate at point b, since its operating point must lie on the ROC curve. Between two imperfect firewalls, the one that has a larger area under its ROC curve is to be preferred, because its operating point can approach closer to point b Note that an alteration of the rule set of a firewall changes its ROC curve.

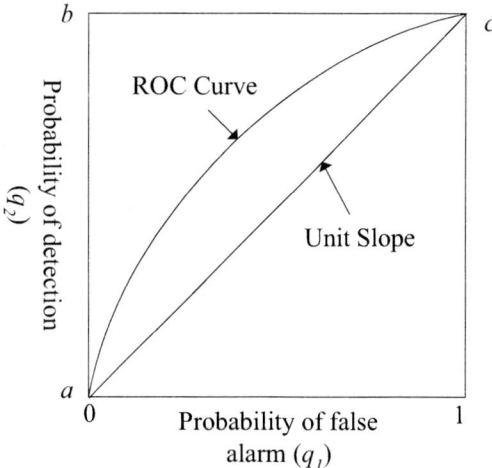

Fig. 2. ROC Curve

Notation:

λ: Mean arrival rate of packets (expected number of packet arrivals per unit time, exponential inter-arrival time assumed)

μ: Mean service rate (expected number of packets processed per unit time, constant in our case)

s: Number of parallel servers (one in our case)

L: Expected number of packets in queue

W: Expected waiting time in system for a packet (sum of waiting time in queue and service time)

π: Proportion of incoming traffic that is benign

q_1:   Probability that a benign packet is blocked by the firewall

q_2:   Probability that an intrusive packet is blocked by the firewall

4   Analysis of the Single Firewall Case

When a single imperfect firewall is in use, the organization needs to determine the point on the ROC curve at which the firewall should operate. This point should be selected to maximize the economic benefit to the organization. In the absence of the firewall, the entry of intrusive packets is unrestricted, and the cost to the organization is C_1 per unit time. Let the total cost be C_2 per unit time in the presence of the firewall. We seek to maximize the net benefit $(C_1 - C_2)$. This quantity must be positive; otherwise the installation of the firewall cannot be justified. Since C_1 does not depend on the firewall characteristics, our job is to minimize C_2. C_2 has the form

$$C_2 = R + D + B + I , \tag{1}$$

where, R = cost of operating firewall (per unit time)

      D = cost of delay at firewall (per unit time)

      B = cost incurred due to blockage of benign traffic (per unit time)

      I = cost incurred due to through passage of intrusive traffic (per unit time)

The first term in the cost function reflects the actual cost of operating the firewall. It can also include the cost of capital for the purchase and installation of the firewall. The second term captures the cost of delay of benign packets queued at the firewall. The third term indicates the cost incurred due to loss of goodwill and consequent loss of business caused by the blockage of benign packets (i.e., by false alarms). The last term gives the cost of theft of informational assets and that of repairing damaged informational assets when intrusive packets get through (i.e., the cost of misdetection). Since the firewall by assumption behaves like an M/D/1 queue, the waiting time W is given by the formula [4] (p 872)

$$W = \frac{2\mu - \lambda}{2\mu(\mu - \lambda)} = \frac{1}{2\mu} + \frac{1}{2(\mu - \lambda)} . \tag{2}$$

Let us consider the traffic entering and leaving the firewall per unit time. Since packets arrive at the rate of λ per unit time, a fraction $\pi\lambda(1 - q_1)$ of the benign traffic is not blocked. Again, a fraction $(1 - \pi)\lambda(1 - q_2)$ of the intrusive traffic gets through.

Thus C_2 becomes

$$C_2 = R + \left[\frac{1}{2\mu} + \frac{1}{2(\mu - \lambda)} \right] \pi\lambda(1 - q_1)K + \pi\lambda q_1 K_1 + (1 - \pi)\lambda(1 - q_2)K_2. \tag{3}$$

In the above equation, the first term is small in value and can be taken to be a constant, although in practice it will depend on the actual firewall in use. In the delay term, only the benign traffic has been considered when computing the cost, since this cost is borne by the genuine users and not by the intruders. K is a conversion factor that converts delay to cost. The third term indicates the cost of blocking benign traffic, and the fourth term indicates the cost incurred as a result of penetration by intruders. K_1 and K_2 are conversion factors. In practice, K_2 is likely to be much larger than K_1, because misdetection often causes irreversible damage, such as when confidential information is compromised. False alarm only causes frustration to genuine users. But the actual number of false alarms can be high because the benign traffic usually makes up a substantial proportion of the total traffic. We expect the first two terms to be much smaller in magnitude than the last two terms.

The relationship between misdetection and false alarm can be defined by an equation of the form $q_1 = q_2^r$, $r > 1$. This is consistent with the shape of the ROC curve in Fig 1. When r increases, the area under the ROC curve is larger, so a higher value of r indicates a firewall of superior performance. We must now minimize C_2 with respect to q_2. We have

$$C_2 = R + \left[\frac{1}{2\mu} + \frac{1}{2(\mu - \lambda)} \right] \pi\lambda(1 - q_2^r)K + \pi\lambda q_2^r K_1 + (1 - \pi)\lambda(1 - q_2)K_2. \tag{4}$$

Differentiating C_2 with respect to q_2 and setting the result to zero, we get

$$q_1^* = \left(\frac{(1 - \pi)K_2}{r\pi \left[K_1 - K\left(\frac{1}{2\mu} + \frac{1}{2(\mu - \lambda)} \right) \right]} \right)^{\frac{r}{r-1}} \tag{5}$$

and

$$q_2^* = \left(\frac{(1 - \pi)K_2}{r\pi \left[K_1 - K\left(\frac{1}{2\mu} + \frac{1}{2(\mu - \lambda)} \right) \right]} \right)^{\frac{1}{r-1}}. \tag{6}$$

Equations (5) and (6) give the values of the quality parameters at the point on the ROC curve at which the total cost C_2 is minimized. By scanning the expressions, we can determine how the optimal operating point shifts on the ROC curve as the parameter values change:

i)   When K_1 increases, both quality parameters decrease in value. This says that when the cost associated with a false alarm is higher, the optimal operating point moves left on the ROC curve to reduce the number of false alarms.

ii) When K_2 increases, the quality parameters increase in value. This says that when the cost associated with misdetection is higher; the optimal operating point moves right on the ROC curve to block more packets and reduce the risk of security breaches.

iii) When K or λ increase, the quality parameters also increase in value. This is because we would want to block more of the traffic to reduce the cost of delay.

iv) An increase in the service rate μ results in a decrease in the values of the quality parameters. This is because when more packets can be serviced, q_1 should be reduced to allow more legal traffic to get through the firewall.

v)   When r increases, the optimal operating point moves left. This can be verified easily by computing the slope of $\ln q_1^*$ with respect to r. But with increase of r, q_2^* might increase or decrease. We illustrate this phenomenon in Fig 3. Suppose Firewall 2 is superior in performance to Firewall 1, *i.e.*, Firewall 2 has a higher value of r. Let the operating point be at point m for Firewall 1. When we switch to Firewall 2, the new operating point can be at point n or at point n^l. The value of q_1^* is smaller at both points, but q_2^* can be smaller or larger.

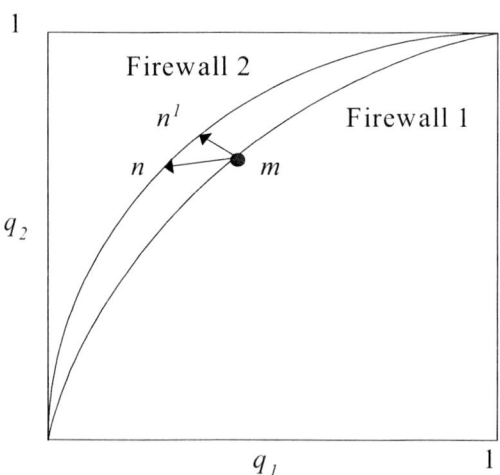

Fig. 3. Shift of Operating Point on Improvement of Firewall Quality

Example: Suppose we want our firewall to operate at the point $q_1 = 0.01$, $q_2 = 0.90$. This means we desire a low rate of false alarms, but are willing to live with misdetection in one out of ten cases. Whether this operating point will be optimal will depend on the parameter values. In particular, since $q_1 = q_2^r$, to achieve the given values of the quality parameters we will need to choose a firewall with $r = 43.71$. It can be readily verified that in the neighborhood of these values, as r increases q_1 will decrease but q_2 will increase.   □

One factor that has been ignored in the foregoing analysis is the dependence of the service rate μ on the quality parameters. It is well known that the service rate is not constant but tends to decline as q_1 increases or q_2 decreases from their ideal values. This dependence can be taken to be of the form $\mu_{actual} = \mu_{opt} \cdot (1 - q_1) \cdot q_2$, where μ_{opt} is the service rate when $q_1 = 0$ and $q_2 = 1$. We should use μ_{actual} in place μ in (3), but that would complicate the subsequent computations considerably.

5   Concluding Remarks

In this paper we have undertaken an economic analysis and derived a mathematical expression for the net benefit that accrues to an organization when firewalls are put in use. For simplicity we have restricted ourselves to the single firewall case. The two quality parameters of the firewall determine the likelihood of false alarm and the likelihood of misdetection. These parameters bear a functional relationship, which can be modeled using the ROC curve. We can then determine the operating point at which the net benefit to the organization is maximized. The quality parameters can be tuned by altering the rule set of the firewall, and if that proves infeasible, by physically changing the firewall. The analysis given above can be extended in the following ways:

a)   Instead of just one firewall, we can take two firewalls into consideration. One of these would be a simple firewall, and the other a complex firewall. The incoming traffic will enter the system at the simple firewall, and the filtered traffic would flow on to the complex firewall. The volume of traffic reaching the complex firewall would depend on the quality parameters of the simple firewall. To determine an optimal operating point, we would have to tune two sets of quality parameters simultaneously.

b)   In a more general setting, assuming just one internal server, there would be one simple firewall and multiple (say $n > 1$) complex firewalls as shown in Fig 1. In this case, to optimize the net benefit, we would have to perform two different tasks: i) to find the best value of n to use; ii) to tune all the $(n + 1)$ sets of quality parameters simultaneously.

References

[1]   Axelsson, S.: The Base-Rate Fallacy and the Difficulty of Intrusion Detection. ACM Transactions on Information and System Security. 3:3 (2000) 186-205

[2]   Canavan, J. E.: Fundamentals of Network Security. Artech House, Norwood MA (2001)

[3]   Cheswick, W. R., Bellovin, S. M., Rubin, A. D.: Firewalls and Internet Security: Repelling the Wily Hacker. Addison-Wesley, Boston MA (2003)

[4]   Hillier, F. S., Lieberman, G. J.: Introduction to Operation Research, (6th Ed). McGraw-Hill, New York NY (2001)

[5]   Lyu, M. R., Lau, L. K. Y.: Firewall Security: Policies, Testing, and Performance Evaluation. Proc 2000 Intl Conf on Computer Systems and Applications (COMPSAC'2000). Taipei Taiwan (2000)

[6]   Menasce, D. A., Almeida, V. A. F.: Capacity Planning for Web Services: Metrics, Models, and Methods. Prentice Hall, NJ (2001)

[7]   Metz, C. E.: Basic Principles of ROC Analysis. Seminars in Nuclear Medicine. 8 (1978) 283-298

[8]   Zweig, M., Campbell, G.: Receiver-Operating Characteristic (ROC) Plots: A Fundamental Evaluation Tool in Clinical Medicine. Clinical Chemistry 39 (1993) 561-577

[9]   Wack, J., Culter, K., Pole, J.: Guidelines on Firewalls and Firewall Policy: Recommendations of the National Institute of Standards and Technology. Computer Security Division, Information Technology Laboratory, National Institute of Standards and Technology (NIST), Gathiersburg MD (2002)

Load Sharing in a Transcoding Cluster

Jiani Guo and Laxmi Bhuyan

Computer Science and Engineering
University of California, Riverside, CA 92521
{jiani,bhuyan}@cs.ucr.edu

Abstract. The growing use of multimedia and e-commerce applications in the Internet presents scalability concerns and new demand for computational resources. A cluster of workstations is especially suitable for such computation-intensive applications. In this paper, we present the design of a transcoding cluster for multimedia applications. Three different load sharing schemes have been implemented. Performance of the system is evaluated in terms of system throughput, out-of-order rate of outgoing media streams and load rebalancing overhead. Based on the experimental results, we present the suitable environments where each load sharing scheme can be applied.

1 Introduction

In this paper, we present a cluster implementation that provides video/audio transcoding service. Multimedia transcoding performs video/audio transformations such as changing the bit-rate, resizing video frames or adjusting the frame resolution etc. Transcoding service is very important for streaming multimedia data in the Internet. It enables the media streams to adapt to the variation of available bandwidth in the transmission channel, and the wide variety of receiver heterogeneity in their inbound network bandwidth, CPU/MEM capacity or display resolution.

Our transcoding cluster consists of a cluster of generic PCs. One PC, called Manager node, acts as the coordinator. Other PCs, called Worker nodes, provide the transcoding service. The Manager node is also connected to a remote PC, which continuously sends media streams to it. A media stream is a complete video/audio segment, like a movie or a conference recording. The remote PC can be regarded as a simplified media server in the practical network. We assume that the media stream data can be divided into a sequence of media units that are ready for independent transcoding. A media unit can be a group of pictures (GOP) of MPEG streams or a FRAME of AVI streams. The Manager node receives these media units from the remote PC, and distributes them to the Workers for transcoding. Each Worker independently processes the media units using the local computing resource and does not require any global stream state. We do not transcode any units on the Manager node. The aims of our implementation are to minimize the processing time for each media unit in the

S. R. Das, S. K. Das (Eds.): IWDC 2003, LNCS 2918, pp. 330–339, 2003.
© Springer-Verlag Berlin Heidelberg 2003

cluster, and to preserve the unit order of outgoing media streams as much as possible.

Load sharing is a critical issue in parallel and distributed systems. A detailed survey of general load balancing algorithms is provided in [ ]. In the network domain, load balancing schemes are particularly adopted to split network service requests among a bunch of servers such as web servers, or distributed cache servers. When the concept of flow is involved, one important goal of any load balancing policies is that the packets belonging to the same flow should be kept in order. For example, in TCP-based flows, out-of-order packets may trigger retransmission or even congestion control, and thus degrade the throughput. In UDP-based flows, such as video/audio transmission, limited receiving buffers may not accommodate the out-of-order packets. This results in higher drop rate, and thus affects the quality of service.

In practice, simple static policies, such as random distribution policy [ ] or modulus-based round robin policy [ ], can achieve satisfactory results. However, the random distribution cannot preserve packet order within a flow if per-flow information is not maintained. Modulus-based round robin policy also has the drawback that all flows are re-mapped if the number of computing nodes is changed. On the other hand, adaptive load balancing policies are usually compli- cated and require prediction of computation time for any incoming requests [ ]. They are difficult to implement and produce increased communication over- head [ ]. Hence, only simple load sharing techniques, such as round robin, are adopted in practice.

In this paper, we implement three kinds of load sharing strategies, self- adjustable scheme, stream-based scheme and feedback-based schemes, in our transcoding cluster. The self-adjustable scheme is first fit that requires no ex- plicit feedback information when scheduling. The stream-based scheme estab- lishes static mapping between streams and Worker nodes. The feedback-based scheme is based on the least load first policy. Guided by the feedback informa- tion obtained from Workers periodically, least load first scheme smartly balances the load among the Worker nodes. However, the scheme may incur substantial overhead to implement the feedback.

We did experiments to test the system performance of our implementation by varying the number of the Worker nodes and load balancing policies. The main performance metrics are out-of-order rate of the streams, inter-arrival time among media units and scalability of the cluster. The paper makes the following contributions:

1. We present three new load sharing strategies for a transcoding cluster. We design and implement a load test mechanism which can periodically feed the load distribution information among the worker nodes back to the manager node to guide load balancing.
2. We implement a transcoding cluster using five linux-based PCs connected over a Gigabit Ethernet, and carry out extensive performance evaluation of the three load sharing algorithms through measurements. The results give

insights into out-of-order properties of the video outputs and scalability of the parallel computation.

The paper is organized as follows: In section 2, we present various load balancing schemes and the feedback mechanism. In section 3, we briefly discuss the experimental framework of the transcoding cluster and the related implementation issues. Experimental results are presented in section 4. Finally, section 5 concludes this paper.

2   Load Balancing

A critical issue in implementing this cluster system is the load balancing strategy, i.e., how to distribute multiple media streams among different computing PCs to achieve the best utilization. A unit based round robin scheduling is adopted in [6]. Using this scheme, we observe that the performance is poor in terms of out-of-order degree for processed units. Hence, we explore other schemes, which are divided into three categories: self-adjustable load balancing scheme, stream-based scheme and the feedback-based load sharing.

2.1   First Fit(FF)

With First Fit, the Manager searches for an available Worker in round robin way. It always chooses the first available one to dispatch a GOP. The way for the Manager to detect if a Worker is available is implementation-related, which we will discuss in section 3.1. In this scheme, the loads among several Workers are naturally balanced to some extent. It is a nice property as far as the processing speed is concerned, since no extra load analyzer is needed to guide scheduling. However, the GOPs of the same stream are most likely to be distributed to different Workers, and thus the delay jitters for each stream at its destination may become severe.

2.2   Stream-Based Mapping(SM)

To preserve the order of computation among media units, as well as to keep the simplicity of the algorithm, we proposed and implemented a stream-based mapping algorithm in [5]. The unit is mapped to a Worker according to the following function:

$$f(c) = c \bmod N \tag{1}$$

where, c is the stream number to which the unit belongs; and N is the total number of Workers in the cluster. Therefore, all the units belonging to one stream will be sent to the same Worker. The Manager always dispatches GOP of stream i to the Worker j, where $j = i\%N$.

However, this scheme works most efficiently for some specific input patterns in a cluster consisting of homogeneous Worker nodes. Assuming there are M streams and N Workers, those specific input patterns satisfy: $M \geq N$ and M is multiple of N.

2.3   Least Load First(LLF)

To improve on the above two schemes and extend the load sharing scheme to a cluster of heterogeneous Workers, it is necessary for the Manager to know the actual workload and processing power of all Workers when scheduling jobs. Thus, we design and implement a feedback mechanism as follows. The Worker periodically reports to the Manager its load information in terms of the maximal possible throughput and CPU utilization. Based on the collected load information, the Manager can calculate the available processing power of all Workers at time t, defined as a vector $(A_1(t), A_2(t), A_3(t), ..., A_N(t))$. Assuming there are N Workers and the monitoring epoch is $Deltat$, $A_i(t)$ is calculated as follows.

Symbol	Definition of the Symbol
a	A real value between 0 and 1
$n_i(t)$	The number of GOPs that are transcoded on the ith Worker during last monitoring epoch
$up_i(t)$	The CPU time used for transcoding task on the ith Worker during last monitoring epoch
$idle_i(t)$	Total CPU idle time of the ith Worker during last monitoring epoch
$u_i(t)$	CPU utilization of the ith Worker during last monitoring epoch
$AU_i(t)$	The smoothed average of the CPU utilization of the ith Worker till time t
$s_i(t)$	The maximal possible throughput of the ith Worker during last monitoring epoch
$AS_i(t)$	The smoothed average of the maximal possible throughput of the ith Worker till time t
$q_i(t)$	The waiting queue length for the ith Worker at the time monitoring is performed

$$u_i(t) = up_i(t)/(up_i(t) + idle_i(t)), \qquad (2)$$
$$AU_i(t) = AU_i(t - \Delta t) \times (1 - a) + u_i(t) \times a, \qquad (3)$$
$$s_i(t) = n_i(t)/up_i(t)/u_i(t), \qquad (4)$$
$$AS_i(t) = AS_i(t - \Delta t) \times (1 - a) + s_i(t) \times a, \qquad (5)$$
$$A_i(t) = AS_i(t) \times \Delta t - q_i(t) \qquad (6)$$

Since the load information varies from time to time, the smoothed average taking the history into account is adopted for the terms. a is a ratio which determines how much the history accounts for the average.

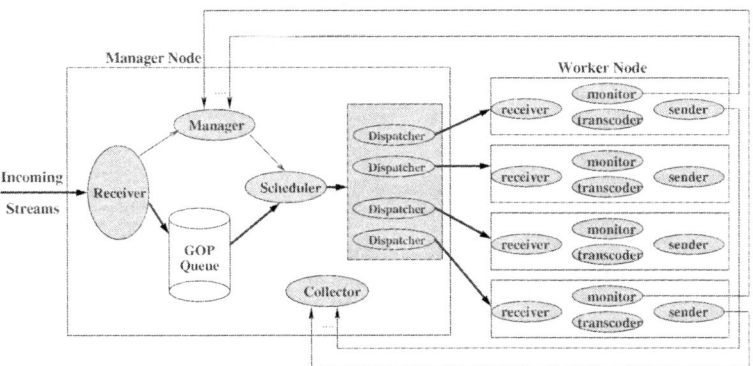

Fig. 1. Computation Model of the Transcoding Cluster

When scheduling GOPs, the Manager always picks the currently least loaded Worker. To achieve this, two pieces of information are maintained for each Worker in each epoch: $A_i(t)$ calculated as above; the number of outstanding requests, i.e., the number of GOPs already dispatched to it and not yet completed. Subtracting the second piece of information from the first one, the scheduler always picks the least loaded Worker to send one GOP, and then add 1 to its outstanding requests.

With this scheme, it is guaranteed that the Workers get the transcoding workload proportional to their capacities. But still, the GOPs of the same stream are possibly to be distributed to different Workers, and thus causes arrival jitters when a stream reaches its destination.

3   Experimental Framework

3.1   Cluster Architecture

Figure 1 demonstrates the computation model of our transcoding cluster. The Worker nodes and the Manager node are connected through a Gigabit Ethernet. The cluster is implemented at the user-level. On the Manager node, five kinds of threads, namely, *receiver*, *scheduler*, *dispatcher*, *collector* and *manager*, are running concurrently. Four threads, *receiver*, *transcoder* and *sender* and *monitor*, reside in the Worker node.

Manager Node The *receiver* receives the GOPs and puts them into the GOP queue. The GOP queue adopts a simple FIFO policy. The *scheduler* fetches units from the GOP queue and puts them into the dispatch queues according to the load balancing policy discussed in section 2. Note that there is a dispatch queue together with a *dispatcher* thread per each Worker node. The *dispatcher* dispatches the GOPs in the dispatch queue to the corresponding Worker node.

The *manager* periodically collects the load statistics information from the Worker nodes and feeds the information to the *scheduler*.

The *collector* collects a brief information about the processed units from the Worker node.

Worker Node The *receiver* receives packets from the Manager and ensemble them into a complete GOP. Once a complete GOP is ready, the transcoder transcodes the GOP. After transcoding, the *sender* sends the GOP to its destination. Once the *receiver* gives the GOP to the transcoder for processing, it requests another GOP from the Manager. Imagine that there are two Workers with different processing power, namely Worker1 and Worker2, and Worker1 is faster than Worker2. The *receiver* thread running on Worker1 will request GOP from the Manager node more frequently than Worker2. Thus, the dispatch queue for Worker1 on the Manager node will be drained faster than that of Worker2, i.e., the dispatch queue for Worker2 will become empty slower than that of Worker1 assuming they contain the same amount of GOPs at the beginning. Therefore, when implementing the FF scheme in our system, by checking if a dispatch queues has a vacancy, the *scheduler* knows if the corresponding Worker is available. The queue length of each dispatch queue is implemented as 2 GOPs in our system.

The *monitor* collects the load statistics information on the Worker node and reports it to the Manager node periodically.

3.2   Hardware Setup

We implement a simplified media server on the remote PC, which continuously sends media stream data to the transcoding cluster. The media streams are short movies encoded in MPEG-1 format. Each GOP of the media stream consists of 15 frames for the playback time of 0.5 second, since the normal playback rate is 30 frames per second (fps). The average GOP size is around 90k. The media server sends streams in round robin fashion. For each stream, the media server splits its GOP into a series of packets, and sends all these packets within 0.3 seconds. Such a round robin scheme reflects the correct scenario where the cluster receives video packets from multiple media severs simultaneously. Although the simple policy does not consider multi-layer encoding or stream error correction encoding used in some commercial applications, it is a reasonable assumption for our measurement purpose.

The transcoding service, provided by each Worker, is derived from a powerful Linux video stream-processing tool implemented by Thomas Ostreich [7]. It can change video compression formats, change the bit-rate, transform color video frames to black/white, and so on. Its current implementation is based on a multi-process and multi-thread model. Because the whole program runs in the user space, invoking the transcoding process consists of context switches and thus affects the performance. In addition, the interface to this service is through files that involves disk I/O operations.

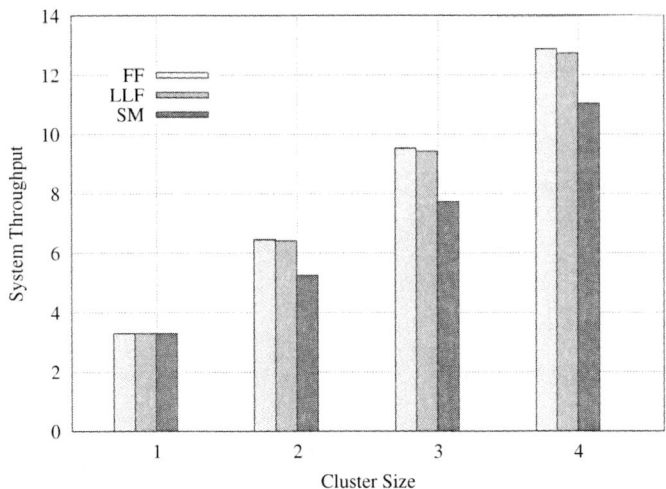

Fig. 2. Scalability of the System Throughput

The remote PC, Manager node and Worker nodes are equipped with Intel Pentium 4 2.53G CPU and 1GB memory. The operating systems is Linux RedHad 9.

4 Performance Evaluation

4.1 System Throughput

The system throughput is measured in terms of GOPs/sec. We care about the scalability of the system throughput as the cluster becomes bigger.

Figure 2 demonstrates the scalability in terms of system throughput for the three different load sharing policies. For FF and LLF, the system throughput scales very well when the number of Worker nodes increases from 1 to 4. As the cluster size increases from 2 to 4, LLF shows slightly lower throughput than FF, due to the load test overhead. However, the overhead is small compared to the time to transcode one GOP, so the throughput is not highly affected. On the other hand, SM avoids dispersing media units of the same stream among different Workers even if a Worker is free. This causes waste of resources, occasional imbalances in load distribution, and reduces the throughput.

4.2 OTI and OFO Rate per Stream

To describe the video quality at the receiver side, we define two metrics as follows.

Metric 1: Output time interval among successive media units of a media stream (OTI per stream) is defined to describe how fast the transcoded media stream can reach the destination. It also gives insights to the arrival jitters.

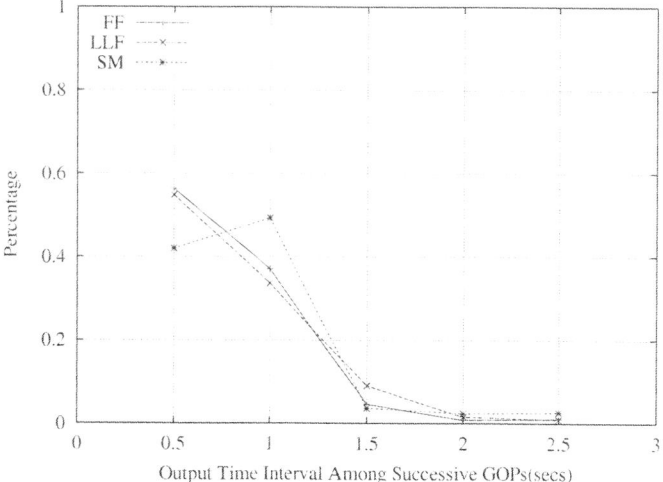

Fig. 3. OTI per Stream

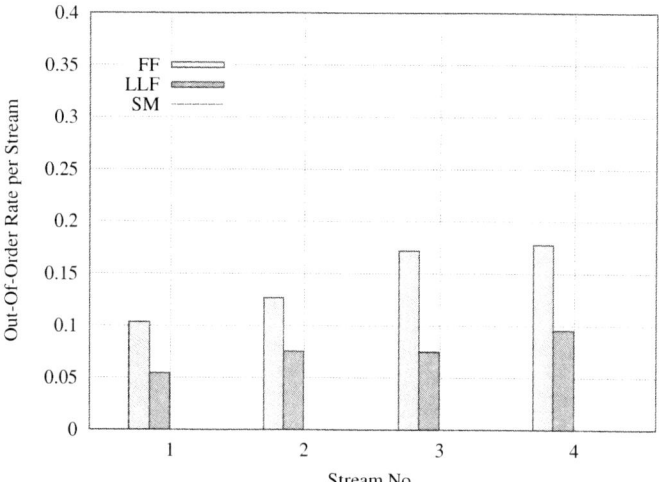

Fig. 4. Out-of-Order Rate per Stream

Metric 2: Out-of-order (OFO) *rate* is used to describe how many GOPs of a stream arrive out-of-order compared to their generating order. Out-of-order occurs when the consecutive GOPs of a stream are transcoded on different Worker nodes. There are two reasons: first, the workload on different Worker nodes is different;; second, different media units consume different computation time.

Figure 3 demonstrates the statistical distribution of OTI per stream for the three load sharing schemes. The corresponding experiment uses 3 homogeneous Worker nodes and 5 media streams. The figure depicts the average OTI per

Table 1. LoadTest Overhead (msecs)

cluster size	1	2	3	4
LoadTest	0.87	1.6	2.3	3.0

Table 2. Load Remapping Overhead (usecs)

cluster size	2	3	4
LoadRemapping	4.2	4.5	4.8

stream. The graph shows that $n\%$ (Y-axis) of the units of the stream arrive within m (X-axis) seconds successively. All the units are classified into 5 categories: the OTI per stream is 0.5 second, 1 second, 1.5 second, 2 second, and 2.5 second, respectively.

FF achieves the best performance in terms of the OTI per stream. As we can see, LLF closely approaches FF although it is a feedback-based mechanism. The reason is that LLF makes more accurate decision when scheduling media units based on the feedback information, which compensate its load test overhead. With SM, media units of the same stream must be processed sequentially on the same Worker node, and thus causes longer delay than the other two schemes.

Figure 4 demonstrates the OFO rate for the same experiment. SM incurs no out-of-order arrival because all the media units of the same stream are processed on the same Worker node sequentially. The largest OFO rate is caused by FF. LLF improves on FF by 50% because of better prediction on the load conditions of the Workers.

4.3   Load Sharing Overhead with LLF

To measure the overhead of using a feedback mechanism to guide LLF, two metrics are defined as follows.

Metric 1: LoadTest Overhead is the average time consumed by the Manager node to poll through all Workers to collect the load statistics information.

Metric 2: Load Remapping Overhead is the time used to set the current loads for each Worker.

As the cluster size increases, there are more Worker nodes in the cluster and the Manager should spend more time to collect information. As shown in table 1, the load test overhead increases roughly proportional to the cluster size.

Load remapping overhead is much smaller than the load test overhead, almost negligible, as shown in table 2. It is because that the remapping overhead is just the operation overhead, which is much less than the network communication overhead involved in the load test.

5   Conclusion

We presented three load sharing policies in a cluster supporting MPEG transcoding service. We found the architecture scales well with the cluster size when accompanied with adequate load distribution policies. There are two main concerns of the system performance. One is the throughput, and the other is the realtime video playout quality. In a system where the media is not intended for playback in realtime, throughput is of most concern. To give the best system throughput, first fit or least load first can be adopted. In a system where realtime playback is expected for the processed streams, a stream-based mapping is better to preserve the unit order of the outgoing media streams.

References

[1] Shirazi, B.A., Hurson, A.R., Kavi, K.M.: Scheduling and load balancing in parallel and distributed systems. IEEE CS Press (1995)   331
[2] Satyanarayanan, M.: Scalable, secure, and highly available distributed file access. IEEE Computer (1990)   331
[3] Katz, E., Butler, M., McGrath, R.: A scalable http server: The ncsa prototype. Computer Networks and ISDN systems **27** (1994) 155–164   331
[4] Zhu, H., Yang, T., Zheng, Q., Watson, D., Ibarra, O., Smith, T.: Adaptive load sharing for clustered digital library servers. Proceedings of the seventh International Symposium on High Performance Distributed Computing (1998) 235–242   331
[5] Guo, J., Chen, F., Bhuyan, L., Kumar, R.: A cluster-based active router architecture supporting video/audio stream transcoding services. Proceedings of the 17th International Parallel and Distributed Processing Symposium (IEEE IPDPS'03), Nice, France (2003)   331, 332
[6] Welling, G., Ott, M., Mathur, S.: A cluster-based active router architecture. IEEE Micro **21** (2001)   332
[7] Ostreich, T.: Linux video stream processing (2002) `http://www.theorie.physik.uni-goettingen.de/~ostreich/transcode/`.   335

A Token Based Distributed Algorithm for Medium Access in an Optical Ring Network

A. K. Turuk, R. Kumar, and R. Badrinath

Department of Computer Science and Engineering
Indian Institute of Technology Kharagpur
Kharagpur-721302
{akturuk,rkumar,badri}@cse.iitkgp.ernet.in

Abstract. In this paper we propose a token based algorithm to access the shared medium in an optical ring network. The algorithm is based on reservation scheme. However, unlike other reservation schemes, which operate on three stages viz. reserve, transmit and release, the proposed scheme operates in two stages and does not explicitly release the resources. The proposed algorithm selects the earliest available data-channel for reservation hence we call it *Earliest Available Channel* (EAC) algorithm. The EAC algorithm operates in a distributed manner. Each node in the network maintains three modules: *send, receive* and *token processing* modules. The algorithm has the capability of handling channel collision and destination conflicts. We study the performance of the algorithm by simulation for bursty traffic modeled by M/Pareto distribution.

1 Introduction

It is widely acknowledged that the rapid growth in demand for bandwidth due to Internet explosion can be satisfied by optical networks, and in particular using the wavelength-division multiplexing (WDM) technology. A single fiber can support hundreds of wavelength channels. With the successful deployment of WDM in core networks, the access networks, viz., local area networks (LANs) and metropolitan area networks (MANs) are bottlenecks. Recently a lot of work has been reported in the literature for the deployment of WDM technology in the access network. Most of the work on LAN reported in the literature employ either a star or a ring as the underlying physical topology of the network. In a LAN the available bandwidth is shared among all the network users. To deal with multiuser access a media access control protocol is needed in such networks. In recent years many media access control protocols have been proposed for WDM-LAN based on star and ring as the underlying physical topology.

There are three well-known access strategy for LAN based on ring topology viz. token ring, slotted ring, and insertion register ring. These have been widely used as local area networks both in commercial system and research prototypes. Ring networks offer several attractive features such as higher channel utilization

S. R. Das, S. K. Das (Eds.): IWDC 2003, LNCS 2918, pp. 340–349, 2003.

and bounded delay. WDM slotted rings are reported in [1],[2],[3],[4]. Synchronization among the slots is a major design factor in slotted rings. Nodes must be synchronized so that a slot starts at the same time on all the wavelength channels in the network at the synchronization point. Synchronization points are the network hub for star topology and the WDM, ADMs for ring topology [5].

Token based WDM ring network is explained in [6],[7]. Unlike *FDDI* rings, authors in [6],[7] discussed multiple tokens in the ring. The number of tokens in the ring, the number of transmitter and receiver that each node is equipped with is equal to the number of data-channels available in the ring.

In this paper, we propose a token based algorithm which we call *Earliest Available Channel* (EAC) algorithm to access the shared medium in a WDM ring network. The algorithm is based on a reservation scheme.

Unlike other reservation schemes that operate in three stages viz. reserve, transmit and release, the EAC algorithm operates in two stages viz. reserve and transmit. In our proposed scheme, reserved resources are not explicitly released. Each node in the network maintains status of it's transmitter, receivers of other nodes and data-channels in the network. Status gives the time at which transmitter, receivers and data-channels are available. Resources (source node transmitter, receiver of destination node and a data-channel) are reserved for a duration which is determined at the time reservation request is made. The duration for which resources are reserved is different for different reservation requests. The reserved resources can be requested for reservation by another node after that period. This does not necessitate the explicit release of reserved resources. Two different nodes can make reservation requests for the same resource during the same cycle of the token but for different times. Transmitter of the source and receiver of the destination are tuned to the same reserved data-channel before communication between them takes place. In other words a lightpath is dynamically established between the source and destination along the reserved data-channel and remains in place until the transmission is completed. Availability of fast tunning lasers makes possible to set up lightpath dynamically.

The EAC algorithm operates in a distributed manner. Each node in the network maintains three modules: *send* module, *receive* module and *token processing* module. A node invokes 'send' module if its *req_made* queue (a queue stores all the reservation requests made by the node) is non-empty. Similarly, *receive* module is invoked if its *req_rec* queue (a queue stores all the transmission requests to the node) is non-empty. *Token processing* module is invoked when a node receives a token. *Req_made* and *req_rec* queues are updated by the *token processing* module. The reservation mechanism is explained in detail in the subsequent sections.

The algorithm has the capability of avoiding channel collision and destination conflicts. We study the performance of the algorithm by simulation for bursty traffic. We compare the performance with another token based algorithm *Multi-Token Inter-Arrival Time* (MTIT) Access Protocol [7]. To the best of

Table 1. Comparison of MTIT with EAC

	MTIT	EAC
Number of Tokens	Equal to the number of data-channel	One
Number of Transmitters and Receivers per node	Equal to the number of data-channel	Two
Fiber-Delay Lines	Exist at every node	No
Channel selection strategy	Selects a free channel	Selects the earliest available channel
Transmission	Simultaneous transmission on each data channel is possible at a node	A single transmission on a data-channel takes place at a node
Header processing	Header is processed at each intermediate node	No processing of header takes place
Packet Removal	Removed by the source	Removed at the destination
Token arrival	Inter-arrival of token at a node may differs	Inter-arrival of token at each node remains same

our knowledge, MTIT is the only token based protocol proposed for optical ring network. A qualitative comparison of MTIT and EAC is given in Table-1.

The rest of the paper is organized as follows. In Section 2 the system model is described. The EAC algorithm is described in Section 3. Simulation results comparing the proposed algorithm with MTIT are reported in Section 4. Finally, some conclusions are drawn in Section 5.

2   System Model

2.1   Assumptions

We consider a WDM ring network with N nodes. The system supports W wavelengths λ_0, λ_1, ..., λ_{W-1}. There are $W - 1$ data channels and one control channel. One of the wavelengths, λ_0, is dedicated to control channel, and rest of the wavelengths are used as data channels. A circuit is established on wavelength λ_0 between every pair of adjacent nodes i and j. The circuit thus established is the dedicated control channel.

Each node is equipped with a *fixed* transmitter/receiver, and a *tunable* transmitter/receiver. The fixed transmitters and receivers are tuned to wavelength, λ_0, to transmit and receive control information between adjacent nodes while tunable transmitters and receivers are tuned to data channels as and when required. For two nodes in the network to communicate, tunable transmitter of the source node and tunable receiver of the destination node must be tuned to the same wavelength (data channel). The system has a *single* token that circulates around the ring on the control channel. The token consists of N sub-fields which we call *slots*; *slot* i is assigned to node i. We define a $TokenPeriod$ (TP) as the period between two successive receives of the token by a node. We calculate TP as $TP = R + N \times p$ where p is the processing delay of token at each node and R is the ring latency. Since TP is same for all nodes in the network, each node gets a fair chance to access the shared medium. Thus, the delay involved is bounded.

A node on receiving the token processes each slot, l, $(0 \leq l < N)$, to update its knowledge about node l in the network. Prior to communication between a pair of nodes, the source must reserve the destination and a data channel. A node reserves the destination and a data channel by writing the control information at it's allotted slot in the token. A node has $N - 1$ buffers for each destination node. Reservation mechanism is explained latter.

2.2   Notations and Definitions

$DAT[i]$: Earliest time at which the node i will be available for receiving

$CAT[i]$: Earliest time at which the data channel i will be available for transmission

τ_d : Destination available time,

τ_c : Channel available time,

t_u : Tuning time of the transmitter/receiver,

t_p : Average propagation delay between source and destination, and

$current\_time$: Time at which an action is taken at a node.

req_made queue : This is the queue that holds the reservation request made by a node. The ith request made by a node is stored at the ith element of the queue. An element of the queue has the following fields : tt – time at which transmitter of the node starts tunning to a data channel, di – identity of destination node to which transmission will take place, dc – wavelength to which the transmitter of the node will be tuned to, td – duration for which transmission will take place.
req_rec queue : This is a sorted queue that holds the reservation request from other nodes for which the current node (here current node is the node that is processing the token) is the destination. For example say, there is a reservation request from node 1, destined to node 5. When node 5 receives the token, reservation request from node 1 is entered in its *req_rec queue*. No other node will make an entry of this request in its *req_rec queue*. Elements of the *req_queue* are same as that of *req_made queue*.
Status : Indicates the status of the node's transmitter ($BUSY/FREE$), and
Finish : Indicates the status of the node's receiver ($BUSY/FREE$).

Control information in a $slot_j(s, d, c, t_c, D)$ are:

    s : value of one indicates node j is requesting for reservation and value of zero indicates no request is made by node j,

    d : identity of the destination node requested for reservation,

    c : identity of the data channel requested for reservation,

    t_c : time at which receiver of the destination, and transmitter of the source starts tunning to data channel c, and

    D : duration of transmission.

3   Algorithm

Each node maintains global status of other nodes indicating the time at which nodes are available for receiving data (the ith index of the vector DAT indicates the time at which node i will be available for receiving data). Similarly, nodes also maintain global status of data channels indicating the time at which data-channels are available for transmitting. Unlike the traditional reservation scheme where resources are reserved only when they are free, our proposed scheme looks ahead to find at what time the required resources will be available and reserves the resources from that point of time. Thus, in our scheme the explicit release of reserved resources is not required. Upon receiving the token, a node first updates its global status DAT and CAT vectors maintained at its node. If its buffers are non-empty, it selects the burst with maximum waiting time and an earliest available data-channel. Nodes then make the reservation request by writing the control information in its allotted slot. Then the token is sent to its adjacent node. When a node receives back the token (i.e., after a period of TP), its reservation request is completed, and all the nodes have recorded the next availability of the requested resources in their global status DAT and CAT vectors. Before transmission, transmitter of the source and receiver of the destination are tuned to the same data-channel. In other words, a circuit (lightpath) is established between the source and Destination, and remains established for the period of transmission.

3.1   EAC Algorithm

Send Module.

    if (req_made queue is Non-empty and $Status = FREE$) **do**

    1 Remove the front element of the req_made queue. Let it be $req\_made(l)$

    2 $Status \leftarrow BUSY$

    3 **if** ($current\_time \geq req\_made(l) \cdot tt$) **do**
        • Start tunning the transmitter to data channel $req\_made(l) \cdot dc$

    4 **if** ($current\_time \geq req\_made(l) \cdot tt + t_u$) **do**
        • start transmitting data to node $req\_made(l) \cdot di$

    5 **if** ($current\_time \geq req\_made(l) \cdot tt + t_u + req\_made(l) \cdot td$) **do**
        • $Status \leftarrow FREE$

    end.

Receive Module.

> **if** (*req_rec* queue is Non-empty and $Finish = FREE$) *do*
> 1 Remove the front element of the *req_rec* queue. Let it be $req\_rec(l)$
> 2 $Finish \leftarrow BUSY$
> 3 **if** ($current\_time \geq req\_rec(l) \cdot tt$) *do*
>   - Start tunning the receiver to $req\_rec(l) \cdot dc$
> 4 **if** ($current\_time \geq req\_rec(l) \cdot tt + t_u$) *do*
>   - Start receiving data
> 5 **if** ($current\_time \geq req\_rec(l) \cdot tt + t_u + req\_rec(l) \cdot td$) *do*
>   - $Finish \leftarrow FREE$

end.

Token Processing Module. When a *node* i receives a token it invokes the token processing module. The following steps are performed to process the received token.

1 Examine $slot_i(s, d, c, t_c, D)$ if $(s = 1)$ *do*
   - $Slot_i(s \leftarrow 0)$
   - Add the request of node i in its *req_send* queue
   - $DAT[d] \leftarrow CAT[c] \leftarrow t_c + t_u + t_p + D$
2 For all $slot_j(s, d, c, t_c, D), j \neq i$ *do*
   - if $(s = 1$ and $t_c + t_u + t_p + D > DAT[d])$ *do*
     - $DAT[d] \leftarrow t_u + t_c + t_p + D$
   - if $(s = 1$ and $t_c + t_u + t_p + D > CAT[c])$ *do*
     - $CAT[c] \leftarrow t_u + t_c + t_p + D$
3 *if* (node i buffers are non-empty) *do*
   - find a burst with maximum waiting time. Let the destination identity of the burst be, say, x.
   - let k be the earliest available channel. That is $k \leftarrow \{m : CAT[m]$ is minimum for $m \leftarrow 1, \cdots W - 1$ }
   - Set $\tau_d \leftarrow DAT[x]$, $\tau_c \leftarrow CAT[k]$
   - Find $\tau \leftarrow max(\tau_d, \tau_c)$. This gives the earliest time at which both the destination node x and the data channel k are available and can be reserved by a node.
   - **if** ($\tau < current\_time + TP)$ $\tau \leftarrow current\_time + TP$
   - Calculate the duration of the transmission D
   - Write the control information in $slot_i(s \leftarrow 1, d \leftarrow x, c \leftarrow k, t_c \leftarrow \tau, D)$
4 Send the token to successor node
   end.

Table 2. Contents of DAT and CAT at the nodes at time t

DAT[0] = 0, DAT[1] = 5, DAT[2] = 7, DAT[3] = 10
CAT[λ_1] = 5, CAT[λ_2] = 7

Table 3. Traffic matrix at time t

Node	0	1	2	3
0		$b(10,4)$	$b(4,3)$	
1	$b(20,4)$		$b(25,3)$	
2		$b(10,3)$		
3	$b(10,5)$			

3.2 Simulation of Algorithm

We illustrate the reservation process with a simulation. We consider a four node ring network. The number of wavelengths is assumed to be three; contents of DAT and CAT vectors at time t are shown in Table-2. Table-3 shows the traffic matrix at time t (in the present example value of $t \geq 5$). The entry $b(x, y)$ corresponding to row m and column n of Table-3 indicates node m has a burst destined to node n. Duration of transmission of the burst is indicated by x, and y indicates the time at which the burst has arrived at node m. We assume the following quanta of values for the parameters: TP is assumed to be 20, t_u to be 2, t to be 6; propagation delay between a pair of adjacent node to be 5, and the processing delay of token at each node is assumed to be negligible.

Let node 0 has the token at time t. Node 0 selects the burst with maximum waiting time i.e, burst destined to node 2, and a earliest available data channel i.e, λ_1. Then it calculates $\tau_d = 7$ ($DAT[2]$), $\tau_c = 5$ ($CAT[\lambda_1]$), $\tau = 7$ (max (τ_d, τ_c). The value of τ is less than $t + TP$ ($\tau < t + TP$) so the value of τ is set to $t + TP$ i.e., 26. Node 0 writes control information in $slot_0$($s = 1, d = 2, c = \lambda_1, t_c = \tau, D = 4$) of the token and sends it to its successor node 1. Node 1 on receiving the token updates the contents of DAT and CAT vectors as shown in Table-4.

Node 1 selects the burst destined to node 2 and the earliest available data-channel λ_2. Assign $\tau_d = 32$, $\tau_c = 7$, $\tau = 32$. When node 1 receives the token, the value of t is updated to $t + 5$ (i.e., the updated value is t + the propagation delay between adjacent nodes which we have assumed to be 5 in our example). Node 1 writes control information in $slot_1$($s = 1, d = 2, c = \lambda_2, t_c = \tau, D = 25$) and sends it to node 2.

Updated values of DAT and CAT vectors at node 2 are shown in Table-5. Request from node 0 and node 1 are entered in the req_rec queue of node 2.

Node 2 selects the burst destined to node 1 and data channel λ_1. It sets the following values: $\tau_d = 5$, $\tau_c = 28$, $\tau = 28$. The value of τ is given by : $\tau < t + TP$ so the value of τ is set to $t + TP$, i.e., 36. Node 2 writes control information in

Table 4. Contents of DAT and CAT at node 1

DAT[0] = 0, DAT[1] = 5, DAT[2] = 32, DAT[3] = 10

CAT[λ_1] = 32, CAT[λ_2] = 7

Table 5. Contents of DAT and CAT at node 2

DAT[0] = 0, DAT[1] = 5, DAT[2] = 59, DAT[3] = 10

CAT[λ_1] = 28, CAT[λ_2] = 59

Table 6. Contents of DAT and CAT at node 3

DAT[0] = 0, DAT[1] = 48, DAT[2] = 59, DAT[3] = 10

CAT[λ_1] = 48, CAT[λ_2] = 59

$slot_2(s = 1, d = 1, c = \lambda_1, t_c = \tau, D = 10)$, and then sends it to node 3. Updated values of DAT and CAT vectors at node 3 are shown in Table-6.

Node 3 selects the burst destined to node 0 and data channel λ_1. It assigns the values : $\tau_d = 0$, $\tau_c = 48$, $\tau = 48$. Node 3 writes control information in $slot_3(s = 1, d = 0, c = \lambda_1, t_c = \tau, D = 10)$ and then sends the token to Node 0.

When node 0 receives the token, and finds its reservation request has been granted it puts it's request in *req_made* queue.

4   Simulation Results

We evaluated the performance of the EAC algorithm through simulation. We included the performance results, in this paper, in terms of throughput and mean packet delay. Throughput is defined as the transmission time divided by transmission time plus scheduling latency. This is a measure of how efficiently data channels are utilized in the network. For simulation, we considered a 10 node optical ring network. The number of wavelengths is 5. Values of other parameters are chosen as follows: capacity of wavelength channel is assumed to the 1 Gb/s, length of the ring is fixed at 100 kms, processing time of token at each node is assumed to be 1 μs, tunning time of transmitter and receiver are assumed to 5 μs. Calculated value of TP is 510 μs.

We consider two cases in our simulation. First, we consider the burst arrival to follow a Poisson distribution; the burst size was increased in (in multiples of 50) for every run of the simulation. Next, we consider the arrival of burst to follow a Poisson distribution and the burst size to follow M/Pareto distribution. In every run of simulation, the number of bursts generated is such that the number of packets generated will be one million.

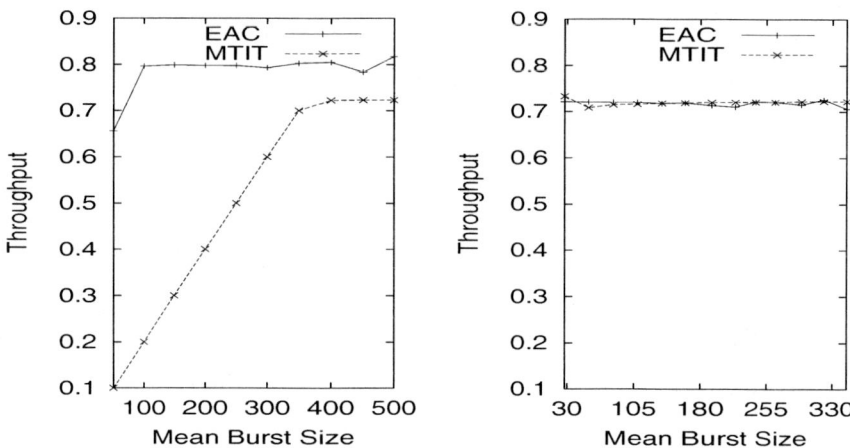

Fig. 1. Burst size *vs.* throughput for (a)fixed burst size, and (b) burst size taken from M/Pareto distribution for five data-channels.

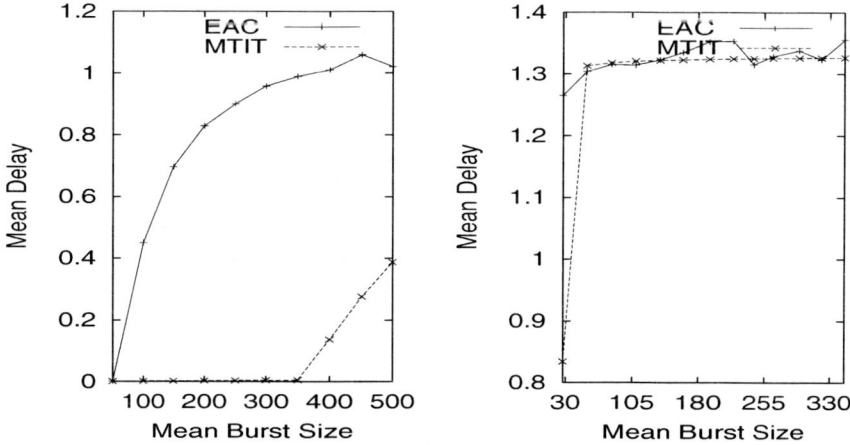

Fig. 2. Burst size *vs.* mean delay for (a) fixed burst size, and (b) burst size taken from M/Pareto distribution for five data-channels

4.1   Burst size *vs.* Throughput

Throughput increases with burst size for both MTIT and EAC as shown in Fig. 1. Wavelength utilization for fixed burst size is better in the case of EAC than MTIT as shown in Fig. 1(a). For burst size taken from M/Pareto distribution, wavelength utilization in EAC is almost identical to that of MTIT (Fig. 1(b)). This is the principal advantage of EAC algorithm that the wavelength utilization is identical with use of reduced resources.

4.2   Burst size *vs.* Mean Packet Delay

We define mean packet delay as the average end-to-end delay experienced by packets. For fixed burst size, delay experienced by packets are more in EAC than MTIT as shown in Fig. 2(a). However, for burst size taken from M/Pareto distribution, delay experienced by packets in both EAC and MTIT are almost identical for larger bursts (Fig. 2(b)).

5   Conclusions

In this paper, we proposed a token based distributed EAC algorithm for medium access in an optical ring network. EAC algorithm is based on a reservation scheme, and has the capability of avoiding channel collision and destination conflicts. We have compared EAC algorithm with another token based MTIT algorithm. MTIT algorithm is not scalable; the number of tokens required is equal to the number of data channels. We found that wavelength utilization is superior in EAC algorithm for bursts of fixed size. The utilization, however, is comparable in both EAC and MTIT algorithms for bursts taken from M/Pareto distribution. Delay experienced by packets are almost identical for both EAC and MTIT when the burst sizes are determined by M/Pareto distribution.

References

[1] Bengi, K., van As H. R.: Efficient QoS Support in a Slotted Multihop WDM Metro Ring. IEEE JSAC, **20** (2002) 216 – 227   341
[2] Marsan, M. A., Bianco, A., Leonard, E., Morabito, A., Neri, F.: All-Optical WDM Multi-Rings with Differentiated QoS. IEEE Commun. Mag., (1999) 58 – 66   341
[3] Marsan, M. A., Bianco, A., Leonardi, F. E., Toniolo, S.: An Almost Optimal MAC Protocol for All-Optical Multi-Rings with Tunable Transmitters and Fixed Receivers. 1997. http://citeseer.nj.nec.com/marsan97almost.html   341
[4] Marsan, M. A., Bianco, A., Abos, E. G.: All-Optical Slotted WDM Rings with Partially Tunable Transmitters and Receivers. http://citeseer.nj.nec.com/34986.html, 1996   341
[5] Spencer, M. J., Summerfield, M. A.: WRAP: A Medium Access Control Protocol for Wavelength- Routed Passive Optical Networks. Journal of Lightwave Technology **18** (2000) 1657 – 1676   341
[6] Fumagalli, A., Cai, J., Chlamtac, I.: A Token Based Protocol for Integrated Packet and Circuit Switching in WDM Rings. IEEE Globecom, Sidney (1998)   341
[7] Fumagalli, A., Cai J., Chlamtac, I.: The Multi-Token Inter-Arrival Time (MTIT) Access Protocol for Supporting IP over WDM Ring Network. Proc. ICC'99 conf. Vancouver, Canada, (1999)   341

An Efficient Technique
for Dynamic Lightpath Establishment
in Survivable Local/Metro WDM Optical Networks

Swarup Mandal[1], Debashis Saha[1], Sougata Bera[2,*],
Shubhadip Ray[2], and Ambuj Mahanti[1]

[1] Indian Institute of Management Calcutta, D.H. Road, Joka, Calcutta:700104
Tel: 91 33 24678300, Fax: 91 33 24678307
{swarup,ds,am}@iimcal.ac.in
[2] Jadavpur University, Calcutta: 700032
Tel: 91 33 24146002
{sougata,shubhadipray}@yahoo.com

Abstract. Providing a reliable and fast connection to a user is one of the challenging issues in a survivable IPover WDM optical network. A dynamic lightpath establishment (DLE) strategy along with a path protection technique may be used to address the issue. In this strategy, a primary lightpath is established when a request for communication from a node comes, and it is released when the communication is over. To protect the primary lightpath, there will also be a backup lightpath. Thus, DLE with path protection is the problem of establishing a pair of primary and backup lightpaths on a call request within a reasonable amount of time. This problem is a variant of the classical Routing and Wavelength Assignment (RWA) problem in the WDM optical network literature. It can be formulated as a combinatorial optimization problem and, in this paper, we have proposed a heuristic search technique, developed based on Best-First Search. We have tested the performance of our technique with respect to the lightpath establishment time (LET) and call blocking probability (CBP) and compared it with the performance of another well-known technique namely Mixed Integer Linear Programming (MILP). We have also studied the performance of our technique with respect to CBP and LET when there are multiple calls to admit. A significant performance gain is noticed in all the above cases.

Keywords: Best-First Search, Lightpath, Survivable WDM Optical Network.

---

* This work is partially supported by the UGC funded major research project (SWOPNET) grant no.F.14-24/2001 (SR-I) and AICTE funded R&D project (OPTIMAN) grant no. 8088/RDII/BOR/R&D (189) /99-2000. Sougata Bera is working in SWOPNET

S. R. Das, S. K. Das (Eds.): IWDC 2003, LNCS 2918, pp. 350-360, 2003.
© Springer-Verlag Berlin Heidelberg 2003

1   Introduction

In the near future, IPover WDM optical networks will be a key technology to satisfy the requirement of large demand of bandwidth and reliable services. In a WDM optical network, a lightpath is established between a pair of source and destination nodes to transmit information. A lightpath consists of an optical channel, or wavelength, between two network nodes that is routed through multiple intermediate nodes [1]. In routing and wavelength assignment (RWA) problem, a prime task is to find such lightpaths. More formally, given the physical topology of a network, we have to find an appropriate logical topology and assign each connection a wavelength. A physical topology is an undirected graph that represents the physical connection of optical fibers among nodes. A logical topology is a directed graph that results when lightpaths connect nodes [2].

The RWA problems, also known as lightpath establishment problems, can be of two types, namely Static Lightpath Establishment (SLE) problem [3] and Dynamic Lightpath Establishment (DLE) problem. In SLE problem, a fixed lightpath is assigned between a pair of nodes, and it remains there even if there is no communication between that pair of nodes. However, in case of DLE, a lightpath between a pair of nodes is established on the fly, and it is released when the communication is over. Again, classical SLE problem assumes that the network is reliable and no fault occurs. But network faults are very common in optical networks. So different researchers have proposed a number of strategies to meet this reliability requirement [3]. These strategies can be broadly classified into two classes, namely *protection schemes* and *restoration schemes*. Restoration schemes (i.e., reactive techniques) dynamically discover a backup lightpath once a failure is detected. Protection schemes (i.e., proactive techniques) set up an edge-disjoint backup lightpath for every primary lightpath. For example in Fig.1, the primary lightpath (1-2, λ_0) is set up for the communication between node 1 and node 2 using the wavelength λ_0.

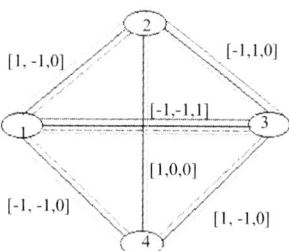

Fig. 1. Network with primary lightpaths $\{(1\text{-}2, \lambda_0), (4\text{-}2, \lambda_0), (4\text{-}3, \lambda_0), (3\text{-}2, \lambda_1), (3\text{-}1, \lambda_2)\}$ and backup lightpaths $\{(1\text{-}3\text{-}2, \lambda_0), (4\text{-}1\text{-}2, \lambda_1), (4\text{-}1\text{-}3, \lambda_0), (3\text{-}1\text{-}2, \lambda_1), (3\text{-}4\text{-}1, \lambda_1)\}$.The bit pattern corresponding to a link indicates the status of wavelengths in the link e.g. the bit pattern of the link E_{23}, [-1,1,0], indicates that the wavelength λ_0 is used for a backup lightpath, λ_1 is used for a primary lightpath and λ_2 is not in use

Now, in the protection scheme, simultaneously an edge disjoint backup lightpath (1-3-2, λ_0) needs to be established for the primary lightpath (1-2, λ_0). Thus, whenever the link 1-2 fails, the call corresponding to the primary lightpath (1-2, λ_0) is served by the backup lightpath (1-3-2, λ_0) without triggering the expensive search process. If there is no further link failure in the path 1-3-2 before the link 1-2 comes up, the call corresponding to the primary lightpath (1-2, λ_0) is fully protected. Again, if there are not many faults in the optical network, the resources used to create the backup lightpath remain idle for most of the time. This leads to a huge loss of revenue, and so sharing of backup lightpaths was proposed to reduce the loss. The sharing of backup lightpaths will increase resource utilization, and, at the same time will not affect the call protection scheme under the single fault tolerant assumption. In Figure 1, the primary lightpaths (4-2, λ_0) and (3-1, λ_2) share the link 1-4 in their backup lightpaths (4-1-2, λ_1) and (3-4-1,λ_1) respectively. In this paper, we have proposed a heuristic search technique namely BDFS for call admission in a survivable WDM optical network with shared backup protection strategy. We have compared the performance of our proposed technique with the performance of the MILP with respect to lightpath establishment time (LET) for single call admission and studied the performance of our technique with respect to call blocking probability (CBP) and LET for multiple call admissions with a network architecture shown in Figure 2.

1.1   Motivation

RWA problem for various different physical topologies has been addressed in [4]. The authors have considered both SLE and DLE. The authors of [5][6] have proposed a solution to the problem of DLE, where they considered dynamic wavelength allocation in WDM ring networks and dynamic channel assignment for WDM optical networks with little or no wavelength conversion. But, most of these researches have not taken network faults into account while solving DLE.

However, faults in WDM networks are a major problem and recently a number of researchers have proposed strategies to handle faults in optical networks [7][8][9][10][11][12]. Different authors have discussed protection schemes in [13] and restoration schemes in [14]. In [15], authors have tried to solve the DLE in a survivable WDM optical network for only a single call admission by using MILP.

In this work, we first improve upon their work in terms of efficiency, and then extend the work to include multiple call admissions simultaneously. So we solve the DLE problem in a survivable WDM optical network under single fault tolerant assumption in the first part of this work. Next we solve the same problem for multiple call admissions, which will be applicable for local or metro optical networks with a centralized reservation scheme [16].

1.2   Organization of Paper

This paper is organized in five sections. Following introduction in Section 1, a mathematical formulation of the problem is presented in Section 2. Section III is devoted to the Solution Methodology using Heuristic Search. Section 4 contains the results and discussion, and Section 5 concludes the paper.

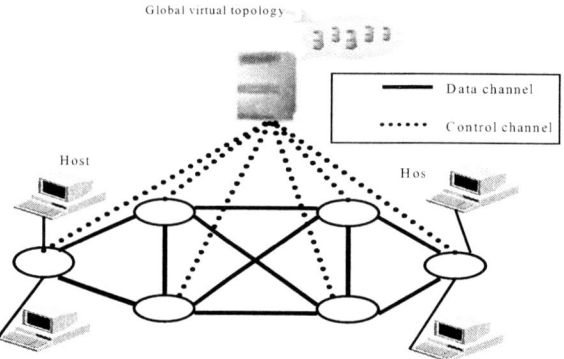

Fig. 2. Upon arrival of a call request, nodes forward the request to a central controller where updated global virtual topology of the network is stored. The controller will perform the calculation of lightpaths and return to the nodes the optimal lightpaths to be used

2   Mathematical Formulation

Let us consider a partially connected WDM optical network of N nodes. Each link in the network is having W number of wavelengths. Let us also consider that, at any instant, there are L calls in progress and C new connections arrive. The problem is to find a pair of wavelength continuous lightpaths (primary and backup) for each of the new calls within a reasonable time-bound. Let us consider the following notations:

PM: Maximum number of possible primary lightpaths considered for each new call

BM: Maximum number of backup lightpaths for each of the P primary lightpaths

E_{rt}:   Represents the link between node r and node t where $1 \leq r, t \leq N$ and $r \neq t$

P_{ij}:   Represents the set of links in a path between node i and node j where $1 \leq i, j \leq N$ and $i \neq j$. $E_{rt} \in P_{ij}$, if the path between node i and node j passes through the node r and node t.

λ_k :   Represents the kth wavelength $1 \leq k \leq W$

$(P_{ij}, \lambda_k)_{mn}$:
    Represents the nth possible primary lightpath with k^{th} wavelength for m^{th} call request where $1 \leq m \leq C$, and $1 \leq n \leq P$

$(P'_{ij}, \lambda_k)'_{mnp}$:
    Represents the pth possible backup lightpath corresponding to the primary lightpath $(P_{ij}, \lambda_k)_{mn}$, where $1 \leq p \leq B$

$(P''_{ij}, \lambda_k)''_z =$
    Represents the already established z^{th} primary lightpath.

$(P'''_{ij}, \lambda_k)'''_z =$
    Represents the already established backup lightpath corresponding to $(P''_{ij}, \lambda_k)''_z$.

S_m :   Set of all possible unique pair of primary and backup lightpaths for m^{th} new connection request where S_{mx} represents the xth pair in S_m , where $0 \le x \le (PM.BM)$. If there is no valid pair of primary and backup lightpaths then the set S_m will be a null set.

$S : S_1 X S_2 X \ldots \ldots X S_C$
    is the total space for a pair of primary and backup lightpaths for C new connection requests.

Each of the pair of primary and backup lightpaths assigned to a new call is subject to the satisfaction of the following constraints:

i.   A primary lightpath $(P_{ij}, \lambda_k)_{mn}$ and its backup lightpath $(P'_{ij}, \lambda_y)'_{mnp}$ must be link disjoint i.e.

$$\text{If } E_{rt} \in P_{ij} \text{ then } E_{rt} \notin P'_{ij} \qquad \text{where } 1 \le k, y \le W \qquad (1)$$

ii.   A primary lightpath $(P_{ij}, \lambda_k)_{mn}$ can not use a wavelength in a link if it is used by another primary lightpath $(P''_{uv}, \lambda_k)''_z$ or a backup lightpath $(P'''_{cd}, \lambda_k)''_z$ i.e.

$$\text{If } E_{rt} \in P_{ij} \text{ then } E_{rt} \notin P''_{uv} \text{ and } E_{rt} \notin P'''_{cd} \text{ where } 1 \le c,d,u,v \le N \qquad (2)$$

iii.   A backup lightpath $(P'_{ij}, \lambda_k)'_{mnp}$ cannot use a wavelength in a link if it is used by a primary lightpath $(P''_{uv}, \lambda_k)''_z$ i.e.

$$\text{If } E_{rt} \in P'_{ij} \text{ then } E_{rt} \notin P''_{uv} \qquad (3)$$

iv.   A backup lightpath $(P'_{ij}, \lambda_k)'_{mnp}$ of the primary lightpath $(P_{ij}, \lambda_k)_{mn}$ may share a wavelength in a link with the already established $(P'''_{cd}, \lambda_k)''_z$ of the already established primary lightpath $(P''_{cd}, \lambda_k)''_z$ if $(P_{ij}, \lambda_k)_{mn}$ is edge disjoint with respect to $(P''_{cd}, \lambda_k)''_z$ (due to assumption of single link fault tolerant network) i.e.

$$\text{If } E_{rt} \in P_{ij} \text{ And } E_{rt} \notin P''_{cd} \text{ then we may have } E_{uv} \in P'_{ij}, P'''_{cd} \qquad (4)$$

Now let us consider a set of pair of primary and backup lightpaths FS which is formed by taking at most one element from each of the S_ms (where FS_m is the mth element of set FS i.e. the pair of primary and backup lightpaths taken from S_m)). Our problem is to find out the FS* from FSs for which CBP will be minimized within reasonable execution time.

3   Solution Methodology Using Heuristic Search

Any combinatorial optimization problem can be formulated as a state space search problem, which can be solved by a heuristic search technique. In this section we are presenting the state space formulation and the description of the algorithm.

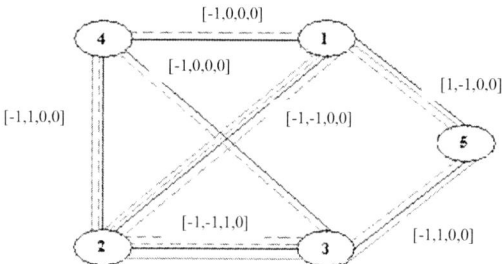

Fig. 3. Network with primary lightpaths $\{(1\text{-}5, \lambda_0), (3\text{-}5, \lambda_1), (3\text{-}2, \lambda_2), (2\text{-}4, \lambda_1)\}$ and backup lightpaths $\{(1\text{-}2\text{-}3\text{-}5, \lambda_0), (3\text{-}2\text{-}1\text{-}5, \lambda_1), (3\text{-}4\text{-}2, \lambda_0), (2\text{-}1\text{-}4, \lambda_0)\}$. The bit pattern corresponding to a link indicates the status of wavelengths in the link e.g. the bit pattern of E_{23}, $[-1,-1,1,0]$, indicates that the wavelength λ_0 and λ_1 are used for backup lightpaths, λ_2 is used for a primary lightpath and λ_3 is not in use

3.1   State Space Formulation

A *state* in the state space shows the assignment of primary and backup lightpaths to the newly requested calls. In the start state, none of the newly requested calls is assigned with a primary lightpath and a backup lightpath. This is indicated by showing all the element of FS as empty. A *goal state* represents a state where either each of the newly requested calls is assigned a pair of primary and backup lightpaths or some of the newly requested calls are assigned pairs of primary and backup lightpaths and no further assignment is possible for the remaining calls. A state transition takes place for a state in level m-1 when a mth newly requested call is assigned with a pair of primary and backup lightpaths from the set of all possible unique pair of primary and backup lightpaths for the mth new call (as defined in S_m). If this assignment is valid with respect to the constraints formulated in equations (1) through (4), then the cost of transition will be zero. On the contrary, if the assignment is not valid with respect to the constraints formulated in equations (1) through (4), then the cost of transition will be one because the new call request will be blocked with respect to this assignment. Now we consider a network with five nodes as shown in the Figure 3 to describe the state space. Each link in the network is having four wavelengths.

For Figure 4 depicts the state space for the network in Figure 3 when there are three simultaneous new call requests between the fallowing pair of nodes: (3,1), (3,2), and (2,4).

In this problem the set S is represented by $\{S_1, S_2, S_3\}$. Again, S_1 is defined by the set, $\{S_{11}, S_{12}, S_{13}, S_{14}\}$, where

$S_{11}=\{(3\text{-}2\text{-}1, \lambda_3), (3\text{-}4\text{-}1, \lambda_0)\}$,
$S_{12}=\{(3\text{-}2\text{-}1, \lambda_3), (3\text{-}5\text{-}1, \lambda_2)\}$
$S_{13}=\{(3\text{-}4\text{-}1, \lambda_1), (3\text{-}2\text{-}1, \lambda_0)\}$
$S_{14}=\{(3\text{-}4\text{-}1, \lambda_1), (3\text{-}5\text{-}1, \lambda_2)\}$.

Similarly:
S_2 is defined by $\{S_{21}, S_{22}, S_{23}\}$, where
$S_{21}=\{(3\text{-}2, \lambda_3), (3\text{-}4\text{-}1\text{-}2, \lambda_1)\}$
$S_{22}=\{(3\text{-}2, \lambda_3), (3\text{-}5\text{-}1\text{-}2, \lambda_2)\}$
$S_{23}=\{(3\text{-}4\text{-}1\text{-}2, \lambda_2), (3\text{-}2, \lambda_0)\}$

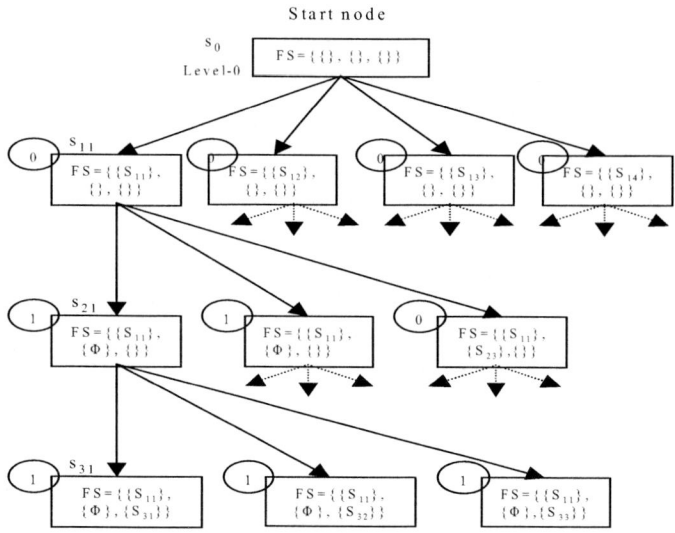

Fig. 4. State Space

Again, S_3 is defined by $\{S_{31}, S_{32}, S_{33}\}$, where

$S_{31}=\{(2\text{-}4, \lambda_2), (2\text{-}1\text{-}4, \lambda_0)\};S_{32}=\{(2\text{-}4, \lambda_2), (2\text{-}3\text{-}4, \lambda_0)\}; S_{33}=\{(2\text{-}1\text{-}4, \lambda_2), (2\text{-}4, \lambda_0)\}.$

The start state s_0 shows that no assignment is made to any call, thus FS_1, FS_2 and FS_3 are shown as empty. Now the first child s_{11} of state s_0 is corresponding to the assignment of S_{11} to the first newly requested call. Since this assignment is valid with respect to the constraints in equation (1) through (4), cost of the state is 0. Similarly the state s_{11} will have its first child as s_{21} corresponding to the decision of assigning S_{21} to the second newly requested call. While checking the validity of this assignment with respect to the constraints, the pair of lightpaths assigned to first call is treated as an already established pair of lightpaths. Under this condition the assignment of S_{21} as the second new call is not valid. Thus, in state s_{21} the newly requested second call is blocked and the cost of s_{21} is one. Again, state s_{21} will have its first child as s_{31} corresponding to the decision of assignment S_{31} to the newly requested third call. Here also we have checked the constraints in a similar way and found it as a valid assignment. Thus, the state s_{31} represents a goal node with cost one.

From the above representation it is clear that the maximum depth of the state space is C and the maximum branching factor of a state is equal to the product (PM.BM). Every leaf node in the state space is a goal state. Thus the state space complexity is $O((PM.BM)^C)$. Here the problem is to find a FS such that the value of the object function (defined in section 2) is minimized.

3.2   The Search Algorithm

Block-Depth-First Search (BDFS) [17] is a search algorithm that is based on a novel combination of staged search and depth-first search [18]. As a result, it has good features of both Best-First and Depth-First Branch-and-Bound (DFBB)[19] techniques and at the same time avoids the bad features of both. Basically the search algorithm comprises two phases namely, forward phase and backtracking phase, as we explain them below.

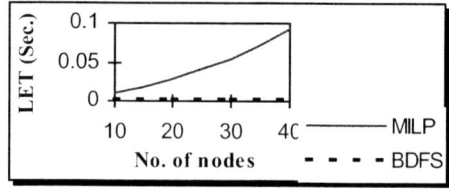

Fig. 5. Plot for LET vs. network size for BDFS and MILP

Forward Phase

In this phase, the algorithm explores the search tree, level by level. All nodes generated are stored in a linear list, LIST. The root node is assigned as level 0 and stored in the LIST. After this the algorithm runs iteratively, working on one level at each iteration. At any level i, it orders nodes using the cost of the node and expands the first node (in a depth first manner) by generating all children at level (i+1). Since in our state space representation of the problem every leaf node in the search tree will represent a feasible solution, the forward phase is bound to terminate with a solution.

Backtracking Phase

This phase basically executes DFBB[19] starting at each unexpanded node. DFBB is performed in the reverse order, i.e., from the last level generated down to level 1.

After the completion of the forward phase, this phase is used to improve the solution. This phase continues until the list, LIST becomes empty to conclude that the solution found is optimal.

4   Results and Discussions

We have tested the performance of BDFS in respect of CBP and LET. We have also compared the performance of BDFS with the performance of MILP [15] for single call admission in networks of varying sizes having a certain call load. We have also studied the CBP and LET as found by BDFS for multiple call admissions under varying network loads in a fixed size network (NSFNET). We further studied the CBP and LET for a network load under varying multiple call admissions in a fixed size network (NSFNET). For this purpose, we coded the algorithms in C and executed them on Borland C++ compiler in Windows environment. Results, as found for different problem instances, are presented below.Figure 5 shows the plot of the results as obtained by MILP and BDFS. The plot is for a single call admission in a network with 50% average wavelength usage (awu). Here, *awu* is defined as the ratio of *total number of wavelengths under use* (total number of 1s and −1s in the bit pattern of all links in the network) to *total number of wavelengths* (number of wavelengths per link times the total number of links).We observe that a significant improvement in LET is achieved by BDFS over that of MILP. As the total number of nodes in the network increases, the LET for BDFS is much better than that of MILP. This is because of the fact that, with the increase in the number of nodes in a network, the number of unique combinations of possible primary and backlight lightpaths for a connection request increases. MILP looks at all the combinations, which meet the constraints in equations (1) through (4). In BDFS, the search process prunes a solution path in the tree whenever it finds a state in the path whose cost is more or equal to the cost of the already found goal. This pruning effect is visibly dominant for a network with large number of nodes.

Figure 6 and Figure 7, show the CBP and LET as found by BDFS for a multiple call admission under varying network loads in NSFNET backbone [15]. It is seen that initially, when the awu is in the range of 30% to 40%, both the plots show a gradual rise after which, there is a steep rise in the CBP. This is because, as the awu per link increases, the goal found in the initial phase will have a high call blocking. As a result, the pruning effect decreases.

Figure 8 and Figure 9 show the CBP and LET found by BDFS for multiple call admission under a network load of 50% awu in the same NSFNET backbone. When the number of calls to be established (simultaneously) is in the range of 5 to 10, the graph shows a gradual linear rise after which, there is an exponential rise in the CBP. As the number of calls increase, the depth of the search tree increases. The goal obtained in the forward phase of the algorithm will have a higher cost thereby resulting in reduced pruning in the backtracking phase.

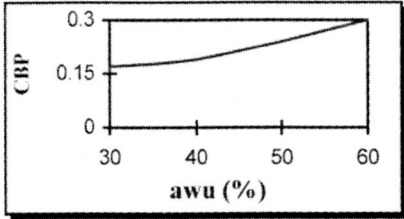

Fig. 6. Plot for CBP vs. awu for multiple call admission

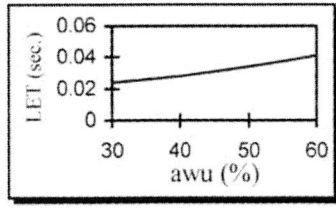

Fig. 7. Plot for LET vs. awu for multiple call admission

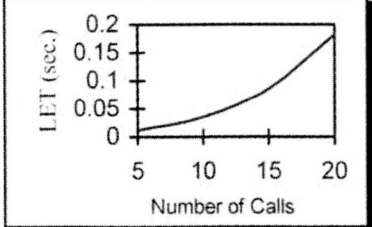

Fig. 8. Plot for Lightpath establishment time vs. Number of calls to be admitted simultaneously

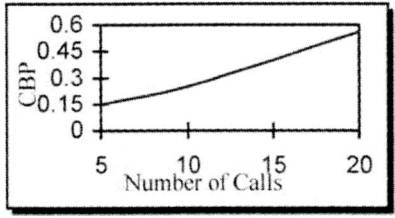

Fig. 9. Plot for CBP vs. number of calls to be admitted simultaneously

5   Conclusion

We have formulated the problem of DLE in a survivable WDM optical network as a
state space search problem and used BDFS to solve it. Experimental results show that
BDFS out-performs MILP for single call admission in networks of varying sizes with
respect to LET. When BDFS is tested for multiple call admission, the CBP and LET
for different problem instances is found to be within acceptable limit.

References

[1]   T.E.,Stern and K.,Bala, Multiwavelength Optical Networks-A Layered
        Approach, Addison Wesley, 1999
[2]   R.,Ramaswami et al. : Optical Networks : A Practical Perspective, Morgan
        Kaufmann Publishers, 1998
[3]   B.,Mukherjee: Optical Communication Networks : McGraw-Hill, 1997
[4]   B. Mukherjee et al., "Some principles for designing a wide-area optical
        network", Proc. IEEE   INFOCOM,94,(Toronto, Canada, Jun. 1994)
[5]   O. Gerstel et al. :"Dynamic wavelength allocation in WDM ring networks",
        IBM Research Report RC 20462 May,1996
[6]   O. Gerstel et al. :"Dynamic channel assignment for WDM Optical Networks
        with little or no wavelength conversion", Proc. 34th Annual Allerton Conf
        (Monticello,11., Oct.1996), pp. 32-43
[7]   A. Fumagalli et. Al, "Survivable networks based on optimal routing and WDM
        self healing rings", Proc. Of IEEE INFOCOM'99.pp. 726-733, March, 1999
[8]   O. Gerstel   and R. Ramaswami, "Optical layer survivability: a services
        perspective", IEEE Communications magazine, vol. 38, no. 3, pp. 104-113,
        March 2000
[9]   M. Alanyal, and E. Ayanoglu, "Provisioning algorithms for WDM optical
        network," IEEE ACM Trans. Net., Vol.7, no.5, Oct. 1999, pp.767-778
[10]   E. Modiano, A. Narula-Tam, "Survivable routing of logical topologies in WDM
        Networks", INFOCOM 2001. Proceedings. IEEE,. Vol. 1, 2001, pp. 348-357
[11]   I. Rubin and J. Ling, "Failure Protection Methods for Optimal Meshed-Ring
        Communication, Networks", IEEE J. Select, Areas Commun,. Vol. 18, No. 10,
        OCTOBER 2000, pp 1950-60
[12]   Eytan Modiano and Aradhana   Narula, "Survivable routing of Logical
        Topologies in WDM Networks, Infocom 2001, Anchorae, Proceedings, IEEE,
        Vol. 1, pp. 348-357, 2001
[13]   S. Ramamurthy and B. Mukherjee, "Survivable WDM Mesh Networks, Part 1-
        protecrion", Proc. of IEEE., INFOCOM'99, pp. 744-751, March, 1999
[14]   R.R.Irashko, and W.D. Grover, "A highly Efficient Path-Restoration Protocol
        for Management of Optica, Network Transport Integrity", IEEE J.Select. Areas
        Commun., Vol 18, No. 5, May 2000. pp.779-794
[15]   Y.P.Aneja, S.Bandopadhyay and A.Jaekel, "An efficient protection scheme for
        WDM Networks using dynamic lightpath allocation", Proc.,HPC Asia, 2002

[16]   D. Saha, and D. Sengupta,*"An optical layer lightpath management protocol for WDM All Optical Network.(AONs)"*, Photonic Network Communications, Kluwer Academ, vol. 2, no. 2, pp. 185-198, 2000

[17]   A. Mahanti, S. Ghosh, A. K. Pal, *"A High Performance Limited-Memory Admissible and Real Time Search Algorithm for Networks"*, Computer Science Technical Report Series, CS-TR-2858, University of Maryland College Park, MD 20742, March 1992

[18]   Nils J. Nilsson, *"Artificial Intelligence: A New Synthesis"*, Morgan Kaufmann Publisher, 1998

[19]   Rao V. N, and Kumar V.," *Analysis of Heuristic Search Algorithms"*, CSci TR 90-40, Institute of Technology, University of Minnesota

Algorithms
for Provisioning Survivable Multicast Sessions
against Link Failures in Mesh Networks*

Narendra K. Singhal and Biswanath Mukherjee

Department of Computer Science
University of California
Davis, CA 95616, USA
{singhaln,mukherje}@cs.ucdavis.edu

Abstract. We investigate new algorithms for efficiently establishing
a multicast session in a mesh network while protecting the session from
a link failure, e.g., a fiber cut in an optical network.
One of the new algorithms, IMPROVED_SEGMENT, protects each segment in the primary tree. The other new algorithm, IMPROVED_PATH,
discovers backup resources for protecting each path, from source to destination, in the tree. We find that IMPROVED_SEGMENT performs
significantly better (around 14% less resource utilization for a typical
wide-area mesh network) than a simple-minded segment-protection algorithm called SEGMENT. For dynamic connection provisioning, IM-
PROVED_PATH is found to perform significantly better than PATH,
and IMPROVED_SEGMENT is found to perform significantly better
than SEGMENT. Among all these algorithms, IMPROVED_PATH is
found to perform the best.
Although we study these algorithms in an optical WDM context, the
approaches are applicable to other contexts as well, such as SONET or
Gigabit Ethernet (GBE).

1   Introduction

Recent advances in networking – particularly high-capacity optical networking
employing wavelength-division multiplexing (WDM) technology [1] – have made
bandwidth-intensive multicast applications such as HDTV, interactive distance
learning, live auctions, distributed games, movie broadcasts from studios, etc.
widely popular [2] [3] [4]. In an event of a fiber cut, all connections traversing in
either direction of the fiber are disrupted and the affected destinations have to be
reached on alternate routes. Most communication networks employ bidirectional
links; hence, all communication links are assumed to be bidirectional in this
study. Because fiber failure is a predominant form of failure in communication
networks, we focus on protecting multicast connections from any fiber cut or
link failure.

* This work has been supported by the US National Science Foundation (NSF) under
  Grant Nos. NCR-95-08239, ANI-98-05285, and ANI-02-07864.

S. R. Das, S. K. Das (Eds.): IWDC 2003, LNCS 2918, pp. 361–371, 2003.

Link failures in a communication network occur often enough to cause service disruption, and they may lead to significant information loss in the absence of adequate backup mechanisms. The loss could be heavy when the failed link in a "light-tree"[1] carries traffic for multiple destinations. A straightforward approach to protecting a multicast tree is to compute a *link-disjoint* [6] backup tree. Two trees are said to be *link disjoint* if their edges don't traverse a common link. Such link-disjoint trees can be used to provide 1+1 dedicated protection [7] where both primary tree and backup tree carry identical bit streams to the destination nodes. When a link fails, the affected destination nodes reconfigure their switches to receive bit streams from the backup tree instead of the primary tree. Pitfalls of this approach include excessive use of resources and at times insufficient availability of network resources to establish link-disjoint trees in a mesh network, leading to the blocking of a large number of multicast sessions [6] [8].

An improvement over the *link-disjoint* approach is "arc disjointedness"[2] for primary tree and backup tree which can significantly reduce network resources reserved for protection from a link failure. Two *arc-disjoint* trees may traverse a common link in opposite directions only, and they can provide 1:1 dedicated protection, i.e., resources for the backup tree are reserved and are used to carry bit streams when any link fails [7]. Earlier work in [9] has exploited the notion of *arc-disjoint* trees to develop algorithms to compute primary and backup trees in an undirected graph. In [4], it is shown how *arc-disjoint* trees can protect a multicast session from any fiber cut in a network even when the link traversed by both primary and backup tree fails. However, *arc-disjoint* trees are not the most efficient solution for protecting a multicast session, and a connection may be blocked due to resource unavailability. In [6] [8], two schemes for segment protection and path protection (called Algorithms SEGMENT and PATH, respectively, in this paper) were proposed. These schemes were found to be efficient than the previous methods based on discovering *link-disjoint* and *arc-disjoint* trees. Because our current study builds upon these schemes (for protecting multicast sessions), we describe them below using simple illustrative examples.

1.1 Segment Protection

In a segment-protection scheme, each segment of the primary tree is protected by discovering a backup segment disjoint to its corresponding primary segment. We define a *segment* as sequence of arcs[3] from the source or from any splitting point (on a tree) to a leaf node or to a downstream splitting point. A destination node is always considered as a *segment end-node* because it is either a leaf node in a tree or is a splitting point where a portion of a signal is dropped locally and the remainder continues downstream (e.g., Drop-and-Continue [DaC] node) [10]. For example, in Fig. 1, the primary tree from source s to destination nodes d_1 and d_2 (shown in dotted lines) occupies the arcs $s \rightarrow u$, $u \rightarrow v$, $v \rightarrow d_1$, and $v \rightarrow d_2$.

---

[1]   In WDM context, a multicast session can be established by a "light-tree" [5].

[2]   We can also refer to "arc disjointedness" as "directed link disjointedness".

[3]   In this study, an arc is an edge in a directed graph.

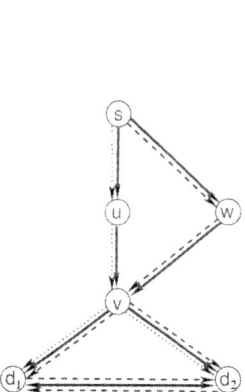

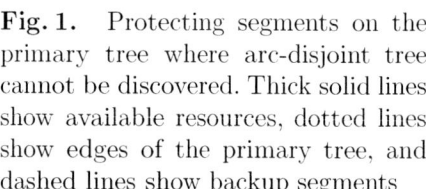

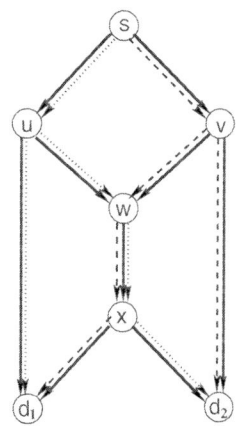

Fig. 1. Protecting segments on the primary tree where arc-disjoint tree cannot be discovered. Thick solid lines show available resources, dotted lines show edges of the primary tree, and dashed lines show backup segments

Fig. 2. Protecting paths on the primary tree where arc-disjoint tree or backup segments cannot be discovered. Thick solid lines show available resources, dotted lines show edges of the primary tree, and dashed lines show backup paths

Here, node v is a *splitting point* and creates three segments, viz., $< s \rightarrow u \rightarrow v >$, $< v \rightarrow d_1 >$, and $< v \rightarrow d_2 >$. Note that, once the primary tree is identified, the splitting points in the tree determine segments on the tree. Observe that, in this figure, it is not possible to find an *arc-disjoint* backup tree because d_1 and d_2 are disconnected from the source after the primary edges ($< s \rightarrow u \rightarrow v >$, $< v \rightarrow d_1 >$, and $< v \rightarrow d_2 >$) are taken out from the graph. However, each of the three segments $< s \rightarrow u \rightarrow v >$, $< v \rightarrow d_1 >$, and $< v \rightarrow d_2 >$ of the tree can be protected by *segment-disjoint* sub-paths $< s \rightarrow w \rightarrow v >$, $< v \rightarrow d_2 \rightarrow d_1 >$, and $< v \rightarrow d_1 \rightarrow d_2 >$, respectively. It is interesting to observe that backup segments are sharing arcs with already-found primary-tree arcs or other backup-segment arcs. Such kind of sharing where backup arcs of a connection share with primary/backup arcs of the same connection is called *self-sharing*. For example, in Fig. 1, although the primary-tree arc $< v \rightarrow d_1 >$ and backup-segment arc $< v \rightarrow d_1 >$ are shown in dotted and dashed lines, they can *share* the same common resource, e.g., a wavelength in the WDM context or a time-slot in the SONET/TDM context. Without loss of generality, we will refer to the WDM context in the rest of the paper. Hence, only one wavelength channel has to be reserved on arcs $< v \rightarrow d_1 >$ and $< v \rightarrow d_2 >$. If one of the links along the primary tree fails, e.g., if the link $< v \longleftrightarrow d_1 >$ fails, d_1 becomes unreachable from the primary tree. The switch at node d_2 will reconfigure such that one of the replicas of the signal coming from the source is dropped locally and the other is forwarded along an already-reserved arc from d_2 to d_1. Readers can verify that other link failures can be handled similarly with appropriate switch reconfigurations.

1.2   Path Protection

In a path-protection scheme, each path in the primary tree from the source node to any destination node is protected by a backup path which is disjoint to its corresponding primary path. Similar to the above *self-shared segment protection* concept, backup paths can share arcs with primary-tree arcs or with already-found backup arcs. For example, in Fig. 1.1, the primary tree (shown in dotted lines) has path $< s \rightarrow u \rightarrow d_1 >$ to d_1 and path $< s \rightarrow u \rightarrow w \rightarrow x \rightarrow d_2 >$ to d_2. These paths are protected by backup paths (shown in dashed lines) $< s \rightarrow v \rightarrow w \rightarrow x \rightarrow d_1 >$ and $< s \rightarrow v \rightarrow d_2 >$, respectively. Note that the backup path to d_1 *self-shares* arc $w \rightarrow x$ with the primary path to d_2 and only one wavelength channel has to be reserved between w and x. Interestingly, this example also shows that a path-protection scheme succeeds where it is not possible to discover an *arc-disjoint* backup tree or to protect each segment.

The problem of setting up a multicast session with protection against any link failure at optimal cost is NP-complete because the subproblem of setting up a minimum-cost tree from source node to all destination nodes (without protection), known as *minimum-cost Steiner tree*, has been shown to be an NP-complete problem [11]. Hence, we develop efficient heuristics to protect multicast connections against any fiber cut (or link failure).

The rest of the paper is structured as follows. Section 2 states the problem for protecting static and dynamic multicast traffic. We describe new algorithms based on segment and path protection schemes in Section 3. Section 4 compares these algorithms through experimental simulations on a typical wide-area mesh network. Finally, Section 5 concludes our paper with a discussion on future work.

2   Problem Description

We are given the following inputs to the problem:

1. A topology $G = (V, E)$ consisting of a weighted directed graph, where V is a set of network nodes (representing IP router(s) connected to a crossconnect), and E is set of links inter-connecting the nodes. Nodes represent IP router(s) connected to a crossconnect and the links correspond to the fibers between nodes. Each link is assigned a weight to represent the cost (one of the components of the cost could be the fiber length) of moving traffic from one node to the other. All lines are bidirectional, with an arc in each direction.

2. In WDM context, the number of wavelength channels carried by each fiber $= W$. When we study the algorithms for establishing only one protected multicast session, we consider $W = 1$, but dynamic provisioning of multiple protected multicast sessions incorporates multiple wavelength channels per

Table 1. Algorithm SEGMENT

Step 1 Create a primary tree using MPH.
Step 2 Identify the segments on the primary tree.
Step 3 Make cost = 0 for the edges along the primary tree.
Step 4 For every primary segment in the tree, repeat Steps 5 through 8.
Step 5 Remove the links along the primary segment.
Step 6 Compute a backup segment arc-disjoint to the primary segment.
Step 7 Update cost = 0 for already-found backup segment.
Step 8 Replace the edges along the primary segment.

fiber. The nodes are equipped with multicast-capable opaque crossconnects which convert the signal from optical to electrical and back to optical (OEO), and hence allow full wavelength conversion [12]. In SONET/TDM context, each link may carry multiple TDM channels.

3. A multicast connection request has to be established while protecting it from any link failure. The connection is *unidirectional* from source to each destination. The group size of the session is denoted by k, which indicates the number of destination nodes that the session serves. If a routing algorithm discovers sufficient resources in a network, the multicast session is established; otherwise, it is blocked. In case of dynamic provisioning, when a new connection arrives, the network graph of available wavelengths on different links, will be composed of directed links because connections arrive and stay for a finite duration before departing.

Our objective is to minimize the network resources (measured in terms of the cost of the links) used for establishing the multicast connections and, hence, to minimize the blocking probability due to resource contention in case of on-line connection provisioning. We achieve this objective by prudently designing protection algorithms. Below, we discuss our algorithms in detail.

3   Multicast Protection Algorithms

Because the *minimum-cost Steiner tree* problem is an NP-complete problem, a heuristic, H, is used to compute the minimum-cost Steiner tree. Although several heuristics exist, we use only one common heuristic, called Minimum-cost Path Heuristic (MPH) [13], to measure the relative performance of the algorithms, in which the closest destination node is selected (on a one-by-one basis) and added to a partially-built tree until all destination nodes are reached. We first present algorithms for segment protection (SEGMENT and IMPROVED_SEGMENT) and then algorithms for path protection (PATH and IMPROVED_PATH).

3.1   Segment-Protection Algorithms

Table 1 describes Algorithm SEGMENT to protect segments of a primary tree created using heuristic MPH. When backup segment is computed for an unprotected segment on the primary tree, the costs of the arcs along the primary tree

Table 2. Algorithm IMPROVED_ SEGMENT

Step 1	Create a primary tree using MPH.
Step 2	Identify the segments on the primary tree.
Step 3	Make cost = 0 for the edges along the primary tree.
Step 4	For every primary segment, repeat Steps 5 through 10.
Step 5	Identify nodes on the path from source to upstream segment end-node (call it set N).
Step 6	Remove the links along the primary segment.
Step 7	Find arc-disjoint backup segments from nodes in set N to downstream segment end-node.
Step 8	Select the backup segment with the least cost.
Step 9	Update cost = 0 for already-found backup segment.
Step 10	Replace the edges along the primary segment.

and of the arcs along already-found backup segments are updated to zero to enhance sharing of the backup segment with already-computed arcs. As a result, the additional cost for computing each new backup segment is minimized.

Table 2 describes Algorithm IMPROVED_SEGMENT. In this algorithm, instead of discovering a backup segment from the upstream segment end-node to the downstream segment end-node, we discover a least-cost backup segment from any upstream node to the downstream segment end-node. We illustrate the advantage of Algorithm IMPROVED_SEGMENT over Algorithm SEGMENT by a simple example. Figure 3 shows the primary tree (in dotted lines) from source s to destination nodes d_1 and d_2 of cost 10 units (4 units for $< s \rightarrow u \rightarrow v >$, 4 units for $< v \rightarrow d_2 >$, and 2 units for $< v \rightarrow d_1 >$). SEGMENT computes the backup segments (shown in dashed lines) - $< s \rightarrow w \rightarrow v >$, $< v \rightarrow d_1 \rightarrow d_2 >$, and $< v \rightarrow d_2 \rightarrow d_1 >$ - to protect the primary segments. The cost of additional edges is 16 (6 from $< s \rightarrow w \rightarrow v >$, 5 from $< d_1 \rightarrow d_2 >$, and 5 from $< d_2 \rightarrow d_1 >$). IMPROVED_SEGMENT computes the backup segment (shown in dash-dot-dot-dot lines) for $< s \rightarrow u \rightarrow v >$ as $< s \rightarrow w \rightarrow v >$, for $< v \rightarrow d_1 >$ as $< v \rightarrow d_2 \rightarrow d_1 >$, and for $< v \rightarrow d_2 >$ as $< w \rightarrow d_2 >$; thus, the cost of additional edges is only 12 units. This example shows that IMPROVED_SEGMENT costs 4 units less than SEGMENT.

3.2   Path-Protection Algorithms

A simple-minded path-protection algorithm, called PATH, is described in Table 3. It is based on the path-protection concept, which protects the path from source to every destination node by computing an optimal disjoint path-pair for every destination node.

An algorithm proposed in [14], known as Suurballe's algorithm, computes a path-pair with least cost between two end-points, if such a path-pair exists.

Table 3. Algorithm PATH

Step 1	For every destination node of the session, repeat Steps 2 and 3.
Step 2	Find an optimal path-pair between the source and destination node.
Step 3	Update cost = 0 for already-found optimal path-pairs.

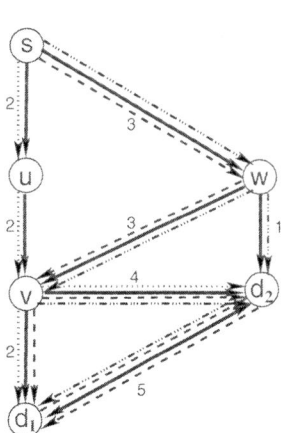

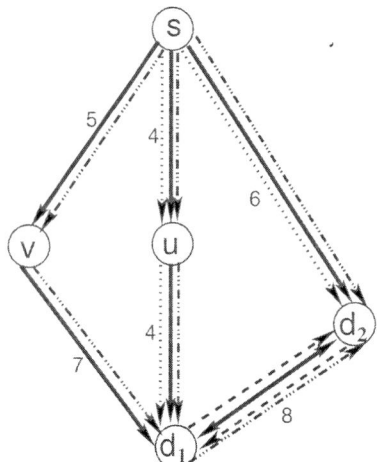

Fig. 3.   Protecting segments on the primary tree (shown in dotted lines) using segment protection algorithms (SEGMENT and IMPROVED_SEGMENT)

Fig. 4.   IMPROVED_PATH in above example performs better than PATH

Once a path-pair is found, the cost of the edges along it are updated to zero to increase sharing of new path-pairs with the already-computed ones and to minimize additional cost.

Table 4 outlines Algorithm IMPROVED_SEGMENT in which first a primary tree is computed and then each primary path in the tree is protected by a backup path. Figure 4 shows the advantage of IMPROVED_PATH due to increased sharing possibility of the backup path with the already-found primary tree. The path-pairs discovered by Algorithm PATH are shown in dash-dot-dot-dot lines and have total cost of 34 units. Algorithm IMPROVED_PATH finds the primary tree (shown in dotted lines), and the backup-path (shown in dashed lines) with a total cost 30 units. Thus, IMPROVED_PATH saves 4 units of resources over PATH.

Table 4.   Algorithm IMPROVED_PATH

Step 1 Create a primary tree using MPH.
Step 2 Make cost = 0 of edges on the primary tree.
Step 3 For every destination node of the session, repeat Steps 4 and 5.
Step 4 Compute a backup path disjoint to primary path from source to destination node.
Step 5 Update cost = 0 for already-found backup path.

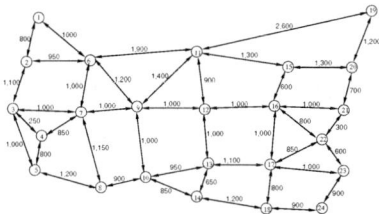

Fig. 5.   A 24-node sample network used in this study

4   Illustrative Examples

We measure the performance of the four algorithms described in Section 3 on the sample wide-area mesh network shown in Fig. 5. (In this figure, the link costs represent fiber lengths between node pairs in kilometers, as shown.) We first investigate the algorithms for static traffic where only one connection request (of varying destination-set size) arrives and then we study them for dynamic traffic where connection requests with finite holding time arrive and then depart.

4.1   Static Traffic

A multicast connection, of destination-set size k, is randomly generated, to compare the algorithms for protecting the connection request. The source node and destination nodes of a connection are uniformly distributed across the network. If a scheme finds enough resources, the connection is established; otherwise, it is blocked. In our first example, we set up a simulation experiment, in which we establish one new connection of specific session-size on the network. We repeat the experiment for 5,000 different connections of the same group size. The size of the multicast group (k) is varied from 1 to 23 (unicast to broadcast).

Figure 6 shows that the average cost of setting up the connections increases with session size for each of the four algorithms. We observe that Algorithm IM-PROVED_SEGMENT performs significantly better than Algorithm SEGMENT (providing an improvement of around 14% in resource utilization for the example network in Fig. 5). We also observe that the new path-protection algorithm, IMPROVED_PATH, performs marginally better than PATH for this example network. Although IMPROVED_PATH precomputes the primary tree which may cause some blocking, it provides increased opportunity for backup paths to share with already-computed primary arcs. On the other hand, PATH computes a path-pair to each destination node which is optimal but has reduced sharing possibility when new path-pairs are computed. These two effects counteract each other resulting in only slightly improved performance for IMPROVED_PATH over PATH for setting up a simple multicast connection. However, when multiple connections need to be set up, IMPROVED_PATH leads to significantly better performance than PATH, as shown below.

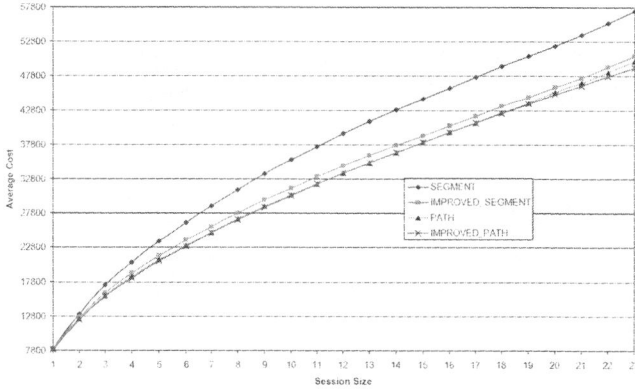

Fig. 6. Cost comparison of proposed algorithms for establishing a multicast session with protection from a link failure

4.2 Dynamic Traffic

In order to study these algorithms in an online environment, we provision multicast connections dynamically on the 24-node sample network shown in Fig. 5. Multicast connections arrive with Poisson distribution and their holding time is negative exponentially distributed. The experiment is run for 50,000 connections of a specific destination size (which varies from 1 to 23) with connection-arrival (network) load of 100 Erlangs and $W = 64$ wavelengths. Figure 7 compares the performance of the four protection algorithms. We observe that, similar to static traffic, Algorithm IMPROVED_SEGMENT performs significantly better than Algorithm SEGMENT. We find that Algorithm IMPROVED_PATH shows marked improvement over Algorithm PATH.

Figure 8 shows the blocking probability for establishing multicast sessions of size 10 using Algorithm IMPROVED_PATH in a network where fibers support different number of wavelength channels ($W = 8, 16, 32$, and 64), for different offered loads. As expected, the graph shows that fibers supporting a larger number of channels have lower blocking probability. Observe that, in this simulation experiment, for limiting blocking probability to 0.1 in a WDM network with $W = 32$, the offered load can be as high as 63 Erlangs. Also, IMPROVED_PATH has better performance among all of these algorithms.

5 Concluding Remarks

We investigated new algorithms, based on segment-protection and path-protection schemes, for protecting multicast sessions from link failures in a communication network. We observed that our new algorithm, IMPROVED_SEGMENT, performs 14% better (in terms of network resource utilization for setting up a single multicast session on a typical mesh network)

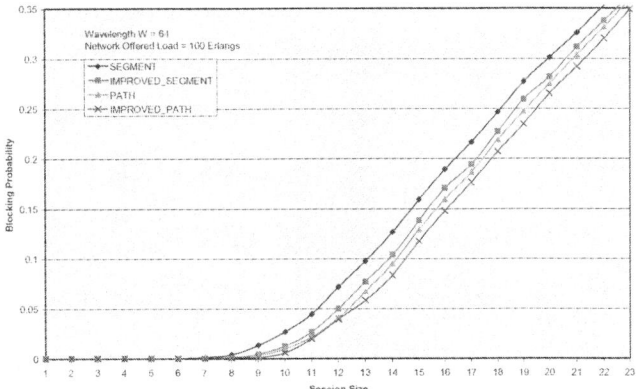

Fig. 7. Blocking probability of various schemes for dynamic provisioning of survivable multicast sessions of sizes from 1 (unicast) through 23 (broadcast)

than the straight-forward segment-protection algorithm, called SEGMENT. Our new path-protection algorithm, IMPROVED_PATH, which first creates a primary tree performs marginally better than the simple path-protection algorithm, called PATH, which computes optimal path-pair to each destination node. Our IMPROVED_SEGMENT algorithm performs significantly better than SEGMENT, and IMPROVED_PATH performs significantly better than PATH in terms of blocking probability. Also, IMPROVED_PATH performs the best among all these algorithms.

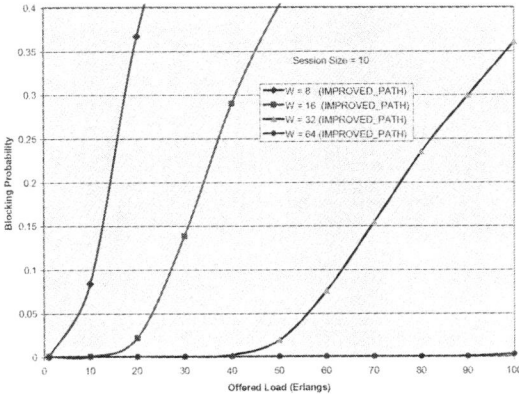

Fig. 8. Blocking probability for establishing multicast sessions of size 10 for fibers supporting different number of wavelength channels (W) at different offered loads

References

[1] B. Mukherjee, *Optical Communication Networks*, McGraw-Hill, July 1997. 361

[2] C. K. Miller, *Multicast Networking and Applications*, Reading, MA: Addison-Wesley, 1999. 361

[3] Y. Sun, J. Gu, and D. H. K. Tsang, "Multicast routing in all-optical wavelength routed networks," *Optical Networks Magazine*, pp. 101–109, July/Aug. 2001. 361

[4] N. Singhal and B. Mukherjee, "Protecting Multicast Sessions in WDM Optical Mesh Networks," *IEEE/OSA Journal of Lightwave Technology*, vol. 21, no. 4, pp. 884–892, Apr. 2003. 361, 362

[5] L. H. Sahasrabuddhe and B. Mukherjee, "Light-Trees: Optical Multicasting for Improved Performance in Wavelength-Routed Networks," *IEEE Communications Magazine*, vol. 37, no. 2, pp. 67–73, Feb. 1999. 362

[6] N. Singhal, L. H. Sahasrabuddhe, and B. Mukherjee, "Dynamic Provisioning of Survivable Multicast Sessions in Optical WDM Mesh Networks," *Technical Digest, OFC, Atlanta, GA, paper TuI5*, Mar. 2003. 362

[7] S. Ramamurthy, L. H. Sahasrabuddhe, and B. Mukherjee, "Survivable WDM Mesh Networks," *IEEE/OSA Journal of Lightwave Technology*, vol. 21, no. 4, pp. 870–883, Apr. 2003. 362

[8] N. Singhal, L. H. Sahasrabuddhe, and B. Mukherjee, "Protecting a Multicast Session Against Single Link Failures in a Mesh Network," *Proc. , IEEE International Conference on Communications, Anchorage, Alaska*, May 2003. 362

[9] M. Medard, S. Finn, R. Barry, and R. Gallager, "Redundant Trees for Pre-planned Recovery in Arbitrary Vertex-Redundant or Edge-Redundant Graphs," *IEEE/ACM Transaction on Networking*, vol. 7, no. 5, pp. 641–652, Oct. 1999. 362

[10] X. Zhang, J. Wei, and C. Qiao, "Constrained Multicast Routing in WDM networks with Sparse Light Splitting," *Proc., IEEE INFOCOM 2000, Tel Aviv, Israel*, vol. 3, pp. 1781–1790, Mar. 2000. 362

[11] S. L. Hakimi, "Steiner's problem in graphs and its implications," *Networks*, vol. 1, no. 2, pp. 113–133, 1971. 364

[12] N. Singhal and B. Mukherjee, "Architectures and Algorithm for Multicasting in WDM Optical Mesh Networks using Opaque and Transparent Optical Cross-Connects," *Technical Digest, OFC, Anaheim, CA, paper TuG8*, Mar. 2001. 365

[13] H. Takahashi and A. Matsuyama, "An approximate solution for the Steiner problem in graphs," *Math. Japonica*, pp. 573–577, 1980. 365

[14] J. W. Suurballe, "Disjoint Paths in a Network," *Networks*, vol. 4, pp. 125–145, 1974. 366

A New Protection Model
to Provide Availability-Guaranteed Service
for Reliable Optical WDM Networks*

Yurong (Grace) Huang[1], Wushao Wen[2],
Jonathan P. Heritage[1], and Biswanath Mukherjee[3]

[1] Electrical and Computer Engineering, Dept. of University of California
Davis, CA 95616, USA
{yrhuang,heritage}@ece.ucdavis.edu
[2] CIENA Corporation
Cupertino, CA 95138, USA
wwen@ciena.com
[3] Computer Science Dept. of University of California
Davis, CA 95616, USA
mukherje@cs.ucdavis.edu

Abstract. Reliability is a crucial concern in high-speed optical networks. A service-level agreement (SLA), which mandates high service availability even in face of network failures, must be met in provisioning reliable connection. In this study, we investigate a generalized connection-provisioning and protection model to provide availability-guaranteed services in a wavelength-division multiplexing (WDM) optical mesh network. A novel link-state model is proposed to represent physical-layer availability and resource status for an optical link. This information can be used by a standard link-state routing protocol to efficiently provide reliable connections. In our proposed link-state-modeling mechanism, current network resource information and the physical components' failure characteristics are aggregated to form a comprehensive dynamic link-state parameter, called link and resource availability (LRA). Based on such a dynamic link-state parameter, we also propose an algorithm for availability-guaranteed connection provisioning. A new generalized protection model is developed through our dynamic LRA-based provisioning. The dynamic LRA-based connection-provisioning algorithm efficiently provides SLA-guaranteed services while significantly improving connection reliability, better utilizing network resources, and achieving fast failure recovery.

* This work has been supported by the US National Science Foundation (NSF) under Grant No. ANI-99-86665.

S. R. Das, S. K. Das (Eds.): IWDC 2003, LNCS 2918, pp. 372-382, 2003.
© Springer-Verlag Berlin Heidelberg 2003

1   Introduction

Wavelength-division-multiplexing (WDM) technology has enabled a transport network to accommodate high-bandwidth telecommunication traffic. In such a WDM mesh network, data traffic from a source to a destination is carried via a wavelength-routing optical channel, called a *lightpath*. Setting up a lightpath for a connection is called *connection provisioning*, which is achieved through a routing protocol and a signaling protocol. A routing protocol, such as Open Shortest Path First (OSPF) [1], is extended to disseminate network topology and resource status information, while a signaling protocol, such as Resource Reservation Protocol (RSVP) [2], is extended and used for establishing, removing, and maintaining connections. Upon arrival of a connection request, the source node uses a routing algorithm for path selection based on its knowledge of the network topology and link-state information. Then, a setup message is sent out to reserve and commit the network resources for the connection request. The link-state parameter used by the routing protocol could be *static* or *dynamic* [3]. The static information does not change in response to connection operations, e.g., establishing or tearing down a connection, whereas the dynamic link-state information varies in response to connection operations so that efficient path selection for new connections can be achieved due to use of the current network state. Therefore, the more accurate the link-state parameters are generated by the network links, the better are the routes chosen by the routing algorithm.

In an optical WDM network, a network-element failure could cause the loss of a large amount of data. With increasing requirements on quality of service (QoS), it is important for connections to survive from failures [3]-[11]. There are several kinds of failures that can disrupt a connection. A *shared risk group* (SRG) [5] is one network component (such as optical amplifier, fiber, and conduit) that could cause multiple channels/connections to be interrupted at the same time when it fails. A link failure occurs when a SRG fails to make the entire link disconnected. A channel failure occurs when the equipment on a single wavelength channel (such as transmitter or receiver) fails [5]. A channel failure only causes a single connection to fail. Moreover, a node could fail due to failure of equipment at the node, e.g., switches. However, redundant equipments are typically used for protection within the node [5]. In this paper, a generalized protection mode is investigated to take both link failure and channel failure into consideration.

A network failure will affect connection reliability which, therefore, is of major concern to both service providers and their customers [3], [6]. A service-level agreement (SLA) that mandates high service availability even in face of network failures must be achieved for reliable connection provisioning. The connection's reliability can be measured via the metric of *availability* which is the probability that a system (component, channel, lightpath, connection, etc.) will be found in the normal operating state at a random time in the future [8].

For a reliable WDM network, certain *protection* or *restoration* (e.g., dedicated/shared, path-based/link-based, etc.) techniques are used to restore traffic whenever a network-element failure occurs [4]. With protection, backup resources are reserved in advance to protect a connection from failure, whereas with restoration, backup resources are allocated on the fly. The reserved backup capacity may be dedicated to protect a connection or may be shared with other connections. For both schemes, there are tradeoffs between network resource utilization and service interruption time [9]. The dedicated-path-protection scheme is simple and fast due to no switch reconfiguration at transit optical crossconnects (OXCs), but more resources need to be reserved for protection. In shared-path-protection scheme, the resource redundancy is reduced to some degree by sharing backup capacity; however, failure recovery could be complex and requires OXC reconfiguration at intermediate nodes [9]. Since our objective here is to improve resource utilization while realizing fast failure recovery via efficient routing, we focus on dedicated-path protection scheme. Efficient routing can also be applied to shared-protection scheme, but it is beyond the scope of this study.

A protection technique typically requires a link-disjoint primary and backup path pair for a connection to survive from a single failure scenario. However, it may not be efficient and fast for a connection to survive from failures when connection provisioning is based on a static link-state parameter that is unaware of the current network state or network failure characteristics. Therefore, in this paper, we propose a novel link-modeling mechanism by which the current network state is, on a timely basis, reflected through a dynamic link-state parameter. The network's free-resource information and network-component-failure characteristics are aggregated to form a comprehensive dynamic parameter, called link and resource availability (LRA). Provisioning based on such a dynamic link-state parameter can automatically balance link loads as well as improve the connection blocking probability. A new generalized protection model is developed through dynamic LRA-based provisioning. In the proposed protection mechanism, the link-disjoint constraint on backup-path computation may not be necessary as long as the customers' availability requirements (defined in the service-level agreement (SLA)) can be satisfied. The dynamic LRA-based connection-provisioning method efficiently provides SLA-guaranteed services while significantly improving connection reliability, better utilizing network resources, and achieving fast failure recovery.

The paper is organized as follows. In Section II, a mechanism for dynamic link-state modeling is proposed, and an example for the computation of link parameter LRA is also presented for illustration. Based on the analysis model of link state, a new generalized protection strategy for availability-guaranteed connection provisioning is proposed in Section III. Illustrative numerical results are discussed in Section IV. Section V concludes our study.

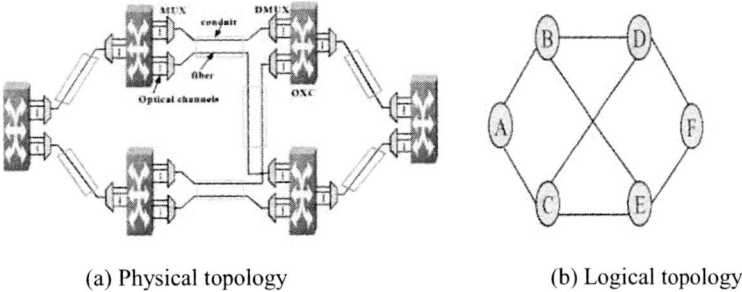

(a) Physical topology   (b) Logical topology

Fig. 1. A sample optical WDM mesh network

2   Link Modelling —
Dynamic Link and Resource Availability (LRA) Analysis

Provisioning reliable connections is a key issue in designing a survivable optical network. The network relies on the link-state parameters distributed by each node to decide on a route for a new connection or to protect a connection from failure. For efficient provisioning to combat network failures, in this section, we present a novel mechanism for link-state modeling which takes into consideration of physical-component's failure characteristics as well as network resource information. The network under study is an optical WDM mesh network which consists of multiple nodes connected by optical fibers in an arbitrary topology. Typically, a logical network topology is used to present the connectivity between network nodes for a physical network layout as shown in Fig. 1.

In our present model of a WDM network, a logical link is an abstraction of one or multiple physical fibers between two nodes, while each fiber supports multiple optical wavelength channels, each of which traverses several physical components (OXC port, MUX/DMUX, amplifier, fiber, conduit, etc.), as shown in Fig. 2. Thus, a logical link is an abstraction of network resources and physical components. In such a network, a failure of any component will affect the network's reliability to some degree. For example, a failure of an OXC port only affects a single wavelength channel, whereas a failure of a SRG (such as a fiber cut) will make multiple channels fail at the same time. Therefore, the distributed link-state database should be able to provide the information of physical components' failure characteristics and the link state should dynamically change according to current network state so that routing and signaling protocols can take physical information into consideration when provisioning a connection. To do so, we aggregate the information of availabilities of the physical components with the free-wavelength-resource information to form a dynamic link-state parameter called link-and-resource availability (LRA). The LRA of a link defines the probability that at least one wavelength channel in this link is available (e.g., free and alive) to accept a new connection. It is used as a link parameter for efficient routing to provide reliable services while balancing network load on each link. The procedure of LRA aggregation is briefly described below and shown in Fig. 3 where a simplified link consisting of six OXC ports and two fibers is used for illustration.

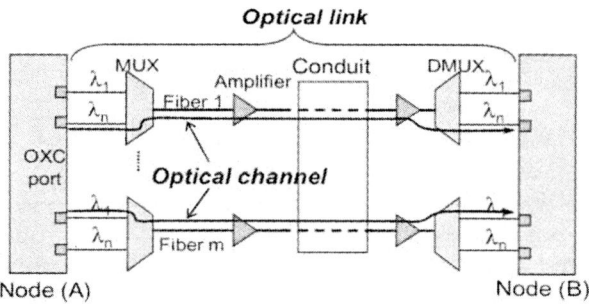

Fig. 2. Example physical configuration of a logical link

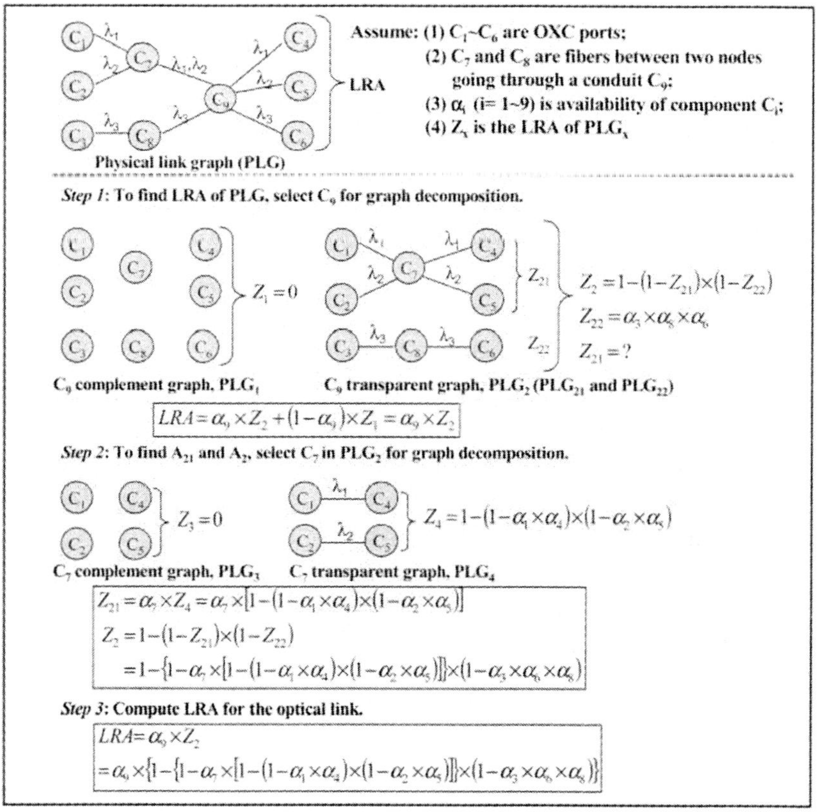

Assume: (1) $C_1 \sim C_6$ are OXC ports;
(2) C_7 and C_8 are fibers between two nodes going through a conduit C_9;
(3) α_i (i= 1~9) is availability of component C_i;
(4) Z_x is the LRA of PLG$_x$

Physical link graph (PLG)

Step 1: To find LRA of PLG, select C_9 for graph decomposition.

$Z_1 = 0$

C_9 complement graph, PLG$_1$

Z_{21}
Z_{22}

C_9 transparent graph, PLG$_2$ (PLG$_{21}$ and PLG$_{22}$)

$Z_2 = 1 - (1 - Z_{21}) \times (1 - Z_{22})$
$Z_{22} = \alpha_3 \times \alpha_8 \times \alpha_6$
$Z_{21} = ?$

$$LRA = \alpha_9 \times Z_2 + (1 - \alpha_9) \times Z_1 = \alpha_9 \times Z_2$$

Step 2: To find A_{21} and A_2, select C_7 in PLG$_2$ for graph decomposition.

$Z_3 = 0$

C_7 complement graph, PLG$_3$

$Z_4 = 1 - (1 - \alpha_1 \times \alpha_4) \times (1 - \alpha_2 \times \alpha_5)$

C_7 transparent graph, PLG$_4$

$$Z_{21} = \alpha_7 \times Z_4 = \alpha_7 \times [1 - (1 - \alpha_1 \times \alpha_4) \times (1 - \alpha_2 \times \alpha_5)]$$
$$Z_2 = 1 - (1 - Z_{21}) \times (1 - Z_{22})$$
$$= 1 - \{1 - \alpha_7 \times [1 - (1 - \alpha_1 \times \alpha_4) \times (1 - \alpha_2 \times \alpha_5)]\} \times (1 - \alpha_3 \times \alpha_6 \times \alpha_8)$$

Step 3: Compute LRA for the optical link.

$$LRA = \alpha_9 \times Z_2$$
$$= \alpha_9 \times \{1 - \{1 - \alpha_7 \times [1 - (1 - \alpha_1 \times \alpha_4) \times (1 - \alpha_2 \times \alpha_5)]\} \times (1 - \alpha_3 \times \alpha_6 \times \alpha_8)\}$$

Fig. 3. Illustrative example for LRA computation process

3   LRA-Based Connection-Provisioning Algorithm

In this section, we propose a dynamic connection-provisioning algorithm in which the routing is based on the dynamic link-state parameter (LRA). In our study, a *connection* could be setup using a single lightpath or a pair of lightpaths (primary path and backup path), if necessary, from a source to a destination. The primary path is used to carry live traffic while the backup path is activated when the primary path fails. While we assume that each node has full wavelength-conversion functionality in our study, it is straightforward to extend it to the case of no wavelength conversion, i.e., enforce the wavelength-continuity constraint where a lightpath should remain on a same wavelength from source to destination.

In our connection-provisioning algorithm, a connection's primary and backup paths may not necessarily be link disjoint as long as the SLA-defined connection-availability requirement can be met. Four rules are used for connection-availability computation in our algorithm, as follows.

Rule 1: Suppose a lightpath from source to destination consists of K hops in series. The lightpath availability, A_L, can be computed as:

$$A_L = \prod_{j=1}^{K} CA_j \text{ where } CA_j \text{ is channel availability of the } jth \text{ hop.}$$

Rule 2: If a connection consists of a single primary path that can satisfy the SLA requirement, the connection availability, A_C, obviously is the availability of the primary path, A_P.

Rule 3: If a connection consists of primary and backup path pairs that are link disjoint, the connection availability, A_C, can be computed as: $A_C = 1 - (1 - A_P) \times (1 - A_B)$ where A_P and A_B are availabilities of the primary path and backup path, respectively.

Rule 4: If a connection cons<ists of primary and backup path pairs that are not link disjoint, it needs to partition the path pair into segments at every junction node which is defined to be a common node on both primary and backup paths. Assume that there are M segments, and the availability of the mth segment is A_m; then, the connection availability, A_C, can be computed as: $A_C = \prod_{m=1}^{M} A_m = \prod_{m=1}^{M} 1 - \left(1 - \prod_{i=1}^{pm} A_{m,p_i} \right) \times \left(1 - \prod_{j=1}^{bm} A_{m,b_j} \right)$

where A_{m,p_i} is the channel availability of ith hop of the mth primary-path segment which consists of pm hops; and A_{m,b_j} is the channel availability of jth hop of the mth backup-path segment which consists of bm hops. The A_m is computed according to the same probability theory used in Rules 1 and 3.

For a connection request, to provide an availability-guaranteed service, the LRA-based algorithm works as follows:

Given: Current network topology with LRA as link weight; and a connection request $R(s, d, SLA_R)$ where $s, d,$ and SLA_R are source node, destination node, and availability requirement of the request (R), respectively.

Step 1: Compute a path, which has maximum multiplication of link LRAs on the route [10], as the primary path for R. If no route resource can be found, block R and STOP. Otherwise, compute primary lightpath's availability using Rule 1. Accept R and STOP if availability of primary path can satisfy SLA_R.

Step 2: To find a backup path: (a) Compute a temporary LRA, LRA_t, for each link along the primary path based on the link state which led to the primary channel reserved in Step 1. (b) For each link along the primary path, reduce LRA_t to $LRA_r = LRA_t * \beta$, where β is a disjoint factor between 0 and 1 and is proportional to the channel availability of this link on the primary path. (c) Find a route, which has the highest multiplication of link LRAs on the route in the temporary topology with LRA_r for the link on the primary path, to be the backup path.

Step 3: If no backup path can be found, block R and STOP. Otherwise, (a) compute overall connection availability of primary path and backup path using Rules 1-4; (b) if the connection availability $< SLA_R$, block R and STOP; otherwise, accept R and update topology by re-computing LRA on each link after resource for R is allocated.

Note that, in Step 2, the link weight is consciously controlled. Instead of removing links on primary path, we reduce these links' LRA by a factor β. β needs to be small enough (e.g., 1% of channel availability in our study) such that link-disjointness is not mandatory but encouraged as long as the connection's availability is guaranteed.

4   Numerical Examples

For availability-guaranteed connection provisioning, we compare the performance of our proposed LRA-based algorithm (DYNAMIC_LRA) with two existing algorithms which are based on static link-state parameters, i.e., STATIC_LEN and STATIC_AVA.

In STATIC_LEN, (a) each link is weighted as its fiber length; (b) primary path is the shortest path (by distance); and (c) backup path is the shortest path after removing each link along the primary path. In STATIC_AVA, (a) weight of each link is an abstract link availability that is an average from all (16 channels in this study) channel availabilities; (b) a path with highest availability is used as primary path; and (c) backup path is the path with highest availability after removing the links along the primary path.

For both cases, backup path is not needed as long as the availability of primary path can satisfy the SLA requirement.

For illustration purposes, in our simulation, we assume the network topology of Fig. 4; availability of fiber and OXC port are uniformly distributed random numbers in the range of 99.95%~99.97% and 99.97~99.99%, respectively; availability requirement for a connection is 99.9%; each fiber supports 16 wavelengths; traffic is dynamic where connection arrivals are Poisson and their holding times are exponentially distributed.

Figure 5 shows the blocking probability vs. network offered load (in Erlangs). It takes into account resource blocking, i.e., a connection cannot find its primary or backup (if necessary) due to resource constraints, and availability blocking, i.e., primary-backup path pair found for a connection cannot satisfy its SLA requirement. The results indicate that significant improvement in blocking can be achieved by our proposed DYNAMIC_LRA algorithm. As shown in Fig. 5, for example, at a load of 120 Erlangs, about 28% improvement in blocking probability is achieved by using

DYNAMIC_LRA. This is because the routing based on dynamic link-state parameter, LRA, can take physical-layer availability into consideration while traffic is automatically balanced to make good use of network resources. Hence more availability-guaranteed connection can be set up by DYNAMIC_LRA.

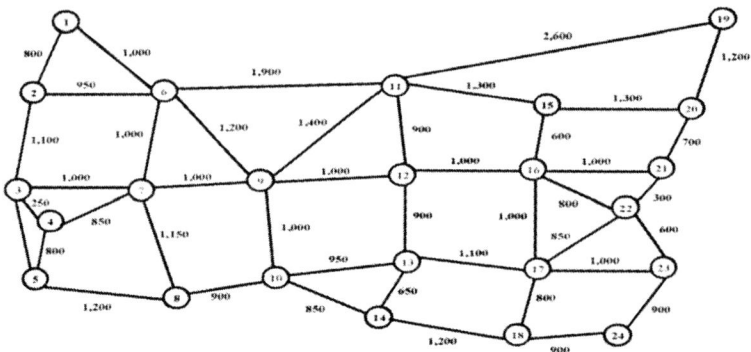

Fig. 4. A sample mesh network with fiber length (in km) marked on each link

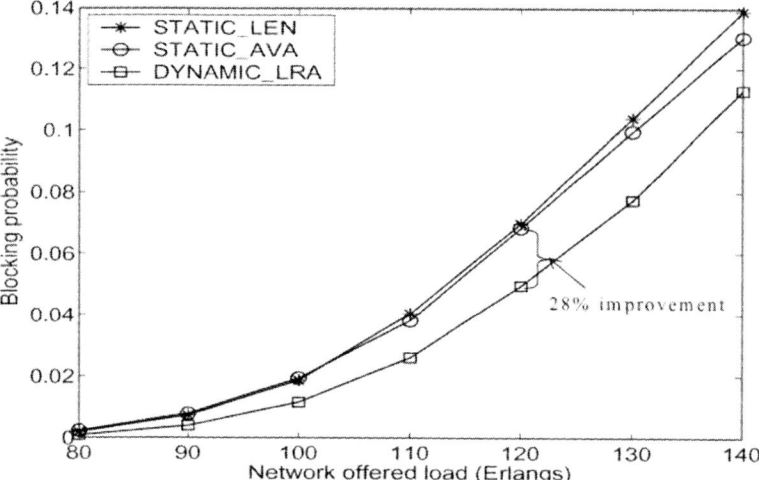

Fig. 5. Connection-blocking probability

Figure 6 shows the percentage of connections that have link-disjoint primary-backup path pairs among all connections that have backup path in DYNAMIC_LRA. We can observe that, in DYNAMIC_LRA, link-disjointedness is preferred even though it is not necessary. In DYNAMIC_LRA, when load increases, the percentage of connections with disjoint primary-backup paths decreases a little bit. This is because some highly-reliable links on primary paths are chosen for backup route. However, such provisioning is achieved under careful control by using the availability constraints to guarantee the connection's availability requirement.

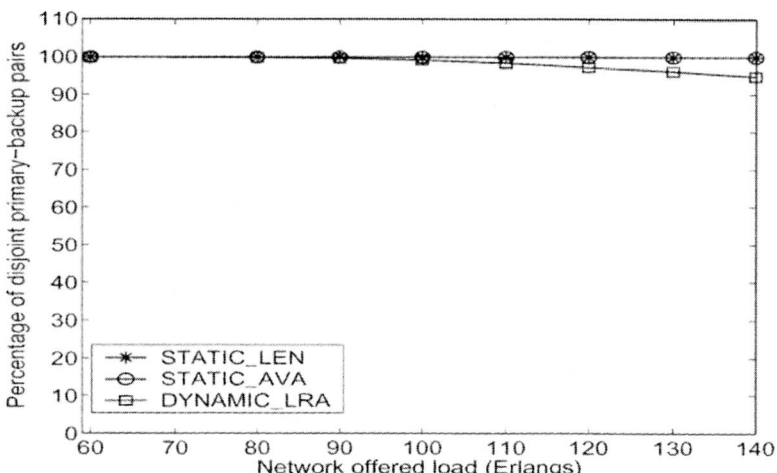

Fig. 6. Percentage of disjoint primary-backup path pairs

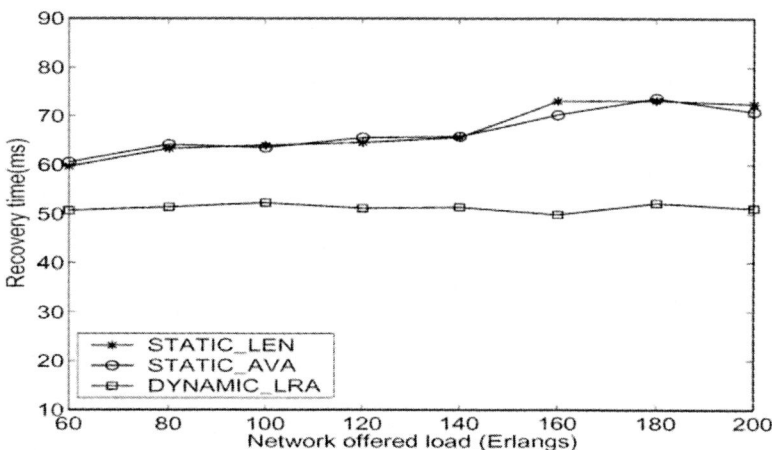

Fig. 7. Average failure recovery time (in ms)

Figure 7 shows the network-wide average failure-recovery time. To evaluate the failure-recovery time, the protection-switching procedure used for study is as follows [11]. When a node detects a failure, it sends an alarm message towards the source node along the primary path. Upon receiving an alarm message, the switching-source node (e.g., a node that initiates switching) sends a request message to the switching-destination node along the backup path. Upon receiving the request message, the switching-destination node sends back a response message along the backup path and the intermediate nodes initiate channel cross-connection. In case of STATIC_LEN and STATIC_AVA, only the source node keeps the primary-path and backup-path information. Hence, the connection's source-destination nodes perform end-to-end switching to the backup path. In case of DYNAMIC_LRA, primary-path or backup-path information is kept at every node the path goes through. Two junction nodes (junction means that the node is on both primary path and backup path), that are closest to the failure in upstream and downstream directions, will perform protection switching.

Note that a connection might be set up using a single primary path which has very low failure possibility. Although connection availability has been taken care of before the connection is set up (due to the fact that only availability-guaranteed connections can be set up), we could try to combat failure for a connection without protection by performing restoration (i.e., dynamic discovery of backup resources).

Here, we examine the network-wide recovery time for protectable failures. To statistically analyze the average network-wide recovery time from a failure, after 100,000 connection arrivals, we freeze the network state and simulate 100,000 single failures on average in the network. The failure could be a fiber failure or channel failure. The simulated failures are generated according to the fiber and channel OXC port's availability. We assume the failure-detection time is 20 μs; the node processing time is 0.4 ms for alarm or request message, and 0.3 ms for response message; the message propagation time is 5 μs/km; and the routing time for restoration is 1 ms.

Our results demonstrate that fast restoration is achieved by using DYNAMIC_LRA as shown in Fig. 7. Furthermore, recovery from channel failures on some highly-reliable links could be very fast because both primary and backup paths of a connection might traverse these links.

5   Conclusion

In this paper, we first proposed a novel link-state-modeling mechanism by which current network resource information and the physical-component's failure characteristics were aggregated to form a comprehensive dynamic link-state parameter, called link and resource availability (LRA). The LRA was analyzed through some analytical models and illustrative examples. Then, based on such a dynamic link-state parameter, we proposed a routing algorithm to provision SLA-availability-guaranteed connections. A new generalized protection model was developed through dynamic LRA-based provisioning. In the proposed protection mechanism, the link-disjoint constraint on backup-path computation may not be necessary as long as the connections' availability requirements can be satisfied. The dynamic LRA-based connection-provisioning approach efficiently provides SLA-guaranteed services while significantly improving connection reliability, better utilizing network resources, and achieving fast failure recovery.

References

[1]   J. Moy, OSPF: Anatomy of an Internet Routing Protocol, Addison-Wesley, May 2000.
[2]   R. Braden et al., "RSVPesource ReSerVation Protocol – Version 1 Functional Specification," RFC2750, IETF, Sep. 1997.

[3]   G. Li, J. Yates, D. Wang, and C. Kalmanek, "Control plane design for reliable optical networks" *IEEE Commun. Mag.*, pp. 90-96, Feb. 2002.

[4]   O. Gerstel and R. Ramaswami, "Optical layer survivability: a services perspective," *IEEE Commun. Mag.*, vol. 38, pp. 104-113, Mar. 2000.

[5]   R. Ramamurthy et al, "Capacity performance of dynamic provisioning in optical networks," *IEEE J. Lightwave Tech.*, vol. 19, no. 1, pp. 40-48, Jan. 2001.

[6]   W. Wen, B. Mukherjee, and S. J. B. Yoo, "QoS based protection in MPLS controlled WDM mesh networks," *Photonic Network Commun.*, vol. 4, no. 3/4, pp. 297-320, 2002.

[7]   Y. Huang, W. Wen, J. Zhang, J. P. Heritage, and B. Mukherjee, "A generalized dynamic protection model for reliable WDM networks," *Proc., ECOC*, to appear, Sept. 2003.

[8]   M. Clouqueur and W. D. Grover, "Availability analysis of span-restorable mesh networks", *J. Select. Areas Commun.*, vol. 20, no. 4, pp. 810-821, May 2002.

[9]   Dacomo, S. D. Patre, G. Maier, A. Pattavina, and M. Martinelli, "Design of static resilient WDM mesh networks with multiple heuristic criteria," *Proc., IEEE INFOCOM*, pp. 1793-1799, 2002.

[10]   J. Zhang, K. Zhu, H. Zang, and B. Mukherjee, "Service provisioning to provide per connection-based availability guarantee in WDM mesh networks," *Proc., OFC*, paper no. FA6, pp. 622-624, March 2003.

[11]   M. Goyal, J. Yates, G. Li, and W. Feng, "Benefits of restoration signaling message aggregation," *Proc., OFC*, paper no. TuI2, pp.203-204, March 2003.

Distributed Connection Control Protocols in All-Optical WDM Networks

Guo Yinghua[1] and Yuan Cong[2]

[1] International Graduate School in ICT, University of Trento
Via Sommarive, 14 I-38050 POVO, Italy
atmgyh@yahoo.com
[2] National Key Lab of ISN, Xidian University
Xi'an, 710071, P.R.China,
congyuan515@yahoo.com

Abstract. In all-optical WDM networks, all-optical connections, *(or Lightpaths)* implement the communication architecture between network entities. The networks must have an efficient connection control mechanism that is able to determine routes and reserve resources in a manner which is scalable and which provides a low degree of blocking for connection requests and also able to support highly dynamic and bursty traffic through the rapid on-demand provisioning of lightpaths. In this paper, we present a distributed lightpath control protocol in which a new resource reservation scheme, SDCR and Conditional One-way data transmission mechanism are deployed. The simulation results show that our protocol outperforms other approaches, regarding the performances of setup delay and network utilization.

1 Introduction

Optical networks have received enormous amount of acceptance in wide-range applications e.g. computer and telecommunication networks because of its capacity of handling the ever-increasing bandwidth demands. In all-optical *(wavelength-routed)* WDM networks, the basic communication mechanism is the all-optical WDM channel, referred as to *lightpaths* [1], which is an optical connection between two nodes in the network and is set up by assigning a dedicated wavelength to it on each link in its path.

Lightpath control has been a critical issue in the evolution of WDM optical networks. In order to establish lightpaths in a wavelength-routed network, we need algorithms and protocols to select routes and assign wavelengths for lightpaths, as well as to reserve network resources. In addition to routing and wavelength assignment *(RWA)* [2,3], we also need provisioning protocols to exchange control information among nodes and to reserve resources along the path.

The provisioning protocols for lightpath establishment in wavelength-routed networks can be either centralized or distributed and have been studied extensively.

S. R. Das, S. K. Das (Eds.): IWDC 2003, LNCS 2918, pp. 383-392, 2003.

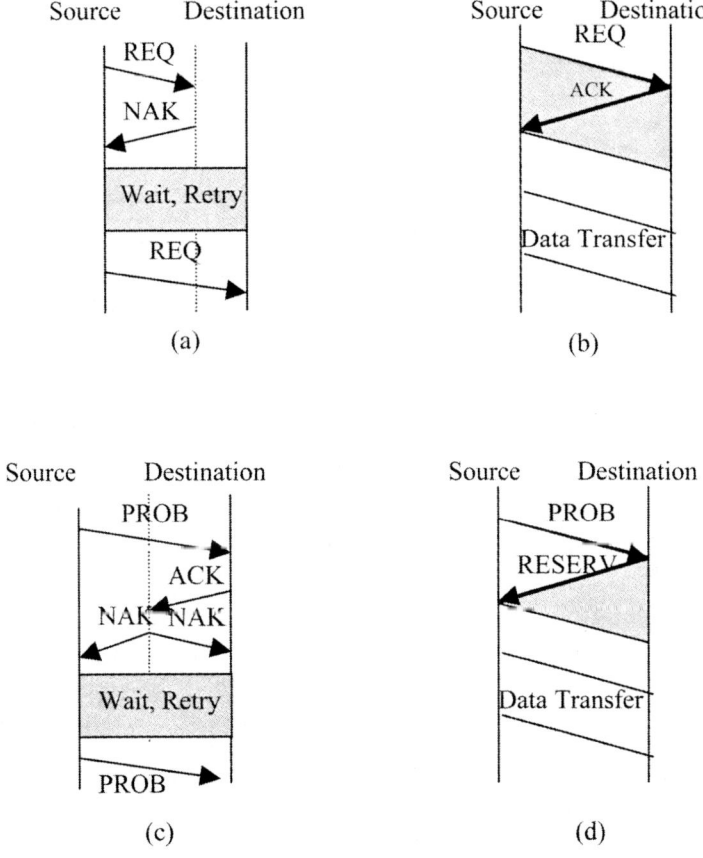

Fig. 1. (a) Unsuccessful SIR (b) Successful SIR (c) Unsuccessful DIR (d) Successful DIR

Some of these investigation focus on the centralized control [2,4], and establishing lightpaths using distributed signaling protocols have been also studied. Surveys of distributed signaling protocols can be found in [5,6,7,8]. In this paper, we propose to introduce a resource reservation mechanism, named *Source and Destination Cooperative Reservation (SDCR)*, which employs two times of reservation attempts initiated by the source node and the destination node respectively. Meanwhile, differing from two current typical data transmission schemes *Two-way* and *One-way*, a more efficient data transmission scheme *Conditional One-way* is proposed, in which the source node starts data transmission on the suggested wavelength T time units after it sends out the reservation message if and only if the source does not receive any NAK message within T time units. We further propose an efficient distributed control protocol for lightpath establishment in wavelength-routed WDM networks.

The rest of this paper is organized as follows. Section 2 briefly reviews various provisioning protocols for setting up lightpaths and introduces the basic arguments for this work. In section 3, with analysis of previous works, we present the proposed lightpath establishment protocol and discuss its significant properties. Section 4 carries out performance evaluation. In section 5, we conclude this paper..

2   Distributed Lightpath Control Protocols

Currently, there are two typical classes of distributed wavelength reservation protocols, namely Source Initiated Reservation (SIR) protocols and Destination Initiated Reservation (DIR) protocols. SIR would be classified as a provisioning method in which reservations are initiated at the source node, and reservations are made in the forward direction. On the other hand, DIR could be classified as a provisioning method in which reservations are initiated by the destination node, and reservations are made in the backward direction. The provisioning scheme in the emerging GMPLS standard is an example of DIR. The basic SIR and DIR schemes are illustrated in Fig.1. In addition, many techniques and variations of the basic scheme (Dropping vs. Holding, Aggressive vs. Conservative, Parallel vs. Sequential, Two-way vs. One-way, etc.) have been developed and a survey of these variations can be found in [7].

A number of works have investigated provisioning protocols for lightpath establishment in wavelength-routed WDM optical networks and some specific provisioning approaches have been proposed [8,9,10]. In [8], a distributed provisioning protocol for managing connections, namely link-state approach is proposed. In this approach, global information including network topology and wavelength usage on each link is available to all nodes. Upon the arrival of a connection request, the source node selects an optimal route to a destination on a given wavelength utilizing the global network information. Then, the node reserves the desired wavelength on each link in the route in order to setup the connection. The limitations of this approach can be briefly described as follows. First, this approach may result in high setup time because the connection isn't be established until signalling traverse two loops from the source to the destination, which is referred as to two-way manner. Second, the connection blocking probability in this approach may be high due to the outdated network information caused by multiple simultaneous connections competition or uncompleted network state update information. Thus, while this approach may be appropriate for a fairly static network in which connections are not changing rapidly, it may not be appropriate for emerging networks in which connections and lightpaths are established more dynamically [11].

In [9], a distributed-routing approach is present. Compared with link-state approach [8], each node in [9] needn't keep the global network information; instead they maintain a routing table for each wavelength, which specifies the next hop and the cost associated with the shortest path to each destination on this wavelength. As a result, routes and wavelength for ligthapths are selected in a distributed fashion, which obtains improvement in blocking probability and scalability to a certain extent. However, in this scheme, control packets still have to travel round trip between the source and the destination before the data transferring, and hence the connection setup time will still be significant in larger networks. Moreover, in both link-state and distributed-routing approaches, wavelengths are reserved in the forward direction to the destination and therefore the problem of bandwidth wasted during reservation period still exists.

In order to reduce bandwidth wasted and make more intelligent routing and wave-length decision, the work [10] presents a distributed lightpath control scheme based on destination routing. In this scheme, upon the arrival of a connection request, the source node sends a PROB message that does not reserve any wavelength to the desti-nation along the shortest path. It is the destination node that decides a route and se-lects a wavelength based on the current network state information it maintains. Finally, the source will begin to transfer data only when it receives ACK message. As a result, the connection setup time is at least twice as much as the propagation delay between the source and the destination. Furthermore, the control overhead will still be signifi-cant in large networks because each node has to maintain global network information and keep this information being updated periodically.

3   Resource Reservation and Data Transmission Schemes

In this section, we firstly present a novel wavelength reservation scheme, *SDCR* (*Source and Destination Cooperative Reservation*), and an efficient data transmission scheme, namely *Conditional One-way* (*by nature, it is one-way if and only if the source's suggested wavelength is in available mode*), which are utilized in our pro-posed protocol. Furthermore, we propose to introduce an efficient provisioning proto-col for lightpath establishment based on above argument. In our considered network, each node maintains global network information that is updated periodically, or when-ever there is a change in information database. The actual route computation and wavelength assignment algorithm used are outside our protocol; several possibilities exist [1,2]. In our protocol, the routing algorithm is *Adaptive Routing* algorithm [3] and we use *First-Fit* scheme for wavelength assignment. For clarity and conciseness, it is assumed that there is no wavelength conversion. However, the work proposed in this paper also can be used in wavelength conversion situation.

In our protocol, a wavelength on a given link can be in one of the following three states: Available, Busy, and *Suggested* that indicates this wavelength has been sug-gested to the destination for a connection on that link. In addition, in order to ex-change information among nodes, some basic kinds of signaling messages are needed as follows: *CAS (Collection And Suggestion)* gathering resource information along a selected route without actually reserving any wavelength and suggesting wavelength which can be used to the destination; *ACK* informing the source that wavelength sug-gestion is successful and occupying this wavelength (*by resetting the state of wave-length from suggested to busy*) on each link along the path; *NAK* confirming that the suggested wavelength has been used by another lightpath and informing the source to stop conditional one-way transmission and wait for response from the destination; *ReWA (Re-Wavelength Assignment)* sent back to the source by the destination to re-assign a wavelength for lightpath when the destination is informed that the wavelength suggestion initiated by source failed.

3.1   Source and Destination Cooperative Reservation

The principle of SDCR that has been implied in lightpath setup procedure can be briefly restated as follows. When a lightpath request arrives, using the global topology and wavelength usage information, the source node *first tries* to select a route and a wavelength, and then sends out the CAS message to suggest this selected wavelength along the route. If this suggestion succeeds, the conditional one-way transmission starts, which is similar to SIR. However, if this suggestion fails, this request won't be blocked like in SIR; instead the destination will *retry* to perform route and wavelength determination to select a route with available wavelength for this request. At this point, it is like DIR. Briefly speaking, our proposed SDCR is a hybrid scheme of SIR and DIR that combines advantages of SIR and DIR, at the same time overcomes their drawbacks.

3.2   Conditional One-Way Transmission

In two-way scheme, the source does not start data transmission until it receives ACK from the destination. As a result, the setup delay is at least twice as much as the propagation delay between the source and the destination. In one-way scheme, the source first sends out REQ to reserve wavelength and then starts data transmission before the ACK comes back. It is clear that, if the reservation fails frequently, the source will transfer much useless data, which wastes the lightpath utilization. Based on this observation, we proposed the conditional one-way transmission scheme in which the source starts data transmission on the suggested wavelength T time units ($T=HP$, *where H is the number of hops for a lightpath, and P is the signaling processing time at one node*) after it sends out the CAS message if and only if the source does not receive any NAK message within T time units.

3.3   Lightpath Setup Procedures

- Upon the arrival of a connection request, the source performs route and wavelength determination procedure to find a route and wavelength for the requested lightpath. Then, the source sends CAS massage that collects resources information and suggests an available wavelength along the selected route.
- When receiving the CAS message, each intermediate node will set this suggested wavelength's state to *Suggested* and forward the CAS message to the next node if this suggested wavelength is *Available*. If this suggested wavelength is either *Suggested* or *Busy*, the node will include a note indicating this suggested wavelength has been occupied by other lightpath into the CAS message for the destination, and then forward the CAS message. At the same time, the intermediate node sends back a NAK message to notify the source the suggestion failure as well as stopping Conditional one-way transmission and waiting for the reply from destination.
- Once the destination node receives the CAS message, if it finds that the suggested wavelength are available on the entire path, it will send back the ACK message along the reverse route. If the suggested wavelength is not available on any link

alone the path, it then re-performs the routing and wavelength determination procedure to decide a route and select a wavelength based on the CAS message and network information. In this case, if a route with an available wavelength can be found, it sends ReWA message to the source. If there are no routes with available wavelengths, the destination will send back NAK message to inform the source that this lightpath setup attempt fails and try it later.

- On the other hand, at the source node, data transmission will start on the suggested wavelength T time units after the source sends out CAS message if and only if the source does not receive any NAK message within T time units. If, however, the source receives any NAK message within T time units, it won't start one-way transmission and will wait for messages from the destination.

3.4   Properties

The proposed protocol in this paper has some significant properties differing from existing distributed lightpath control approaches.

1. Lower Blocking Probability
   The source node can select a route and wavelength with high success probability since they maintain the global network topology and wavelength usage information. At this point, the performance equals to SIR. However, even if this *first-try* fails, the destination will retry instead of blocking this request like DIR. Hence, differing from the SIR and DIR in which there is only one reservation attempt initiated by either the source or the destination, the scheme "source first-try, destination retry" is able to improve the success probability of lightpath establishment.

2. Wide Applied Ranges
   Our proposed protocol can be used in wide ranges from static to quite dynamic networks. In static network, most of the blocking will be due to insufficient resources. At this point, the SIR is not an appropriate scheme for lightpath establishment in static networks because the wavelength is wasted during the reservation period in the forward direction. Unlike SIR, SDCR does not reserve any wavelength in the forward direction, and the suggested wavelength can be also suggested by other lightpath, which decreases the wavelength waste in the lightpath setup period. On the other hand, in dynamic network, blocking due to outdated information may become an increasingly significant component of the overall connection blocking probability, and DIR is not suitable for lightpath establishment in such situation. However, the SDCR is also suitable for this case because it suggests wavelength sooner than DIR with high success probability, leaving less opportunity for other lightpath requests to interfere with the current lightpath request.

4   Performance Evaluation

In this section, we evaluate our proposed protocol comparing with two current approaches, Link-state approach [8] and Distributed-routing approach [9], in terms of lightpath setup delay and network utilization via simulation of the mesh network [6] in PM (Path Multiplexing) scenario. In our simulations, we assume: 1) The simulated mesh network has 16 nodes and 25 links (the number on the links represents link distances in tens of kilometres and the number of wavelengths on each link is 8, Fig. 2), 2) New connections will arrive according to the Poisson process with arrival rate λ, 3) The destination of each connection is uniformly distributed, 4) Connection duration has an exponential distribution with mean 100 ms, 5) Signaling message processing time at a node, P, is 10 μ s; the time to configure across-connect, C, is 500 μ s; the average propagation delay between two nodes D is 14.7ms, and the average hop distance is H_a = 2.28.

Lighpath setup delay is the time required to establish a lightpath once a lightpath request arrives. When the network traffic load is light and there are no reattempts, the Average Setup Time (AST) can be formulated AST = 2D + (2 H_a +1) P + C = 30.2 ms for Link-state approach and Distributed-routing approach, which are plot in Fig.3. From Fig.3, when the network load is light, we observe that the setup times in Link-state approach and Distributed-routing approach are fairly close to this bound (30.2 ms), but the setup time in our proposed protocol is significant low. This is due to the fact that, under low network load, the first-try of wavelength reservation initiated by the source succeeds with quite high probability, which in fact is one-way transmission scheme. It is clear that the setup time in one-way scheme is less than the one in two-way scheme. As the load increases, since the first-try will fail frequently, the transmission scheme used in our proposed protocol is essentially two-way scheme, in which the setup delay is at least twice as much as the propagation delay between the source and the destination. Therefore, the setup delay difference between Distributed-routing approach and our protocol is not significant.

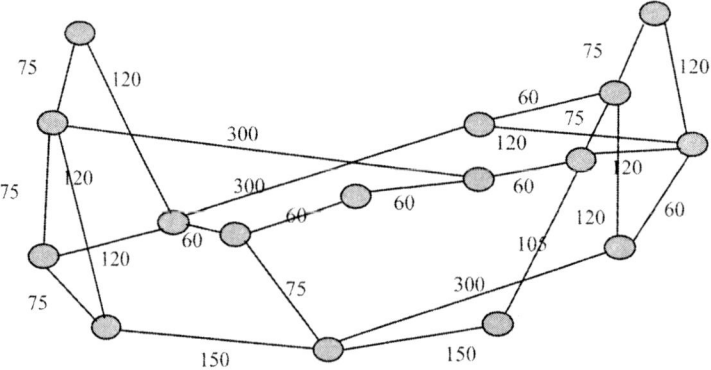

Fig. 2. A nationwide backbone network

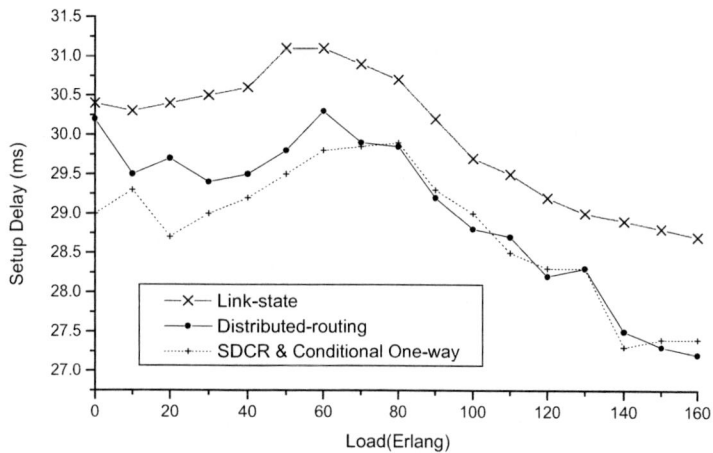

Fig. 3. Lightpath Setup Delay vs. Load

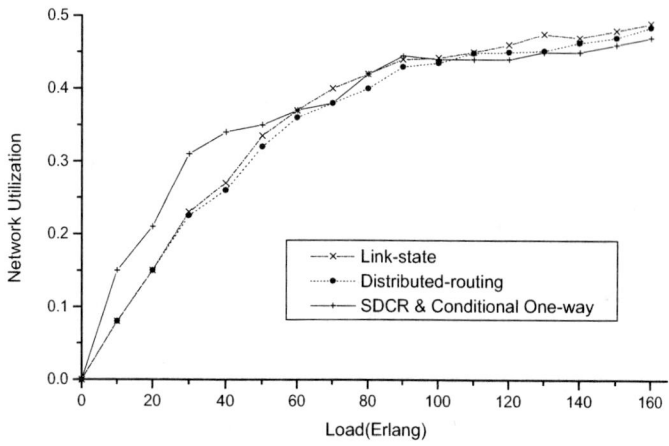

Fig. 4. Network Utilization vs. Load

From Fig.4, we can see that networks reach the saturable state where the network utilization is around 50 percent for a load of 160 Erlangs. We also find that our protocol's performance in term of network utilization is different under different network load. Under light network load situation, our protocol obtain higher network utilization than link-state and distributed-routing approaches because the overhead in our protocol is smaller than overhead in SIR, even in DIR when the *first-try* of suggestion initiated by the source succeeds. However, under heavy network load, the *first-try* of wavelength suggestion fails frequently, which causes that lightpath establishment must be implemented by the *retry* initiated by the destination. As a result, the overhead in our protocol is similar to overhead in link-state and distributed-routing, in turn, the difference of network utilization in the three approaches is not significant in heavy load situation.

Furthermore, we are aware that, by maintaining global network state information, each node can make more intelligent decision on route selecting and wavelength reservation; however, this obtaining of global information also results in high control overhead and increase in blocking probability caused by outdated information due to propagation delay in large networks. High control overhead may be one drawback of our proposed protocol. On the other hand, if each node only maintains its local information (e.g. its outgoing link and wavelength usage in those links), the control overhead and probability of outdated information can be decreased; but, there is a high possibility that nodes will make wrong decision on route selecting and wavelength reservation. Hence, we postulate that there is an equilibrium value for the range in which each node collects the network state information and exchanges control messages. In the following work, one issue is to determine the optimal range. In this range, each node does not maintain the global network state information any longer; instead, every node obtains only the wavelength usage information from its neighboring nodes within this optimal distance. Also, when information updating, each node only sends pertinent information to the nodes within this area instead of broadcasting.

5   Conclusion

In this paper, we proposed a new wavelength reservation scheme, *SDCR* and an efficient data transmission scheme, *Conditional One-way*. Based on these two schemes, we proposed an efficient provisioning protocol for lightpath establishment, which can be used in both static and dynamic wavelength-routed WDM networks. We compared our proposed protocol with Link-state approach and Distributed-routing approach and simulated their performance in a mesh network. Simulation results shows that, regarding setup delay and network utilization, our protocol performs better than other approaches, especially in light load situation.

References

[1]   I. Chlamtac, A. Ganz, and G. Karmi, "Lightpath communications: An approach to high-bandwidth optical WAN's," *IEEE Trans. Commun.,* vol. 40, pp. 1171–1182, July (1992).

[2]   R. Ramaswami and K.N. Sivarajan. "Optimal routing and wavelength assignment in all-optical networks". In *Proceedings of IEEE Infocom*, pages 534–543, June (1994).

[3]   H. Zang, J.P. Jue, and B. Mukherjee, "A Review of Routing and Wavelength Assignment Approaches for Wavelength-routed Optical WDM Networks," Optical Networks Magazine, vol. 1,no. 1, pp.47-60, Jan.(2000)

[4]   A. Aggarwal et al. "Efficient routing and scheduling in optical networks," *Proc. of the ACM-SIAM Symp. On discrete algorithms (SODA '93)*, (1993), pp.412-423

[5]   X. Yuan, R. Melhem, R. Gupta et al., "Distributed Control Protocols for Wavelength Reservation and Their Performance Evaluation," *Photonic Network Communications*, vol. 1, No.3, pp. 207-218, (1999).

[6]   H. Zang, J.P.Jue, L. Sahasrabuddhe et al., "Dynamic Lightpath Establishment in Wavelength-Routed WDM Networks," IEEE communication magazine, vol.39, no.9, pp.100-108, Sept. (2001)

[7]   Yousong Mei and Chunming Qiao, "distributed control schemes for dynamic lightpath establishment in WDM optical networks", *Proc. Of 2000 Workshop on optical networks*, Richardson, Texas, Feb.(2000).

[8]   R.Ramaswami and A. Segall. "Distributed Network Control for Optical Network". IEEE/ACM Transactions on Networking, vol.5, no.6, pp. 936-943, Dec. (1997).

[9]   H. Zang, L. Sahasrabuddhe, J.P. Jue, S. Ramamurthy, and B. Mukherjee, "Connection Management for Wavelength-Routed WDM Networks," Proceedings, IEEE Globecom '99, Rio de Janeiro, Brazil, vol. 2, pp. 1428-1432, December (1999).

[10]   Jun Zheng; Mouftah, H.T, " Distributed lightpath control based on destination routing for wavelength-routed WDM networks," Global Telecommunications Conference, 2001. GLOBECOM '01. IEEE, Volume: 3 , (2001) Page(s): 1526 -1530.

[11]   Jue, J.P. Gaoxi Xiao, "An adaptive routing algorithm for wavelength-routed optical networks with a distributed control scheme," *Computer Communications and Networks, 2000. Proceedings. Ninth International Conference* on, (2000).

Author Index

Lecture Notes in Computer Science

For information about Vols. 1–2834
please contact your bookseller or Springer-Verlag